# ELECTRICAL CRAFT PRINCIPLES

## 4th Edition

## Volume 2

## JOHN WHITFIELD

The Institution of Electrical Engineers

Published by: The Institution of Electrical Engineers, London,
United Kingdom

First published 1974
© 1974: Peter Peregrinus Ltd.

2nd Edition 1980
© 1980: Peter Peregrinus Ltd.

3rd Edition 1989
© 1989: Peter Peregrinus Ltd.

4th Edition 1995
© 1995: The Institution of Electrical Engineers

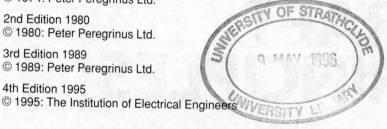

The Institution of Electrical Engineers,
Michael Faraday House,
Six Hills Way, Stevenage,
Herts. SG1 2AY, United Kingdom

While the author and the publishers believe that the information and
guidance given in this work is correct, all parties must rely upon their own
skill and judgment when making use of it. Neither the author nor the
publishers assume any liability to anyone for any loss or damage caused
by any error or omission in the work, whether such error or omission is
the result of negligence or any other cause. Any and all such liability is
disclaimed.

The moral right of the author to be identified as author of this work has
been asserted by him/her in accordance with the Copyright, Designs and
Patents Act 1988.

**British Library Cataloguing in Publication Data**

A CIP catalogue record for this book
is available from the British Library

ISBN 0 85296 833 7

Printed in England by Short Run Press Ltd., Exeter

# ELECTRICAL CRAFT PRINCIPLES

## Volume 2

# Contents

# Preface to the Fourth Edition

This book has been on sale for the best part of 20 years through three editions, and the continuing demand for it has resulted in the need for this, the fourth edition.

Important changes have taken place since the appearance of the third edition in 1988. The extremely important 16th Edition of the IEE Wiring Regulations has become a British Standard, *BS 7671*. There have also been changes in the syllabuses of the examining bodies responsible for the certification of electrical craftsmen, as well as in the pattern of education and training which they are required to follow.

The increased 'hands-on' nature of the training is to be welcomed, as is the widening of syllabuses to include more electronics. Multiple-choice questions are now in wide use, and a selection of this type of test has been provided in each chapter of this book. I must admit to having doubts about presenting the reader with faulty information in the wrong answers to the multiple-choice questions, but am sure that a sensible approach, including the analysis of each wrongly answered question, will ensure that no harm results.

When originally written, the first volume of this book covered Part II of the craft syllabuses for electricians. Changes have meant that this fourth edition now covers part of Part II (the balance will be found in Volume 1) as well as the 'C' certificate course. No attempt has been made to deal with the practical training aspects, which require a much more detailed analysis than is possible in these volumes.

The volume will be found suitable as the main textbook or for providing essential background learning for a wide range of courses, which include:

City & Guilds 2320 Electrical and Electronic Craft Studies
City & Guilds 2360 Electrical Installation Work
City & Guilds 2210 Telecommunication and Electronics Mechanics
City & Guilds 2380 16th Edition of the IEE Wiring Regulations
City & Guilds 2710 Telecommunications Technicians
City & Guilds 2900 Essential Electricity

The author believes that the material included in this book, in addition to being suitable for craft students, is suitable for much of the early study in the BTEC courses leading to qualifications in electrical and electronic engineering.

Norwich                                                         J. F. Whitfield

# Acknowledgments

None of the work in this book can be claimed as original, and the author acknowledges with thanks the many teachers, writers and students who collectively have been responsible for his approach to the subject. In particular, he extends grateful thanks to

- Mr. Nat Hiller, B.Sc.(Eng), C.Eng., F.I.E.E., the Editor, for his continued support and assistance
- his wife, for her patient help and understanding
- Robin Mellors-Bourne, IEE Books Managing Editor, and Fiona MacDonald, Production Editor, for their help and support
- the following individuals and organisations, who have given valuable assistance with information and illustrations:

Avo Ltd.
BICC Ltd.
Crompton Parkinson Ltd.
Evershed and Vignoles Ltd.
Ferranti Ltd.
GE Lighting Ltd.
GEC Machines Ltd.
GEC Transformers Ltd.
Gould Instrument Division
Laurence, Scott & Electromotors Ltd.
Joseph Lucas Ltd.
Marconi Instruments Ltd.
Mullard Ltd.
Parsons Peebles Ltd.
Robin Electronics Ltd.
Standard Telephones & Cables Ltd.
Tektronix UK Ltd.
Thorn EMI Instruments Ltd.
H. Tinsley & Co. Ltd.
Weir Electrical Instrument Co. Ltd.

# Symbols and abbreviations

## 1 Terms

| Term | Symbol or abbreviation |
|---|---|
| alternating current | AC |
| area | $a$ or $A$ |
| armature conductors, number of | $Z$ |
| armature current | $I_A$ |
| armature resistance | $R_A$ |
| brush voltage drop | $V_B$ |
| capacitance | $C$ |
| charge (electric) | $Q$ |
| coefficient of utilisation | $U$ |
| copper loss (power) | $P_c$ |
| current: steady or RMS value | $I$ |
| instantaneous value | $i$ |
| maximum value | $I_m$ |
| mean or average value | $I_{av}$ |
| direct current | DC |
| eddy-current loss (power) | $P_E$ |
| electric charge | $Q$ |
| electromotive force | EMF |
| steady or RMS value | $E$ |
| instantaneous value | $e$ |
| energy | $W$ |
| field current | $I_F$ |
| frequency | $f$ |
| frequency, resonant | $f_r$ |
| hysteresis loss (power) | $P_H$ |
| illuminance | $E$ |
| impedance | $Z$ |
| induced EMF | $E$ |

| Term | Symbol or abbreviation |
|---|---|
| induced EMF, instantaneous | $e$ |
| inductance, mutual | $M$ |
| inductance, self | $L$ |
| iron loss (power) | $P_I$ |
| length | $l$ |
| line current | $I_L$ |
| line voltage | $V_L$ |
| luminous flux | $\Phi$ (phi) |
| luminous intensity | $I$ |
| magnetic flux | $\Phi$ (phi) |
| magnetic flux density | $B$ |
| magnetising force | $H$ |
| magnetomotive force | MMF |
| maintenance factor | $M$ |
| number of turns | $N$ |
| parallel paths, number of | $a$ |
| permeability of free space | $\mu_0$ (mu) |
| permeability, relative | $\mu_r$ (mu) |
| permittivity of free space | $\epsilon_0$ (epilson) |
| permittivity, relative | $\epsilon_r$ (epilson) |
| phase angle | $\phi$ (phi) |
| phase current | $I_p$ |
| phase voltage | $V_p$ |
| poles, pair of | $p$ |
| potential difference | PD |
| steady or RMS value | $V$ |
| instantaneous value | $v$ |
| maximum value | $V_m$ |
| mean or average value | $V_{av}$ |
| across a resistive component | $V_R$ |

| Term | Symbol or abbreviation | Term | Symbol or abbreviation |
|---|---|---|---|
| power, average value | $P$ | rotor speed (r/min) | $N$ |
| instantaneous value | $p$ | rotor speed (r/s) | $n$ |
| power factor | PF | secondary current | $I_2$ |
| primary current | $I_1$ | secondary EMF | $E_2$ |
| primary EMF | $E_1$ | secondary turns | $N_2$ |
| primary turns | $N_1$ | secondary voltage | $V_2$ |
| primary voltage | $V_1$ | slip | $s$ |
| quantity of electric | | speed (r/min) | $N$ |
| charge | $Q$ | speed (r/s) | $n$ |
| reactance, capacitive | $X_c$ | synchronous speed (r/min) | $N_s$ |
| reactance, inductive | $X_L$ | synchronous speed (r/s) | $n_s$ |
| resistance | $R$ | time | $t$ |
| revolutions per minute | r/min | time, periodic | $T$ |
| revolutions per second | r/s | torque | $T$ |
| root mean square | RMS | velocity | $v$ |
| rotational velocity | $\omega$ (omega) | wavelength | $\lambda$ (lambda) |

# SI units

If we are to measure physical, mechanical and electrical quantities, we must use a system of units for the purpose. All units, no matter how complex, are based on a number of basic units. The system of units adopted in much of the world is the SI (an abbreviation of Système International d'Unités):

| Physical quantity | Name of unit | Symbol |
| --- | --- | --- |
| length | metre | m |
| mass | kilogram | kg |
| time | second | s |
| electric current | ampere | A |
| temperature | kelvin | K |
| luminous intensity | candela | cd |

Imperial units (such as the inch or the pound) are not recognised in this volume.

## Multiple and submultiple units

There are many examples in practical electrical engineering where the basic units are of inconvenient size.

**Multiple units** are larger than the basic units.

The prefix **meg** or **mega** (symbol M) means **one million times**.

For instance,      1 megavolt (1 MV) = 1 000 000 volts

and      1 megohm (1 MΩ) = 1 000 000 ohms

The prefix **kil** or **kilo** (symbol k) means **one thousand times**.

For instance,      1 kilovolt (1 kV) = 1 000 volts.

**Submultiple units** are smaller and are decimal fractions of the basic units.

The prefix **milli** (symbol m) means **one-thousandth of**.

For instance,      $1 \text{ milliampere } (1 \text{ mA}) = \dfrac{1}{1000} \text{ ampere}$

The prefix **micro** (symbol $\mu$, the Greek letter 'mu') means **one millionth of**.

For instance, $\qquad 1 \text{ microhm } (1 \ \mu\Omega) = \dfrac{1}{1\ 000\ 000} \text{ohm}$

The prefix **nano** (symbol n) means **one thousand millionth of**.

For instance, $\qquad 1 \text{ nanosecond } (1 \text{ ns}) = \dfrac{1}{1\ 000\ 000\ 000} \text{second}$

The prefix **pico** (symbol p) means **one million millionth of**.

For instance, $\qquad 1 \text{ picovolt } (1 \text{ pV}) = \dfrac{1}{1\ 000\ 000\ 000\ 000} \text{volt}$

Two words of warning are necessary concerning the application of these extremely useful prefixes. Firstly, note the difference between the symbols M and m. The ratio is one thousand million! Secondly, always convert a value into its basic unit before using it in an equation. If this is done, any unknown value in the equation can be found in terms of its basic unit.

# 2   Units

| Unit | Symbol | Unit or |
| --- | --- | --- |
| ampere | A | electric current |
| ampere-turn | At | magnetomotive force |
| ampere-turn per metre | At/m | magnetising force |
| centimetre | cm | length |
| candela | cd | luminous intensity |
| coulomb | C | quantity of electric charge |
| cycle per second (hertz) | Hz | frequency |
| degree Celsius | °C | temperature |
| farad | F | capacitance |
| henry | H | self or mutual inductance |
| hertz (cycle per second) | Hz | frequency |
| joule | J | work or energy |
| kilogram | kg | mass |
| kilovolt-ampere | kVA | apparent power |
| kilowatt | kW | power |
| kilowatt-hour | kWh | energy |
| lumen | lm | luminous flux |
| lux | lx | illuminance |
| metre | m | length |
| metre-newton (joule) | m-N | work or energy |
| newton | N | force |
| newton metre | Nm | torque |

| Unit | Symbol | Unit of |
|---|---|---|
| ohm | Ω (omega) | resistance, reactance or impedance |
| radians per second | rad/s | rotational velocity |
| reactive kilovoltamperes | kVAr | reactive power |
| reactive voltamperes | VAr | reactive power |
| second | s | time |
| tesla (weber per square metre) | T | magnetic flux density |
| volt | V (or U) | PD and EMF |
| voltampere | VA | apparent power |
| watt | W | power |
| weber | Wb | magnetic flux |
| weber per square metre (tesla) | T | magnetic flux density |

# 3  Multiples and submultiples

| Prefix | Symbol | Meaning | |
|---|---|---|---|
| meg- or mega- | M | one million times | or $\times 10^6$ |
| kil- or kilo- | k | one thousand times | or $\times 10^3$ |
| milli- | m | one thousandth of | or $\times 10^{-3}$ |
| micro- | $\mu$ (mu) | one millionth of | or $\times 10^{-6}$ |
| nano- | n | one thousand millionth of | or $\times 10^{-9}$ |
| pico- | p | one million millionth of | or $\times 10^{-12}$ |

## Chapter 1

# Capacitance and capacitors

## 1.1  Introduction

Capacitors are included in much of the equipment that the modern electrical craftsman is called on to understand and to install. They range in size from the miniature unit included in a fluorescent-lamp starter switch, to reduce broadcast interference, to the large tank-enclosed capacitors for industrial power-factor correction. When used in a direct-current circuit, a capacitor will form a block to steady current, but when it is connected to an AC supply a current appears to pass through it. An understanding of the theory and construction of capacitors thus becomes necessary before dealing with alternating-current systems in subsequent chapters.

The reader should appreciate that the word 'condenser' is often incorrectly used in place of 'capacitor'.

## 1.2  Electric charge and electric fields

Consider a pair of flat, conducting plates, arranged parallel to each other as in Figure 1.1 and separated by an insulator, which may simply be air. Each plate, as an electrical conductor, will contain large numbers of mobile negatively charged electrons. If the plates are connected to a DC supply,

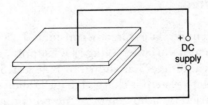

**Figure 1.1   A charged-plate system**

negative electrons will be attracted from the upper plate to the positive pole of the supply, and will be repelled from the negative pole onto the lower plate. The upper plate will become positively charged owing to its shortage of electrons, whereas the surplus electrons on the lower plate will give it a negative charge. The difference in the polarity of charge between the plates means that a potential difference exists between them, the flow of electrons dying away and ceasing when the PD between the plates is the same as the supply voltage.

When in this condition, the plates are said to be 'charged', and an electric field exists between them. This electric field is, like the magnetic field, quite invisible, but can be detected by the effect it has on a charged body placed in it, the body being attracted to the positive plate if negatively charged, and vice-versa. The field is represented by imaginary 'lines of electric flux' as shown in Figure 1.2.

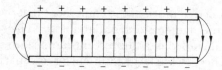

**Figure 1.2   Electric field due to charged plates of Figure 1.1**

If the plates are disconnected from the supply, and are connected together through a resistor, the surplus of electrons on the negative plate will flow through the resistor to the positive plate, satisfying its shortage of electrons. This electron flow constitutes a current, which will die away and cease as the charges on the plates reduce. The current flowing in the resistor will liberate heat energy showing that energy was stored in the electric field. The charge stored in an electric field can be measured in coulombs, that is the current which flows to charge the system multiplied by the time for which it flows. In practice, the charge that can be stored by a parallel-plate system depends on the plate area, the distance between the plates, the PD between them and the nature of the insulating material that separates them.

## 1.3   Capacitance

Assume that the system of plates already considered is connected in circuit with a two-way switch, a variable-voltage DC supply, and an instrument called a 'ballistic galvanometer', as shown in Figure 1.3. When the switch is in position 1, the plate system is charged to a PD shown by the voltmeter, and which can be varied by adjustment of the potential divider. With the switch in position 2, the charge on the plates will produce a current through the ballistic galvanometer, which measures the quantity of charge passing through it, thus measuring the charge stored on the plates. This process

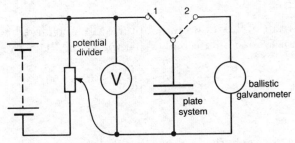

**Figure 1.3    Circuit to indicate nature of capacitance**

is repeated a number of times, and each time the ratio (quantity of charge stored)/(PD between the plates) is found to be constant. This constant is called the '**capacitance**' (symbol $C$) of the plate system, which is called a '**capacitor**', and is denoted by the symbol used for the plate system in Figure 1.3. The capacitance of a capacitor is measured in '**farads**' (symbol F).

Thus
$$C = \frac{Q}{V}$$

where      $C$ = capacitance (F)

$Q$ = stored charge in the capacitor (C)

$V$ = PD between the capacitor plates (V)

Previously, it was shown that the unit of electric charge is the coulomb, which is also the ampere second, so

$$Q = It$$

where      $Q$ = stored charge in a capacitor (C)

$I$ = charging current to capacitor (A)

$t$ = time for which charging current passes (s)

Note that a capacitor has a capacitance of one farad if it can store one coulomb when the PD between its plates is one volt.

## Example 1.1

A steady current of 15 A flows into a previously uncharged capacitor for 10 seconds, when the PD between the plates is 10 000 V. What is the capacitance of the capacitor?

$$C = \frac{Q}{V} = \frac{I \times t}{V} = \frac{15 \times 10}{10\ 000}\ \text{farads} = 0 \cdot 015\ \text{F}$$

In practice the farad is usually too large, and the microfarad (μF), which is one millionth of a farad, is widely used.

## Example 1.2

A steady current of 2 A flows into a previously uncharged 8 μF parallel-plate capacitor for one millisecond. What will be the PD between its plates?

$$V = \frac{Q}{C} = \frac{I \times t}{C} = \frac{2 \times 0 \cdot 001}{8 \times 10^{-6}} \text{ volts} = 250 \text{ V}$$

## Example 1.3

A 2 μF capacitor is charged so that the PD between its plates is 1000 V. For how long could this capacitor provide an average discharge current of 1 mA?

$$Q = CV = It$$

$$t = \frac{CV}{I} = \frac{2 \times 10^{-6} \times 1000}{0 \cdot 001} \text{ seconds} = 2 \text{ s}$$

It should be understood that the charging or discharging currents of a capacitor are not normally steady values, both starting at a high level and falling at a decreasing rate to zero (see Section 1.10).

## 1.4   Dielectric breakdown

The insulating material which occupies the space between the plates of a capacitor is called the 'dielectric'. Insulating materials have very few free electrons available to form a current under normal conditions but, if they are subjected to an intense electric field, electrons may be torn from their atoms and give rise to a current through the insulator. If this happens, the insulation properties may be destroyed, and 'dielectric breakdown' is said to have taken place.

Although this breakdown can occur in capacitors, it is often more serious if it occurs in the insulation of current-carrying conductors. The strength of an electric field can be expressed in terms of the PD applied across a certain thickness of insulation and this value is known as the 'average potential gradient', usually measured in kilovolts per millimetre. The potential gradient at which the insulator fails is called its 'dielectric strength', which is a measure of the excellence of the insulation.

## Example 1.4

The impregnated-paper dielectric of a parallel-plate power capacitor is $0 \cdot 5$ mm thick, and has a dielectric strength of 5 kV/mm. If the capacitor concerned has an applied PD of $1 \cdot 8$ kV, what is the average potential gradient in the paper? At what applied voltage could the capacitor be expected to break down?

$$\text{average potential gradient} = \frac{\text{applied voltage (kV)}}{\text{dielectric thickness (mm)}}$$

$$= \frac{1 \cdot 8}{0 \cdot 5} \text{ kilovolts per millimetre} = 3 \cdot 6 \text{ kV/mm}$$

likely breakdown voltage (kV) = dielectric strength (kV/mm)
$$\times \text{ dielectric thickness (mm)}$$

$$= 5 \times 0 \cdot 5 \text{ kilovolts} = 2 \cdot 5 \text{ kV}$$

Table 1.1 lists some insulators, giving typical applications and dielectric strengths. The figure for the latter may vary due to factors such as temperature and purity, so values given are approximate only.

**Table 1.1: Dielectric strength and application of common insulators**

| Material | Dielectric strength | Application |
|---|---|---|
| | kV/mm | |
| Air | 3 | General |
| Bakelite | 20–25 | Plugs, machinery etc. |
| Bitumen | 14 | Cable-sealing boxes |
| Glass | 50–120 | Overhead-line insulators |
| Mica | 40–150 | Commutators, hot elements, capacitors etc. |
| Micanite | 30 | Machines |
| Impregnated paper | 4–10 | Power cables, capacitors |
| Paraffin wax | 8 | Capacitors etc. |
| Porcelain | 9–10 | Fuse carriers, overhead-line insulators etc. |

## 1.5 Capacitors in parallel and in series

Capacitors can be connected in parallel or in series, and in either case it is important to know the equivalent capacitance of the group.

Consider three capacitors of capacitances $C_1$, $C_2$ and $C_3$, respectively, connected in parallel as shown in Figure 1.4. When the group is connected to a supply $V$, the capacitors will each store a charge, and we will refer to these charges as $Q_1$, $Q_2$ and $Q_3$, respectively. The total stored charge $Q_T$ will be the sum of the individual charges. Thus

$$Q_T = Q_1 + Q_2 + Q_3$$

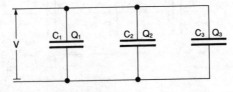

**Figure 1.4   Capacitors in parallel**

But $Q = CV$. The supply voltage $V$ applies to all capacitors, so if $C_T$ is the equivalent capacitance of the group,

$$C_T V = C_1 V + C_2 V + C_3 V$$

Dividing by $V$,

$$C_T = C_1 + C_2 + C_3$$

Thus the equivalent capacitance of a group of parallel-connected capacitors is found by adding together the capacitances of the individual capacitors. By connecting in capacitors in this way, a unit of high capacitance can be made up from a number of smaller capacitors.

## Example 1.5

Capacitors of capacitance 2 μF, 5 μF, 8 μF and 10 μF, respectively, are connected in parallel to a 200 V DC supply. Calculate:

(*a*) the equivalent capacitance of the group
(*b*) the total charge, and
(*c*) the charge on each capacitor:

(*a*)   $C = 2 + 5 + 8 + 10$ microfarads $= 25$ μF
(*b*)   $Q_T = C_T V = 25 \times 10^{-6} \times 200$ coulombs $= 5000 \times 10^{-6}$ C $= 5$ mC
(*c*)   $Q_2 = C_2 V = 2 \times 10^{-6} \times 200$ coulombs $= 0 \cdot 4$ mC
   $Q_5 = C_5 V = 5 \times 10^{-6} \times 200$ coulombs $= 1$ mC
   $Q_8 = C_8 V = 8 \times 10^{-6} \times 200$ coulombs $= 1 \cdot 6$ mC
   $Q_{10} = C_{10} V = 10 \times 10^{-6} \times 200$ coulombs $= 2$ mC.

Check:

$$Q_T = Q_2 + Q_5 + Q_8 + Q_{10}$$

$$= 0 \cdot 4 + 1 + 1 \cdot 6 + 2 \text{ millicoulombs} = 5 \text{ mC}$$

Now consider the three capacitors connected in series as in Figure 1.5. The supply voltage is $V$, and the PDs across the individual capacitors are $V_1$, $V_2$ and $V_3$, respectively. Since the capacitors are all in series, the same charging current must apply to each for the same time, so each capacitor will have the same charge of, say, $Q$ coulombs.

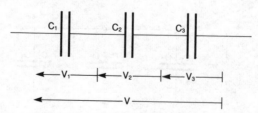

**Figure 1.5   Capacitors in series**

Now
$$V = V_1 + V_2 + V_3$$

But $V = Q/C$, so, if $C_T$ is the equivalent value of the series-connected capacitors,

$$\frac{Q}{C_T} = \frac{Q}{C_1} + \frac{Q}{C_2} + \frac{Q}{C_3}$$

Dividing by $Q$,

$$\frac{1}{C_T} = \frac{1}{C_1} + \frac{1}{C_2} + \frac{1}{C_3}$$

## Example 1.6

Calculate the equivalent capacitance of 5 µF, 10 µF and 30 µF capacitors connected in series.

$$\frac{1}{C_T} = \frac{1}{5} + \frac{1}{10} + \frac{1}{30} = \frac{6 + 3 + 1}{30} = \frac{10}{30}$$

Therefore
$$C_T = \frac{30}{10} \text{ microfarads} = 3 \text{ µF}$$

This example makes it clear that the equivalent capacitance of a string of series-connected capacitors is less than that of the smallest individual unit. The advantage of series connection is that each capacitor has a PD of less than the total applied voltage, so that low-voltage capacitors may be connected in series to form a higher-voltage unit. The supply voltage, however, is only shared equally between the capacitors if they all have equal value.

## Example 1.7

The series-connected capacitors of Example 1.6 are connected to a 240 V DC supply. Calculate the charge on each capacitor, and the potential difference across each.

From Example 1.6, $C_T = 3$ µF

Total stored charge, $Q = C_T V = 3 \times 10^{-6} \times 240$ coulombs = 720 µC

Since the capacitors are connected in series, the charge on each is the same as the total charge, i.e. 720 µC.

The PD across the 5 µF capacitor,

$$V_5 = \frac{Q}{C_5} = \frac{720 \times 10^{-6}}{5 \times 10^{-6}} \text{ volts} = 144 \text{ V}$$

and similarly for the other two capacitors:

$$V_{10} = \frac{Q}{C_{10}} = \frac{720 \times 10^{-6}}{10 \times 10^{-6}} \text{ volts} = 72 \text{ V}$$

$$V_{30} = \frac{Q}{C_{30}} = \frac{720 \times 10^{-6}}{30 \times 10^{-6}} \text{ volts} = 24 \text{ V}$$

Check:

$$V_5 + V_{10} + V_{30} = V$$

$$144 \text{ V} + 72 \text{ V} + 24 \text{ V} = 240 \text{ V}$$

Note that the lowest-value capacitor has the highest PD across it. This means that, if the capacitors are similarly constructed, they must all be rated at the highest voltage. For this reason, capacitors are not often connected in series unless all are of the same capacitance.

## 1.6  Capacitance of a parallel-plate capacitor

The construction of capacitors will be considered in Section 1.8, where it will be shown that there are many practical alternatives to the pair of parallel plates shown in Figure 1.1; calculation of the capacitance of a system becomes difficult, however, if the plates are not flat and parallel.

The capacitance of a parallel-plate capacitor depends on a number of factors, which are:

### (i) Plate area

Figure 1.6 shows that increasing the plate area $A$ of a capacitor is similar to connection of extra capacitors in parallel. This increases capacitance, so $C \propto A$.

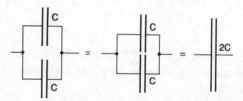

**Figure 1.6   Capacitance increases with increasing plate area**

### (ii) Plate spacing

Figure 1.7 shows that an increase in the spacing of the plates $d$ is similar to connection of additional capacitors in series. This reduces capacitance, so $C \propto 1/d$.

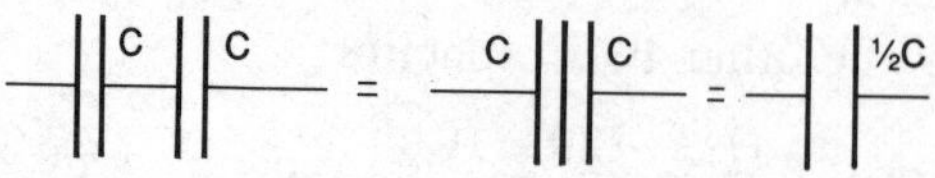

**Figure 1.7   Capacitance reduces with increasing plate spacing**

## *(iii) Relative permittivity of the dielectric*

Different dielectric materials have differing abilities to store electrical energy in the form of an electric field. This ability is reflected in the relative permittivity of the material, which is given the symbol $\epsilon_r$ ($\epsilon$ is Greek 'epsilon'). Table 1.2 gives the relative permittivities of some dielectric materials.

**Table 1.2: Relative permittivity ($\epsilon_r$) for some dielectrics**

| Dielectric | Relative permittivity |
|---|---|
| Air | 1 |
| Bakelite | 4·5–5·5 |
| Glass | 5–10 |
| Mica | 3–7 |
| Impregnated paper | 2 |
| Polystyrene | 2·5 |
| Porcelain | 6–7 |

Capacitance depends directly on relative permittivity.

Thus, $$C \propto \epsilon_r$$

## *(iv) Number of plates*

To reduce the size of a capacitor for a given capacitance, the interleaved construction shown in Figure 1.8 is sometimes used. The capacitor shown has 11 plates, and ten dielectrics. It consists, effectively, of ten capacitors in parallel. Thus, if the arrangement has $N$ plates,

$$C \propto (N-1)$$

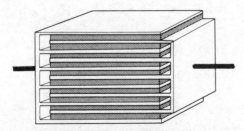

**Figure 1.8   Interleaved-plate capacitor**

Taking all the factors into account, we have

$$C \propto \frac{A\epsilon_r(N-1)}{d}$$

It can be shown that a capacitor having two plates ($N-1=1$) each 1 m square and separated by an air dielectric 1 m thick would have a capacitance of $8 \cdot 85 \times 10^{-12}$ F. This is called the permittivity of free space ($\epsilon_o$), and has a constant value of $8 \cdot 85 \times 10^{-12}$.

$$C = \frac{\epsilon_o \epsilon_r A(N-1)}{d}$$

where        $C$ = capacitance (F)

$\epsilon_0$ = a constant, $8 \cdot 85 \times 10^{-12}$

$\epsilon_r$ = the relative permittivity of the dielectric

$A$ = the area of each plate (m$^2$)

$N$ = the number of plates

$d$ = the thickness of each dielectric (m)

## Example 1.8

A capacitor with interleaved plates (Figure 1.8) has 11 plates, each 120 mm square, separated by mica sheets 1 mm thick. Calculate the capacitance of the arrangement, assuming the relative permittivity of the mica is 4.

$$C = \frac{\epsilon_o \epsilon_r A(N-1)}{d}$$

$\epsilon_o = 8 \cdot 85 \times 10^{-12}$

$\epsilon_r = 4$

$A = 120 \text{ mm} \times 120 \text{ mm} = 14\,400 \times 10^{-6} \text{ m}^2$

$\qquad\qquad\qquad\qquad\qquad = 14 \cdot 4 \times 10^{-3} \text{ m}^2$

$N = 11$, so $N - 1 = 10$

$d = 1 \text{ mm} = 10^{-3} \text{ m}$

$$C = \frac{8 \cdot 85 \times 10^{-12} \times 4 \times 14 \cdot 4 \times 10^{-3} \times 10}{10^{-3}} \text{F}$$

$\qquad = 5100 \times 10^{-12} \text{ F}$

$\qquad = 0 \cdot 0051 \text{ } \mu\text{F}$

## Example 1.9

What will be the area (to the nearest square metre) of each of the two plates of a parallel plate capacitor which has a dielectric of polystyrene 2 mm thick and a capacitance of 1 μF? From Table 1.2, the relative permittivity of polystyrene is 2·5.

$$C = \frac{\epsilon_o \epsilon_r A(N-1)}{d}, \text{ so } A = \frac{Cd}{\epsilon_o \epsilon_r (N-1)}$$

$$A = \frac{1 \times 10^{-6} \times 2 \times 10^{-3}}{8 \cdot 85 \times 10^{-12} \times 2 \cdot 5 \times 1} \text{ m}^2$$

$$= 90 \cdot 4 \text{ m}^2$$

Each plate will thus have to be about the size of a tennis court, and this is unacceptably large.

## Example 1.10

A paper capacitor has two parallel plates, each of effective area $0 \cdot 2$ m$^2$. If the capacitance is 7080 pF calculate the effective thickness of the paper if its relative permittivity is 2.

$$C = \frac{\epsilon_o \epsilon_r A(N-1)}{d}, \text{ so } d = \frac{\epsilon_o \epsilon_r A(N-1)}{C}$$

$$d = \frac{8 \cdot 85 \times 10^{-12} \times 2 \times 0 \cdot 2 \times 1}{7080 \times 10^{-12}}$$

$$= \frac{3 \cdot 54}{7080} \text{ m}$$

$$= 0 \cdot 5 \times 10^{-3} \text{ m}$$

$$= 0 \cdot 5 \text{ mm}$$

## 1.7 Energy stored in a capacitor

If a charged capacitor is short-circuited, the sharp 'crack' made by the arc produced by discharge current is an indication that energy was previously stored in the capacitor. This energy was stored in the electric field set up by the capacitor.

To produce an expression for the stored energy, assume that a capacitor is charged by a constant current of $I$ amperes so that the potential difference across its plates builds up at a uniform rate to $V$ volts after $t$ seconds (see Figure 1.9). Thus, a current of $I$ amperes has passed to the capacitor for $t$ seconds, so the charge on the capacitor is

$$Q = It$$

During this time, the average PD across the capacitor has been $V/2$ volts. The power applied to charge the capacitor (power = voltage × current),

$$P = V/2 \times I \text{ watts}$$

and the energy provided (energy = power × time)

$$W = V/2 \times I \times t \text{ joules}$$

But          $I \times t = Q$ and $Q = C \times V$

Therefore     $W = V/2 \times C \times V$ joules

$$W = \frac{CV^2}{2} \text{ joules}$$

where          $W$ = energy stored in the electric field (J)

$C$ = capacitance of field system (F)

$V$ = potential difference across field system (V)

## Example 1.11

Calculate the energy stored in a 10 μF capacitor when the PD across its plates is (*a*) 10 V, (*b*) 100 V, (*c*) 1000 V.

$$W = \frac{CV^2}{2}$$

(*a*)          $W = \dfrac{10 \times 10^{-6} \times 10^2}{2}$ joules = 0·5 mJ

(*b*)          $W = \dfrac{10 \times 10^{-6} \times 100^2}{2}$ joules = 0·05 J

(*c*)          $W = \dfrac{10 \times 10^{-6} \times 1000^2}{2}$ joules = 5 J

The energy stored by a capacitor is quite small. For example, the energy stored by a 10 μF capacitor with a PD of 1000 V would be used by a 100 W lamp in a twentieth of a second.

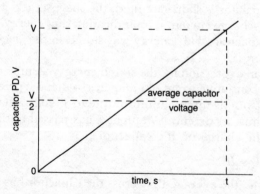

**Figure 1.9    Increase in potential difference across a capacitor charged with a constant current**

## Example 1.12

What is the capacitance of a capacitor which stores energy of 2 J for a potential difference of 240 V?

$$W = \frac{CV^2}{2} \text{ so } C = \frac{2W}{V^2}$$

$$C = \frac{2 \times 2}{240 \times 240} \text{ farads} = 6 \cdot 94 \times 10^{-5} \text{ farads or } 69 \cdot 4 \text{ } \mu F$$

## Example 1.13

Calculate the potential difference between the terminals of a 10 μF capacitor if it must store an energy of 0·2 J.

$$W = \frac{CV^2}{2} \text{ so } V^2 = \frac{2W}{C}$$

$$V^2 = \frac{2 \times 0 \cdot 2}{10 \times 10^{-6}} = 4 \times 10^4$$

Therefore $\qquad V = \sqrt{(4 \times 10^4)} \text{ volts} = 2 \times 10^2 \text{ volts} = 200 \text{ V}$

Although the energy stored in a capacitor is small, it is capable of providing a shock to a person touching the terminals, or of causing a fire. A charged capacitor will tend to lose its charge after the charging source is removed by leakage through the dielectric but, to ensure that this process is reasonably fast, a discharge resistor is connected across the terminals of power capacitors. Such a resistor must have a high value, or it will offer a low-resistance path for current from the charging source, but must be low enough to discharge the capacitor within a reasonable time. A typical value would be 10 MΩ.

## 1.8 Construction of capacitors

Capacitors are manufactured with many differing forms of construction, but always the principle is that of two or more plates separated by an insulating dielectric. Often the plates are not flat, as has been suggested so far in this chapter. The usual need when designing a capacitor is to produce a unit of the smallest possible physical size for a given capacitance and voltage rating, since in many cases the capacitor is a part of other apparatus whose size is limited. For example, a typical transistor radio receiver may contain 50 or more capacitors. If these were the flat-plate type considered in Example 1.9 (with a plate area of over 90 m² for a capacitance of 1 μF), the radio would hardly be portable.

In Section 1.6 we have shown that

$$C = \frac{\epsilon_o \epsilon_r A (N-1)}{d}$$

$\epsilon_o$ is a constant, and cannot be changed, but to build a capacitor with higher capacitance we could do one or more of the following:

(1) use a dielectric with a higher relative permittivity;
(2) increase the area of each plate;
(3) use more plates connected in parallel; and
(4) reduce the thickness of the dielectric.

The dielectric used will often be dictated more by its physical properties rather than by its relative permittivity. Glass and porcelain have higher relative permittivities than impregnated paper and polystyrene, but are inflexible and difficult to use, particularly in small sizes.

An increase in plate area makes an unwieldy unit which will not fit into apparatus, although the size of a flat-plate capacitor can be reduced by interleaving the plates as shown in Figure 1.8. Another method very often used to give a large plate area in a capacitor of small dimensions is to use flexible plates which are wound into a tight roll. Long strips of flexible metal foil are separated by a strip of flexible dielectric, such as waxed paper or polystyrene. A further dielectric strip is placed on top of the arrangement, which is then rolled up as shown in Figure 1.10. The resulting capacitor is often sealed in a metal can to exclude moisture, or is mounted on a rack with other units and submerged in an oil-filled tank to give a high-capacitance unit. Such units are used for industrial power-factor correction (see Section 5.12). The purposes of the oil are to ensure that moisture is excluded from the arrangement and to remove heat by convection. Unfortunately, the oil is flammable, and a number of fires in industrial and commercial premises have been traced back to the escape of oil from capacitors. Some years ago the problem was thought to have been solved by the introduction of polychlorinated benzine (PCB) as a filler to replace the oil, because it is virtually nonflammable and nondegradable. However, PCB has been found to be highly toxic, and as it is nondegrading its disposal

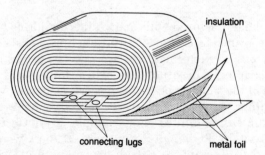

**Figure 1.10   Construction of a rolled-foil capacitor, opened to show construction**

is very difficult. Its use to fill capacitors ceased some time ago, but very large numbers of PCB-filled units are still in operation. It is MOST IMPORTANT that old capacitor units which contain PCB are treated with the greatest care, and that the specialist advice of the Health and Safety Executive is sought concerning disposal. UNDER NO CIRCUMSTANCES should such units be thrown away or buried. Care must also be taken not to come into contact with the material, and medical advice should be sought if such an event occurs.

The metal-foil-paper-dielectric method is also used for capacitors intended for installation in electronic equipment (see Figure 1.11).

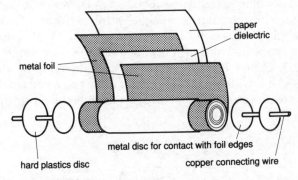

**Figure 1.11    Tubular rolled-foil capacitor, opened to show construction**

There are many methods of capacitor construction, particularly for small capacitors used as components of electronic apparatus. Figure 1.12 shows three examples of such capacitors. It has been shown that a capacitor having

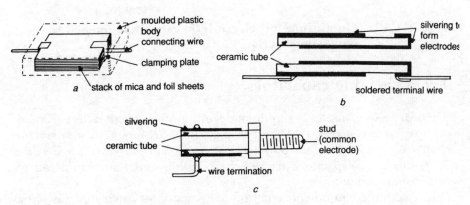

**Figure 1.12    Capacitors for electronic work**
    (*a*) Stacked-foil mica-dielectric type
    (*b*) Tubular ceramic-dielectric type
    (*c*) Metallised ceramic-dielectric standoff type

a very thin dielectric will have a high capacitance for its size, and this applies to electrolytic capacitors. Two plates of aluminium are separated by a chemical conductor in paste or liquid form, called an **electrolyte**, such as aluminium borate. A current passed between the plates through the electrolyte produces a very thin layer of aluminium oxide on one of the plates, which, being a good insulator, becomes the dielectric. The plate having the oxide layer must be used as the positive connection, or leakage current will destroy the oxide, and thus destroy the capacitor. Electrolytic capacitors have their polarity clearly marked and this polarity must be followed. Incorrect polarity will result in an increasing leakage current which will cause the electrolyte to boil, building up pressure so that the capacitor bursts. It follows that electrolytic capacitors must never be used where the applied voltage is alternating. The electrolytic type of capacitor has comparatively low voltage rating, cannot be used on AC supplies unless two units are connected back-to-back, deteriorates fairly rapidly if unused, and has much higher leakage currents than other types. Despite these disadvantages, it is widely used because of its low cost and small size. Most electrolytic capacitors are manufactured on the rolled-foil principle as shown in Figure 1.13.

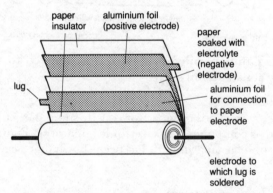

**Figure 1.13   Aluminium-foil electrolytic capacitor, opened to show construction**

## 1.9   Variable capacitors

There are few examples in power engineering where continuous variation of capacitance is necessary. Where changes in power requirements demand changes in capacitance for power-factor correction purposes (see Section 5.12) blocks of capacitors are switched, giving step changes instead of continuous variation.

In electronic equipment, continuous variation of capacitance is often necessary, as in the tuning unit of a radio set. Such capacitors are multiplate air-spaced types, one set of plates (the rotor vanes) being free to move within the second set of plates (the stator vanes). Since the effective plate area is that of the plate overlap, moving the rotor vanes out of the stator vanes will reduce capacitance, and vice-versa (see Figure 1.14). Where increased variable capacitance is required, two or more sets of plates may be mounted on a common shaft and connected in parallel.

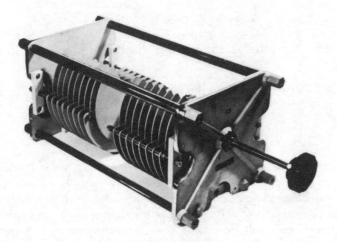

**Figure 1.14   Air-spaced variable capacitor**

In some circuits, variation of capacitance is needed only to adjust values during initial setting-up procedures and, in these cases, 'trimming' capacitors are used. These are usually small units, and in some cases may be miniature versions of the variable-plate type described above. Other trimmer-type capacitors rely on variation of plate spacing, sometimes involving compression of polythene dielectrics.

## 1.10   Charge and discharge curves

In Section 1.2 we considered a pair of parallel plates (Figure 1.1) connected to a DC supply. We pointed out that electrons move to one plate, and away from the other, charging the capacitor so that an electric field exists between the plates. The movement of electrons is an electric current, and will continue until the potential difference between the plates is equal to the supply voltage, when it will cease, leaving the capacitor charged.

The current flowing will not have a steady value, but will be changing continually. Consider the arrangement of Figure 1.15. A capacitor $C$ is connected in series with a resistor $R$ to a DC supply of $V$ volts through a switch. The potential difference across the capacitor at any instant will be considered as $v$ volts, and the current at any instant as $i$ amperes. If the capacitor is initially discharged, as soon as the switch is placed in position

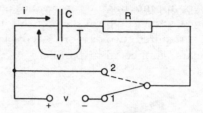

**Figure 1.15   Circuit for charge and discharge of a capacitor**

1 the supply voltage applies to the circuit. The current at this instant will be given by

$$i_1 = \frac{V}{R}$$

This current begins to charge the capacitor, so that the voltage $v$ builds up across its plates. Since the positive pole of the supply attracts electrons from the left-hand plate, this plate becomes positively charged, and the polarity of the capacitor voltage $v$ opposes that of the supply voltage $V$. The effective circuit voltage is thus $V - v$, and the currrent at any instant will be given by

$$i = \frac{V - v}{R}$$

As $v$ increases, $V - v$ reduces, and drives a reducing current through the circuit. The reducing current charges the capacitor at a reducing rate, so that $v$ builds up less quickly. The result is that the current, which is initially large, dies away at an ever-decreasing rate as shown in Figure 1.16, which also shows the corresponding curves for capacitor voltage $v$ and the charge on the capacitor $q$. When $v = V$, the capacitor is fully charged, and the current has died away completely.

The curves shown often occur in science and are called natural curves of decay (current) or of growth (voltage and charge). Because the values

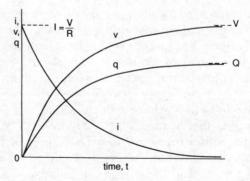

**Figure 1.16   Capacitor-charging curves**

of current, voltage and charge only change for a short time before becoming steady, they are called '**transients**'.

The charging action described is reversed when the switch (Figure 1.15) is placed in position 2. Initially $v = V$, and $i_1 = V/R$, but this discharging current reduces the capacitor voltage $v$, and generally $i = v/R$. The current thus follows a similar decay curve to that of the capacitor voltage in the charging condition, with current dying away as $v$ falls to zero as shown in Figure 1.17.

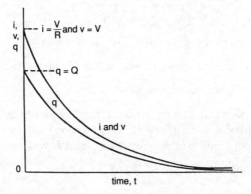

**Figure 1.17   Capacitor-discharging curves**

The time taken for the charging and discharging actions to take place depends on the time constant of the circuit, which is equal to $C \times R$. Since the current gets smaller at a decreasing rate, the exact instant at which the transient is complete cannot be exactly defined. However, this time is quite short for most practical cases, being virtually completed in five time constants.

## 1.11   Applications of capacitors

Capacitors have many important applications, almost all of them depending on the ability of the capacitor to store energy which can later be returned to the circuit from which it was taken. Unfortunately, we need to do a little more preparatory work before the action of capacitors in many circuits can be fully understood, but such applications are considered later in this book. Examples are:

power-factor correction — Section 5.12

smoothing circuits        — Section 6.5

resonant circuits         — Sections 4.5 and 4.7.

# 1.12   Construction of capacitor charge and discharge curves

The mathematical formulas from which the curves can be drawn are too complex at this stage, but it is still useful to be able to construct curves, even if they are approximate.

All constructions are based on the **time constant**, which can be defined as the time it would take for the voltage or current to reach its final value if it continued to change at its initial rate. For capacitance circuits, the time constant (here given the symbol $\tau$) is:

$$\tau = C \times R$$

where      $\tau$ is the circuit time constant in seconds

$C$ is the circuit capacitance in farads (note the unit)

$R$ is the circuit resistance in ohms

It will be appreciated that the time taken (the transient time) for the values of voltage and current to reach their final (steady) values is dependent on the circuit time constant. Higher values of $C$ and $R$ will extend the transient time.

## Example 1.14

A 22 μF capacitor and a 68 kΩ resistor are connected in series. Calculate the time constant of the combination.

$$\tau = C \times R = 22 \times 10^{-6} \times 68 \times 10^{3} = 1496 \times 10^{-3} = 1 \cdot 496 \text{ s}$$

To construct, for example, a decay curve, first calculate the circuit time constant $\tau$. Next, draw a graph with a vertical axis equal to the circuit voltage or current and a horizontal axis of time up to five times the circuit time constant ($5\tau$). From the initial value of voltage or current, draw a straight line down to the final value for a time of $\tau$ seconds. In fact, the voltage or current will not follow a straight-line relationship to its final value, but will begin to curve as soon as it starts to change. For this method, we must assume that it follows the straight-line relationship for a definite time which we are free to choose. The shorter the chosen time, the longer the graph will take to construct, but the more accurate it will be.

Mark the chosen time on the first straight line, then add a second straight line from this point to reach the final value after $\tau$ seconds. Next, choose a time for which you will assume that the rising voltage remains linear after the first chosen time, then draw a third straight line from this point to reach the final value after $\tau$ seconds. Continue the process of drawing lines until the curve is complete. Figure 1.18 shows the curve for a current decay typical of a $CR$ circuit. Too few construction lines are drawn to give a good curve, but the figure shows how the curve is constructed.

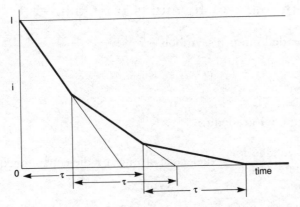

**Figure 1.18   Partial construction of current-decay curve**

Figure 1.19 shows a fuller construction for a voltage-growth curve, such as the PD across a capacitor as it is charged. The process is the same as that illustrated in Figure 1.18 but assumes a shorter time for straight-line change and hence more construction lines. As the curve becomes more horizontal, you will probably choose longer time intervals between your construction lines.

It must be appreciated that the resulting curve is not absolutely accurate, but it becomes more so as the time intervals for which voltage is assumed to change at a constant rate are reduced.

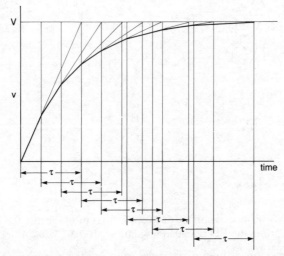

**Figure 1.19   Full construction of voltage-growth curve**

## 1.13   Summary of formulas for Chapter 1

(See text for definition of symbols)

$$Q = It \quad I = \frac{Q}{t} \quad t = \frac{Q}{I}$$

Average potential gradient $= \dfrac{V}{d}$

Breakdown voltage for a parallel-plate system $=$ dielectric strength $\times$ dielectric thickness
   (V)                              (V/m)                         (m)

Capacitor charge and voltage:

$$Q = CV \quad C = \frac{Q}{V} \quad V = \frac{Q}{C}$$

For capacitors in parallel:

$$C_T = C_1 + C_2 + C_3 + \ldots$$

For capacitors in series:

$$\frac{1}{C} = \frac{1}{C_1} + \frac{1}{C_2} + \frac{1}{C_3} + \ldots$$

Parallel-plate capacitor:

$$C = \frac{\epsilon_o \epsilon_r A(N-1)}{d}$$

$$\epsilon_o = \frac{Cd}{\epsilon_r A(N-1)} \qquad \epsilon_r = \frac{Cd}{\epsilon_o A(N-1)} \qquad A = \frac{Cd}{\epsilon_o \epsilon_r A(N-1)}$$

$$d = \frac{\epsilon_o \epsilon_r A(N-1)}{C} \qquad N = \frac{Cd}{\epsilon_o \epsilon_r A} + 1$$

Energy stored in a capacitor:

$$W = \frac{CV^2}{2} \qquad C = \frac{2W}{V^2} \qquad V = \sqrt{\frac{2W}{C}}$$

## 1.14   Exercises

1  A previously uncharged capacitor takes a current of 25 mA for 20 ms, at the end of which time the PD between its plates is 100 V. What is its capacitance?

2  A 20 μF capacitor is fully discharged, and is then charged by a current of 5 mA for 1 s. What will be the PD between its plates?

3 For how long must a charging current of 1 A be fed to a 2 μF capacitor to raise the PD between its plates by 1000 V?

4 Define in your own words the following terms:
(i) dielectric, (ii) capacitance, and (iii) capacitor.

5 A 200 μF capacitor is fully charged from a 400 V DC supply. What is the value of the constant current which will completely discharge the capacitor in 10 ms?

6 Two flat plates are arranged so as to be 5 mm apart. If the PD between them is 2 kV, what is the potential gradient between them?

7 The average potential gradient between the parallel plates of a capacitor is 3·5 kV/mm. If the plates are 0·8 mm apart, what PD exists between the plates?

8 If an electric field set up between two parallel plates with a PD of 250 V has an average potential gradient of 100 V/mm, how far apart are the plates?

9 (a) Describe with the aid of a sketch the construction of a small capacitor. Give one instance of the use of such a capacitor and state its function in the circuit.
(b) What charge is carried by a 10 μF capacitor when connected to a 240 V DC supply?

10 Name one type of capacitor and describe its construction.

11 (a) Describe with the aid of a sketch the construction of a 'paper' capacitor.
(b) Give one example of the use of this capacitor in a practical circuit, and give reasons for its use.
(c) Calculate the resulting capacitance if two capacitors of values 6 μF and 10 μF are connected (i) in series, (ii) in parallel.

12 (a) Name THREE types of capacitor.
(b) Describe ONE of them, with the aid of a sketch.
(c) Give one instance of the use of such a capacitor, and state its function.

13 What values of capacitance would be obtained if two capacitors of 0·1 μF and 0·2 μF were connected (a) in series, and (b) in parallel?

14 What values of capacitance would be obtained if two capacitors of 0·2 μF and 0·3 μF, respectively, were connected (a) in series, and (b) in parallel?

15 What is the equivalent capacitance of a bank of four capacitors of 5 μF, 7 μF, 10 μF and 16 μF, respectively, if they are connected (a) all in series, and (b) all in parallel?

16 Three capacitors of values 6 μF, 8 μF and 10 μF, respectively, are connected (i) in series, and (ii) in parallel. Calculate the resultant capacitance in each case.

17 Calculate the capacitance of a parallel-plate capacitor having two plates, each 20 mm by 10 mm, and separated by a dielectric 0·855 mm thick having a relative permittivity of 3.

18 How many plates has a parallel-plate capacitor having a capacitance of 0·02 μF, if each plate is 50 mm square and each dielectric is 0·442 mm thick, with a relative permittivity of 4?

19 A parallel-plate capacitor is made up of 13 plates, each 100 mm × 150 mm, interleaved with mica for which $\epsilon_r = 6$. The capacitance of the arrangement is 5780 pF. Calculate the thickness of each mica sheet.

20 Calculate the energy stored in a 12 μF capacitor when the PD across its plates is 240 V.

21 A capacitor stores 2 J of energy for a potential difference of 400 V. What is its capacitance?

22 What must be the PD across a 100 pF capacitor if it is to store energy of 2 μJ?

## 1.15   Multiple-choice exercises

1M1   A device designed to be capable of storing electric charge is called a:
   (*a*) condenser             (*b*) inductor
   (*c*) capacitor             (*d*) plate system.

1M2   The unit of capacitance is the:
   (*a*) henry               (*b*) line of electric flux
   (*c*) volt                (*d*) farad.

1M3   If a previously uncharged capacitor reaches a potential difference of 250 V when a steady current of 15 mA flows to it for 2·5 s, its capacitance is:
   (*a*) 150 μF        (*b*) 15 F        (*c*) 94 μF        (*d*) 667 F.

1M4   The material between the plates of a capacitor is called the:
   (*a*) insulator           (*b*) dielectric
   (*c*) separator           (*d*) charge storing material.

1M5   When the potential difference across a capacitor becomes so great that current is forced to flow between the plates, the result is known as:
   (*a*) failure             (*b*) an explosion
   (*c*) dielectric breakdown  (*d*) dielectric loss.

1M6   When three capacitors of individual capacitances of 8 μF, 1·2 μF and 820 nF are connected in parallel, the capacitance of the combination will be:
   (*a*) 829·2 nF        (*b*) 9·2 μF        (*c*) 0·46 μF        (*d*) 10·03 μF.

1M7   The effective capacitance of the circuit shown in the Figure below is:

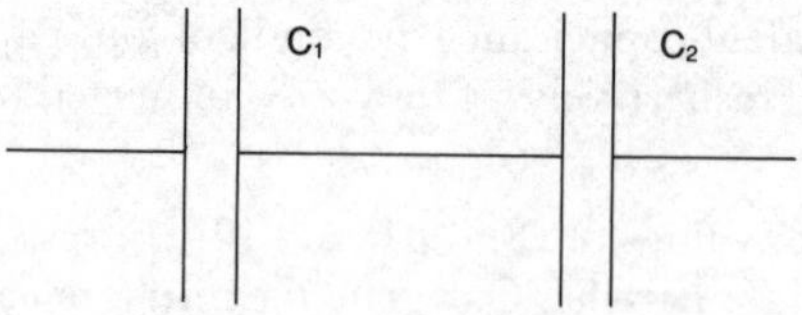

**Figure 1.20   Diagram for Exercise 1M7**

   (*a*) $\dfrac{C_1 \times C_2}{C_1 + C_2}$          (*b*) $C_1 + C_2$

   (*c*) $\dfrac{C_1 + C_2}{C_1 \times C_2}$          (*d*) $\dfrac{1}{C_1} + \dfrac{1}{C_2}$

1M8  The capacitance of the network shown in the figure is:
  (*a*) 5 µF  (*b*) 4·1 µF  (*c*) 2 µF  (*d*) 1 µF.

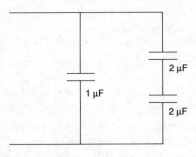

**Figure 1.21  Diagram for Exercise 1M8**

1M9  The ability of a dielectric material to store electric energy is indicated by its:
  (*a*) dielectric strength  (*b*) thickness
  (*c*) permeability  (*d*) relative permittivity.

1M10 The formula relating the capacitance ($C$) of a capacitor to its plate area ($A$), dielectric thickness ($d$), relative permittivity of the dielectric ($\epsilon_r$) and the permittivity of free space ($\epsilon_o$) is:

  (*a*) $C = \dfrac{\epsilon_o \times A}{\epsilon_r \times d}$  (*b*) $C = \dfrac{\epsilon_o \times \epsilon_r \times A}{d}$

  (*c*) $C = \dfrac{\epsilon_o \times d \times \epsilon_r}{A}$  (*d*) $C = \dfrac{A}{\epsilon_r \times \epsilon_o \times d}$

1M11 A rolled-foil capacitor has a capacitance of 56 nF and is made with a dielectric whose thickness is 0·5 mm and whose relative permittivity is 1·8. If the permittivity of free space is $8·85 \times 10^{-12}$, the effective plate area of the capacitor will be:
  (*a*) 0·45 m$^2$  (*b*) 1·76 m$^2$  (*c*) 176 cm$^2$  (*d*) 45 cm$^2$.

1M12 The formula for the energy ($W$) stored in a capacitor of capacitance $C$ if it has a potential difference of $V$ is:

  (*a*) $W = 2CV$  (*b*) $W = \dfrac{C^2 V}{2}$

  (*c*) $W = \dfrac{2}{CV^2}$  (*d*) $W = \dfrac{CV^2}{2}$

1M13 The potential difference across the terminals of a 1·2 µF capacitor when it has stored energy of 86·4 µJ is:
  (*a*) 36 V  (*b*) 2·4 kV  (*c*) 12 V  (*d*) 1·92 V

1M14 A capacitor is constructed of two long strips of aluminium foil, each with a layer of thin polystyrene lying on top of it. The two strips are then laid on top of each other and rolled tightly. The resulting capacitor is known as the:
  (*a*) rolled-foil type  (*b*) variable type
  (*c*) electrolytic type  (*d*) air-spaced type.

1M15  If an electrolytic capacitor is connected in circuit with the wrong polarity, the likely result is that:
(*a*) the effective capacitance will increase
(*b*) there will be no noticeable result
(*c*) the capacitor will burst
(*d*) the circuit time constant will be reduced.

1M16  When a capacitor is discharged, the graph of current against a base of time is called:
(*a*) a step function          (*b*) a natural curve of growth
(*c*) a constant function      (*d*) a natural curve of decay.

1M17  A resistor is connected in series with a 47 nF capacitor, resulting in a time constant of $0 \cdot 846$ ms. The resistor value is:
(*a*) 18 k$\Omega$          (*b*) $3 \cdot 98$ $\mu\Omega$      (*c*) $55 \cdot 5$ k$\Omega$      (*d*) 1 M$\Omega$

1M18  The most common application of capacitors in power systems is for:
(*a*) providing lagging currents    (*b*) voltage-stabilisation systems
(*c*) power-factor correction       (*d*) rectifying alternating currents.

*Chapter 2*

# Inductance and inductors

## 2.1   Introduction

In Chapter 9 of 'Electrical Craft Principles', Volume 1, the effect of electromagnetic induction was introduced. It was pointed out that any change in the magnetic flux linking a conductor would result in an electromotive force (EMF) being induced in it. This change of flux can be as a result of movement of the magnetic system or the conductor, when the effect is referred to as dynamic induction; or the change of flux may be due to a change of current in the coil producing it, when the effect is known as static induction.

The induction effect is of the greatest importance in electrical engineering, and will be considered in more detail in this chapter as a basis for later work.

## 2.2   Induced EMF

An electromotive force, or EMF, drives the electric current and is measured in volts. Such a force can be produced in several ways, but by far the most important of these is that of induction by means of a changing magnetic field.

Any conductor, or system of conductors, which is subjected to a changing magnetic flux will have an EMF induced in it. We often say that such an EMF is due to 'flux cutting' when a conductor is moved across a magnetic field, or when the field moves across the conductor. Since movement is involved, such an EMF is said to have been produced by 'dynamic induction'. This is the principle of the electric generator, which will be described in more detail in Chapter 8 of this book.

An electromotive force can also be produced without any physical movement at all, by changing the magnetic flux produced by an electromagnet. This method is often referred to as a change in 'flux linkage', indicating that there has been a change in the amount of magnetic flux 'linking' with, or passing through, a conductor or coil. Owing to the fact that the conductors concerned remain stationary, this method is called 'static induction'; the effects of static induction are very important in electrical engineering, since machines such as transformers (see Chapter 7) rely on them. The principles of static induction are considered in this chapter.

It is obviously of importance to be able to calculate what EMF will be induced in a conductor by a changing magnetic flux. In fact, the EMF depends not on the amount of flux change, but on the rate at which it changes. A small amount of magnetic flux changing very rapidly can induce a greater EMF than a large change of flux taking place very slowly.

The unit of magnetic flux, the **weber** (symbol Wb), was given its value quite deliberately so that flux changing at the rate of one weber per second will induce an EMF of one volt into a conductor linked or cut by this change.

Thus
$$e = \frac{\Phi_2 - \Phi_1}{t}$$

where    $e$ = average EMF induced (V)

$\Phi_2$ = final value of magnetic flux (Wb)

$\Phi_1$ = initial value of magnetic flux (Wb)

$t$ = time taken for flux to change from $\Phi_1$ to $\Phi_2$ (s)

When a coil with a number of turns is subjected to the changing magnetic flux, each turn will have an induced EMF equal to the rate of change of flux in webers per second. Since the turns are in series, the total EMF across the coil terminals will be the EMF per turn multiplied by the number of turns.

Thus
$$e = \frac{(\Phi_2 - \Phi_1)N}{t}$$

where            $N$ = the number of turns on the coil

## Example 2.1

A magnetic flux of 250 mWb linking a coil of 10 000 turns is switched off and falls to zero in 0·5 s. Calculate the average EMF induced.

$$e = \frac{(\Phi_2 - \Phi_1)N}{t} = \frac{(250 \times 10^{-3} - 0)10\ 000}{0·5} \text{ volts}$$

$$= \frac{2500}{0·5} \text{ volts} = 5000 \text{ V or 5 kV}$$

## Example 2.2

The magnetic flux linking a 200-turn coil on a transformer increases from 0·1 Wb to 0·4 Wb in 0·3 s. Calculate the average EMF induced.

$$e = \frac{(\Phi_2 - \Phi_1)N}{t} = \frac{(0 \cdot 4 - 0 \cdot 1)200}{0 \cdot 3} \text{ volts}$$

$$= \frac{0 \cdot 3 \times 200}{0 \cdot 3} \text{ volts} = 200 \text{ V}$$

## Example 2.3

The magnetic flux in a coil of 100 turns rises from zero to a steady value in $0 \cdot 8$ s, during which time an average EMF of 25 V is induced. Calculate the steady value of magnetic flux.

Since $$\Phi_1 = 0, \ e = \frac{\Phi_2 N}{t} \quad \text{and} \quad \Phi_2 = \frac{et}{N}$$

Thus $$\Phi_2 = \frac{25 \times 0 \cdot 8}{100} \text{ webers} = 0 \cdot 2 \text{ Wb}$$

## Example 2.4

The magnetic flux of Example 2.3 falls to zero, inducing an average EMF of 10 V as it does so. How long does the flux take to collapse?

$$\Phi_2 = 0 \quad \text{so} \quad e = \frac{-\Phi_1 N}{t} \quad \text{and} \quad t = \frac{-\Phi_1 N}{e}$$

Thus $$t = \frac{-0 \cdot 2 \times 100}{10} \text{ seconds} = -2 \text{ s}$$

The idea of the magnetic flux falling to zero in negative time is clearly ridiculous — this would mean that the flux reached zero before it started to fall! Obviously there is something wrong here, and this will now be explained.

## 2.3   Lenz's law

Imagine a coil of wire connected to a small torch battery through a switch. As soon as the switch is closed, current will pass through the coil and, since the coil is a solenoid, will begin to set up a magnetic field. The increasing magnetic flux will pass through the coil itself, and since the flux is changing, will induce an EMF into the coil. If this EMF had a direction assisting that of the battery, it would drive more current through the coil, setting up more flux and inducing still greater EMF. This effect would be cumulative, and before long the current in the coil would generate so much heat that the conductors of the coil would melt.

A little thought will convince the reader that this state of affairs is ridiculous. If this actually happened, it would be impossible to increase the current in almost any conductor system without its suddenly growing large enough to destroy the current-carrying conductors. We know from observations of domestic electrical apparatus that this is not so. Furthermore, we would be getting the energy used to heat the coil from nowhere, since such a large amount of energy could not come from a torch battery.

This can lead us to only one conclusion: that the EMF induced in the coil does not assist the battery in driving current into the coil, but opposes the battery EMF and tries to prevent the current from rising. This basic law of nature is known as **Lenz's law**, which states that 'the direction of an induced EMF is always such as to oppose the effect producing it'. Thus, EMF induced by the magnetic flux change owing to an increasing current will oppose that current and try to prevent its increase. Similarly, the EMF induced owing to a reducing current will assist that current and try to keep it flowing. The EMF cannot stop the change in current because, if it did so, there would be no magnetic flux change and the EMF would cease to exist. It does, however, slow the change in current.

## 2.4   Mutual inductance

Consider two coils placed side by side as indicated by Figure 2.1*a*. The left-hand coil is connected to a DC supply (in this case a battery) by means of a switch, while the right-hand coil is connected to a centre-zero indicating instrument. As long as the switch is open, there will be no current in the left-hand coil, and hence no effect at all on the right-hand coil. As soon as the switch is closed, the current in the left-hand coil will begin to rise. For reasons already discussed in Section 2.3, the current will not rise instantaneously to its full value. Whilst it increases, the current in this coil will set up an increasing magnetic flux, some of which will pass through the right-hand coil, this increase or change in flux inducing an EMF in this coil, the resulting current causing deflection of the indicating instrument (see Figure 2.1*b*). In practice, the current is likely to rise fairly rapidly, giving a sudden 'kick' of the instrument needle, which will quickly return to the zero position. The reason for the return to zero is that only while the magnetic flux is changing will the EMF be induced. As soon as the current in the left-hand coil reaches a steady value, the magnetic flux will also become steady, and as there is no change of flux in the right-hand coil, no EMF will be induced (see Figure 2.1*c*).

If the switch is opened, the current in the left-hand coil will again fall to zero, but again the change in current will take time. While the magnetic flux is reducing, an EMF will again be induced in the right-hand coil, giving a deflection on the instrument (see Figure 2.1*d*). The deflection will be in the opposite direction to that occurring when the current was increasing, because the EMF induced will have different polarity for increasing and for decreasing magnetic flux. When the current has fallen to zero, there will be no magnetic flux, no change of flux and no EMF, so the needle of the instrument will return to the zero position.

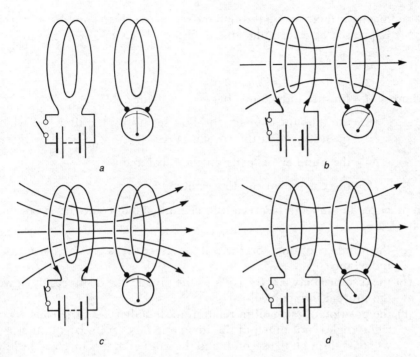

**Figure 2.1  Mutual inductance between coils**

This effect of an EMF being induced in one coil when the current changes in a second coil placed close to it is called mutual induction, and mutual inductance (symbol M) is said to exist between the coils. The unit of mutual inductance is the **henry** (symbol H), and can be defined in two ways:

(*a*) The mutual inductance between two coils is one henry when current changing in the first coil at a rate of one ampere per second induces an EMF of one volt in the second coil. Expressed as a formula, this definition is

$$e = \frac{M(I_2 - I_1)}{t}$$

where  $e$ = average EMF induced in the second coil (V)

$M$ = mutual inductance between the coils (H)

$I_1$ and $I_2$ are the initial and final currents, respectively, in the first coil (A)

$t$ = the time for the current to change from $I_1$ to $I_2$ (s)

(*b*) The mutual inductance between two coils is one henry when a current of one ampere in the first coil produces a flux linkage of one weber turn in the second. Weber turns are found by multiplying the magnetic flux passing through the second coil by its number of turns. Expressed as a formula, this definition becomes

$$M = \frac{\Phi_2 N_2}{I_1}$$

where   $M$ = mutual inductance between the coils (H)

$\Phi_2$ = the amount of magnetic flux set up by the first coil which passes through the second (Wb)

$N_2$ = the number of turns on the second coil

$I_1$ = the current in the first coil (A)

In practice, a number of factors affect the mutual inductance between coils. These are:

(a) the numbers of turns on the coils — more turns give higher mutual inductance;
(b) the distance between the coils — the greater the distance, the lower the mutual inductance;
(c) the position of one coil in relation to the other — if the coils are at right angles, not much of the magnetic flux set up by the first coil will pass through the second, and the mutual inductance will be low; and
(d) the presence of a magnetic circuit — if both coils are wound on an iron core, this will not only increase the magnetic flux set up, but will channel almost all of it through the second coil, considerably increasing the mutual inductance.

The alternative symbols for two coils between which mutual inductance exists are shown in Figure 2.2.

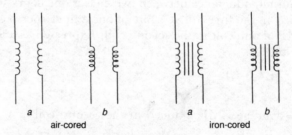

**Figure 2.2   Symbols for mutual inductors**
      Preferred symbols are those marked *a*

## Example 2.5

Two adjacent air-cored coils each have 500 turns. When a steady current of 5 A passes through one coil, the magnetic flux linking with the second is 15 mWb. Calculate the mutual inductance between the coils.

$$M = \frac{\Phi_2 N_2}{I_1} = \frac{15 \times 10^{-3} \times 500}{5} \text{ henrys} = 1 \cdot 5 \text{ H}$$

## Example 2.6

The current in the first coil of Example 2.5 is switched off and falls to zero in $0 \cdot 3$ s. Calculate the average EMF induced in the second coil while the current is changing.

$$e = \frac{M(I_2 - I_1)}{t} = \frac{1 \cdot 5(0 - 5)}{0 \cdot 3} \text{ volts} = \frac{-7 \cdot 5}{0 \cdot 3} \text{ volts}$$

$$= -25 \text{ V}$$

The negative sign indicates that the EMF obeys Lenz's law, and opposes the current change.

## Example 2.7

Two air-cored coils have a mutual inductance of $0 \cdot 4$ H, and one coil of 1200 turns has a linking flux of 5 mWb when a steady current flows in the other. Calculate the value of the steady current.

$$M = \frac{\Phi_2 N_2}{I_1} \quad \text{so} \quad I_1 = \frac{\Phi_2 N_2}{M} = \frac{5 \times 10^{-3} \times 1200}{0 \cdot 4} \text{ amperes} = 15 \text{ A}$$

## Example 2.8

When the current in the coil system of Example 2.7 is increased to 25 A, an average EMF of 80 V is induced in the 1200-turn coil. Calculate the time taken for the current to change.

$$e = \frac{M(I_2 - I_1)}{t} \quad \text{so} \quad t = \frac{M(I_2 - I_1)}{e}$$

$$t = \frac{0 \cdot 4(25 - 15)}{80} \text{ seconds} = \frac{0 \cdot 4 \times 10}{80} \text{ seconds} = 0 \cdot 05 \text{ s}$$

## Example 2.9

The induction coil of an internal-combustion engine has two windings. The first is connected to the battery, and carries a current of 2 A when the contact points are closed. The second winding has 15 000 turns, and is linked by a magnetic flux of 2 mWb when the current of 2 A flows in the first. Calculate the mutual inductance between the windings.

When the contact points open, the current in the first winding falls to zero in 2 ms. Calculate the average EMF induced in the second winding while the current is changing.

$$M = \frac{\Phi_2 N_2}{I_1} = \frac{2 \times 10^{-3} \times 15\,000}{2} \text{ henrys} = 15 \text{ H}$$

$$e = \frac{M(I_2 - I_1)}{t} = \frac{15(0 - 2)}{2 \times 10^{-3}} \text{ volts} = \frac{-15 \times 2}{2 \times 10^{-3}} \text{ volts}$$

$$= -15\,000 \text{ V or } -15 \text{ kV}$$

Example 2.9 illustrates one very practical example of how mutual inductance can be used. A rapidly changing current in one winding induces a very high EMF in the other, and this EMF is fed to a sparking plug via the distributor system. A typical ignition coil of this type is shown in Figure 2.3. Another very important application of mutual inductance is the transformer, which is considered in more detail in Chapter 7.

**Figure 2.3    Cutaway view of motor-vehicle ignition coil showing primary and secondary windings**

Since iron-cored magnetic systems are subject to an effect called saturation, the amount of magnetic flux set up will not be directly proportional to the magnetising current. Since the value of mutual inductance depends on the ratio of magnetic flux and current, it follows that it cannot have a constant value, and will become smaller as saturation occurs. The saturation effect will be explained in Chapter 3. The mutual inductance between fixed coils not containing magnetic materials will, however, be constant.

## 2.5   Self inductance

We saw in Section 2.4 how a changing magnetic flux, set up by one coil, passed through a second coil and induced an EMF. The magnetic-flux change applies to both coils. In many cases, not all of the flux set up by the first coil will link with the second, so the change of flux in the first coil will be greater than that in the second. Thus, if a changing flux provides an EMF in the second coil, it must also do so in the first.

This induction of an EMF in a coil, due to a change of current in it resulting in a change of magnetic flux linkage, is called 'self inductance', given the symbol L. The unit of self inductance, like that of mutual inductance, is the henry, symbol H. Like the unit of mutual inductance, the henry as a unit of self inductance can be defined in two ways.

(a) The self inductance of a coil is one henry when a rate of change of current of one ampere per second in the coil induces an EMF of one volt in it. Expressed as a formula, this definition is

$$e = \frac{L(I_2 - I_1)}{t}$$

where   $e$ = average induced EMF (V)

$L$ = self-inductance of system (H)

$I_1$ and $I_2$ are the initial and final currents, respectively, in the system (A)

$t$ = time taken for current change from $I_1$ to $I_2$ (s)

(b) The self inductance of a coil is one henry if a current of one ampere in the coil produces within it a flux linkage of one weber turn. Weber turns are found by multiplying the magnetic flux linking the coil by its number of turns. Expressed as a formula, this definition is

$$L = \frac{\Phi N}{I}$$

where          $L$ = self-inductance of system (H)

$\Phi$ = magnetic flux linking the system (Wb)

$N$ = number of turns

$I$ = current carried by the system (A)

The factors affecting the self inductance of a conductor system are:

(a) the number of turns — more turns gives higher self inductance;
(b) the way the turns are arranged — a short thick coil will have higher self inductance than a long, thin one; and
(c) the presence of a magnetic circuit — if the coil is wound on an iron core, the same current will set up a greater magnetic flux and the self-inductance will be higher.

Standard circuit symbols for an inductor are shown in Figure 2.4.

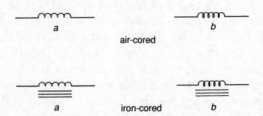

**Figure 2.4   Sybols for inductors**
Preferred symbols are those marked *a*

## Example 2.10

A coil of 1000 turns sets up a magnetic flux of 15 mWb when carrying a current of 30 A. What is its self inductance?

$$L = \frac{\Phi N}{I} = \frac{0 \cdot 015 \times 1000}{30} \text{ henrys} = 0 \cdot 5 \text{ H}$$

## Example 2.11

A supply connected to an air-cored coil of self inductance $0 \cdot 2$ H causes the current to increase from zero to 4A in $0 \cdot 08$ s. Calculate the average EMF induced in this time.

$$e = \frac{L(I_2 - I_1)}{t} = \frac{0 \cdot 2 \times (4 - 0)}{0 \cdot 08} \text{ volts} = \frac{0 \cdot 2 \times 4}{0 \cdot 08} \text{ volts} = 10 \text{ V}$$

## Example 2.12

An iron core is placed inside the coil of Example 2.11. When the coil is connected to the same supply, the current takes 6 s to reach a steady value of 4 A, the average EMF induced during this time being 20 V. Calculate the new self inductance of the coil.

$$e = \frac{L(I_2 - I_1)}{t} \quad \text{so} \quad L = \frac{et}{I_2 - I_1} = \frac{20 \times 6}{4} \text{ henrys} = 30 \text{ H}$$

## Example 2.13

A coil is wound with 400 turns and has a self inductance of 2 H. What current must flow to set up a flux of 30 mWb?

$$L = \frac{\Phi N}{I} \quad \text{so} \quad I = \frac{\Phi N}{L} = \frac{0 \cdot 03 \times 400}{2} \text{ amperes} = 6 \text{ A}$$

## Example 2.14

A coil has a self inductance of 3 H and a resistance of 8 $\Omega$. Calculate the current when the coil is connected to a 12 V DC supply. The supply is switched off and the current falls to zero in 0·2 s. Calculate the average EMF induced.

The self inductance of the coil will tend to reduce the rate of growth of current (see Section 2.7), but will not affect its final steady value.

Therefore Ohm's law applies, and

$$I = \frac{V}{R} = \frac{12}{8} \text{ amperes} = 1 \cdot 5 \text{ A}$$

$$e = \frac{L(I_2 - I_1)}{t} = \frac{3(0 - 1 \cdot 5)}{0 \cdot 2} \text{ volts} = \frac{-3 \times 1 \cdot 5}{0 \cdot 2}$$

$$= -22 \cdot 5 \text{ V}$$

The negative sign indicates that Lenz's law applies (see Section 2.3) and any induced EMF will always oppose the current change producing it. If the current rises, setting up more linking magnetic flux, the induced EMF opposes the voltage driving the current and tries to prevent the change. If the current falls, reducing the linking flux, the induced EMF assists the current and tries to maintain it.

This effect of opposition results in the induced EMF being called a 'back EMF'. It will always oppose change, but can never prevent it. If it did so, there would be no change in magnetic flux, and the EMF would cease to exist. A DC circuit containing high self inductance will always make current change a slow process. For this reason, iron-cored coils of high self inductance, called 'chokes', are connected in any DC circuit where variations of current are to be smoothed out. Figure 2.5 shows a typical smoothing choke. The use of the choke as a smoothing component will be discussed more fully in Chapter 6. Iron-cored coils do not have a fixed magnetic-flux-to-current ratio, and thus have a variable self inductance, which tends to fall as the current increases and flux approaches the level of magnetic saturation.

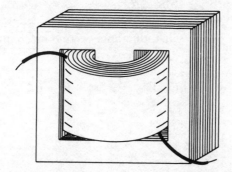

**Figure 2.5   Choke for smoothing in power-rectifier circuit**

## 2.6   Energy stored in a magnetic field

Energy is required to set up a magnetic field, and this energy will be stored in it. The energy will not be lost when the current setting up the magnetic field is switched off or otherwise reduced, but will be returned to the circuit in terms of the current driven by the EMF induced by the collapsing field.

To produce an expression for the energy stored in a magnetic field, consider a coil of self inductance $L$ henrys with a supply of $V$ volts connected to it. Assume that this voltage causes a final current of $I$ amperes, which is reached $t$ seconds after the instant of switching on. In practice, the growth of current would not be a steady one but, for the sake of simplicity, let us assume that it rises evenly as shown in Figure 2.6. The average current will be $I/2$ amperes, so the power provided while the field is set up will be $P = V\,I/2$ watts. Since energy = power × time, the energy provided to the magnetic field will be $W = V\,I\,t/2$ joules.

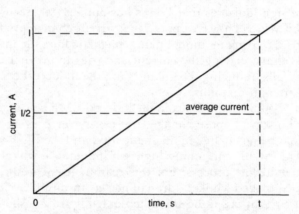

**Figure 2.6   Current/time graph for proof of expression for energy stored in magnetic field**

In Section 2.5 we used the expression

$$e = \frac{L(I_2 - I_1)}{t}$$

or $\qquad$ voltage $= \dfrac{\text{self inductance} \times \text{current change}}{\text{time}}$

In this case, this would be $\quad V = \dfrac{LI}{t} \quad$ so $\quad Vt = LI$.

Thus

$$W = \frac{Vt \times I}{2} = \frac{LI \times I}{2}$$

or

$$W = \frac{I^2 L}{2}$$

where   $W$ = energy stored in the magnetic field (J)

   $L$ = self inductance of the system providing the magnetic field (H)

   $I$ = current in the system (A)

## Example 2.15

Calculate the energy stored in the magnetic field of a coil of self inductance $0 \cdot 2$ H when carrying a steady current of 10 A.

$$W = \frac{I^2 L}{2} = \frac{10^2 \times 0 \cdot 2}{2} \text{ joules} = 10 \text{ J}$$

## Example 2.16

When carrying a steady current of 4 A, the energy stored in the magnetic field of a coil is $0 \cdot 4$ J. Calculate the self inductance of the coil.

$$W = \frac{I^2 L}{2}$$

so   $$L = \frac{2W}{I^2} = \frac{2 \times 0 \cdot 4}{4 \times 4} \text{ henrys} = 0 \cdot 05 \text{ H or 50 mH}$$

## Example 2.17

The energy stored in the magnetic field of a 120 mH inductor is $0 \cdot 24$ J. Calculate the steady current in the inductor.

$$W = \frac{I^2 L}{2}$$

so   $$I^2 = \frac{2W}{L} = \frac{2 \times 0 \cdot 24}{120 \times 10^{-3}} = 4$$

$$I = \sqrt{I^2} = \sqrt{4} \text{ amperes} = 2 \text{ A}$$

## 2.7   Growth and decay curves

In practice, an inductor will consist of a coil of wire, and the wire will have resistance. This resistance can be shown separately from the inductance as in Figure 2.7, although in fact the two are mixed together, each turn on a coil contributing to both its self inductance and its resistance.

At the instant of switching on a DC supply to a circuit (by closing the switch of Figure 2.7 in position 1), the rate of rise of current will be high, and this will result in a high induced EMF. Obeying Lenz's law, this EMF will oppose the current and reduce the rate at which it rises, so that the graph of increasing current with time will be as shown in Figure 2.8. This is called a 'curve of natural growth'.

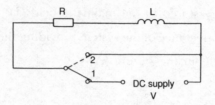

**Figure 2.7**   **Circuit for growth and decay of current in an inductor**

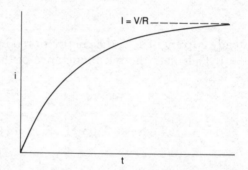

**Figure 2.8**   **Current/time graph after closing switch to position 1 in circuit of Figure 2.7 (curve of natural growth)**

If the current is switched off by changing the switch in Figure 2.7 from position 1 to position 2, the energy in the magnetic field will be transferred in the form of a current circulating in the closed loop provided. This current will die away, and will fall to zero when all of the energy stored in the magnetic field has been converted to energy in the form of heat in the resistance of the coil. In practice, there are difficulties in operating the switch, but these will be considered in Section 2.9. The decaying current in the closed circuit will be driven by the induced EMF. As the rate of current change gets less, the EMF will reduce, slowing the rate of current change still further. The graph of the reducing current with time is shown in Figure 2.9 and is called a 'curve of natural decay'.

The growth and decay rates are both dictated by the 'time constant' of the circuit, which is $L/R$. To take an example, the growth or decay of current in a $0 \cdot 5$ H coil of resistance 25 $\Omega$ would be virtually completed within about one tenth of a second of switching on or off (about five time constants).

It should be noted that the final steady direct current in an inductive circuit depends only on its resistance and the applied voltage. The self inductance of the circuit will, by inducing an EMF, slow the current change, but will have no effect on the current when it becomes steady. This current obeys Ohm's law, so that $I = V/R$.

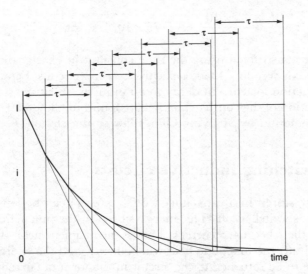

**Figure 2.9    Full construction of a current-decay curve**

## 2.8    Construction of inductor charge and discharge curves

The mathematical formulas from which the charge and discharge curves can be drawn are too complex at this stage, but it is still useful to be able to construct these curves, even if they are approximate.

All constructions are based on the **time constant**, which can be defined as the time it would take for the voltage or current to reach its final value if it continued to change at its initial rate. For inductive circuits, the time constant (here given the symbol $\tau$) is

$$\tau = \frac{L}{R}$$

where             $\tau$ is the circuit time constant in seconds

$L$ is the circuit inductance in henrys

$R$ is the circuit resistance in ohms

It will be appreciated that the time (the transient time) taken for the values of voltage and current to reach their final (steady) values is dependent on the time constant. Higher values of $L$ and/or lower values of $R$ will increase the transient time.

### Example 2.18

A 82 mH inductor and a 47 k$\Omega$ resistor are connected in series. Calculate the time constant of the combination.

$$\tau = \frac{L}{R} = \frac{82 \times 10^{-3}}{47 \times 10^3} = 1 \cdot 75 \times 10^{-6} \text{ s} \quad \text{or} \quad 1 \cdot 75 \; \mu\text{s}$$

Curves are constructed using the time constant in exactly the same way as explained in Section 1.12 for capacitive/resistive circuits. Figure 2.9 shows the construction for current decay in an inductive/resistive series circuit; an explanation of the construction will be found in Chapter 1. The initial current $I$ is found by applying Ohm's law to the circuit $(I = V/R)$.

## 2.9   Switching inductive circuits

The energy stored in a capacitor (see Chapter 1) can be left in it when the supply is switched off. The energy stored in the magnetic field of an inductor is there because a current provides the magnetic field. If the current is switched off, the magnetic field will collapse, and the energy stored in it will have to be returned to the circuit in the form of current driven by the induced EMF.

If an attempt is made simply to switch off the DC supply to a highly inductive circuit, this is likely to result in a sudden interruption of current. The current will fall very quickly, resulting in a very large induced EMF, which will either cause an arc across the switch contacts or will break down the insulation between adjacent turns on the coil. The energy from the magnetic field will be aided by the supply voltage in maintaining an arc, which is likely to cause severe damage to the switch.

The circuit of Figure 2.7 would seem to provide a solution, since it has a discharge path for current from the inductor. In practice, however, the switch must break from 1 before making contact at 2 to prevent short circuiting the supply. This break can damage coil insulation as described above, or, more likely, will set up an arc from contact 1 to the switch blade,

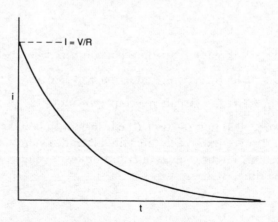

**Figure 2.10    Current/time graph after closing switch to position 2 in circuit of Figure 2.7 (curve of natural decay)**

which will 'pull' it to contact 2, thus providing mains voltage between contacts 1 and 2 across the arc. This will maintain the arc, resulting in severe damage to the switch and possibly a fire.

The circuit of Figure 2.11, with a permanently connected discharge resister $R_D$, would appear to offer a solution, but in practice a correct choice for the value of the resistor is difficult. If $R_D$ is small, the rate of change of decaying current will be small, with low induced EMF and no risk of breaking down coil insulation. However, with the switch closed, this low-value resistor will be directly across the supply, drawing a heavy current and getting very hot in the process. A high-value discharge resistor will reduce the power loss with the switch closed, but will increase the rate of current decay when the switch is opened, possibly to the extent where the induced EMF breaks down coil insulation.

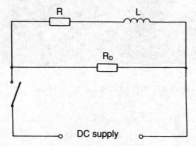

**Figure 2.11   Inductive circuit with discharge resistor**

The conflicting requirements of low discharge resistance when the switch is open and high resistance when it is closed can be met by the use of a special voltage-sensitive resistor. This is a semiconductor device having a very high resistance at normal supply voltage. When the switch is opened, the EMF induced in the coil increases, and applies to the special resistor, which reduces sharply in value to provide a safe and slow discharge-current path. Figure 2.12 shows such a device connected in circuit, and Figure 2.13 is a photograph of some voltage-sensitive resistors.

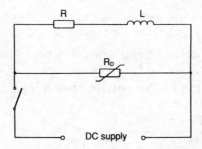

**Figure 2.12   Inductive circuit with non-linear discharge resistor**

**Figure 2.13    Selection of nonlinear resistors**

Another method of providing a low-resistance discharge path for an inductor fed from a DC supply is to connect a diode in parallel with it as in Figure 2.14. The diode has very low resistance in one direction (the direction pointed by the arrow on its symbol for conventional current) and very high resistance in the other. The diode offers very high resistance to supply voltage, but a very low-resistance path for decaying current. When used in this way the device is called a 'flywheel diode' or 'free-wheel diode'.

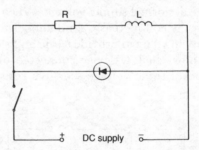

**Figure 2.14    Inductive circuit with flywheel diode**

## 2.10   Summary of formulas for Chapter 2

See text for definitions of symbols.

$$e = \frac{(\Phi_2 - \Phi_1)N}{t} \qquad t = \frac{(\Phi_2 - \Phi_1)N}{e} \qquad N = \frac{et}{\Phi_2 - \Phi_1}$$

$$\Phi_2 = \Phi_1 + \frac{et}{N} \qquad \Phi_1 = \Phi_2 - \frac{et}{N}$$

Mutual inductance:

$$M = \frac{\Phi_2 N_2}{I_1} \qquad I_1 = \frac{\Phi_2 N_2}{M} \qquad \Phi_2 = \frac{MI_1}{N_2} \qquad N_2 = \frac{MI_1}{\Phi_2}$$

Mutually induced EMF:

$$e_2 = \frac{M(I_2 - I_1)}{t} \qquad t = \frac{M(I_2 - I_1)}{e_2} \qquad M = \frac{e_2 t}{I_2 - I_1}$$

$$I_2 = I_1 + \frac{e_2 t}{M} \qquad I_1 = I_2 - \frac{e_2 t}{M}$$

Self inductance:

$$L = \frac{\Phi N}{I} \qquad I = \frac{\Phi N}{L} \qquad \Phi = \frac{LI}{N} \qquad N = \frac{LI}{\Phi}$$

Self-induced EMF:

$$e = \frac{L(I_2 - I_1)}{t} \qquad t = \frac{L(I_2 - I_1)}{e} \qquad L = \frac{et}{I_2 - I_1}$$

$$I_2 = I_1 + \frac{et}{L} \qquad I_1 = I_2 - \frac{et}{L}$$

Energy stored in a magnetic field:

$$W = \frac{I^2 L}{2} \qquad L = \frac{2W}{I^2} \qquad I = \sqrt{\frac{2W}{L}}$$

## 2.11 Exercises

1 The supply to a coil of 1250 turns is switched on, the magnetic flux taking 0·2 s to reach a steady value of 80 mWb. Calculate the average EMF induced during this period.

2 A transformer secondary winding of 200 turns has a linking flux which falls from 30 mWb to 10 mWb in 0·01 s. Calculate the average EMF induced.

3 The magnetic flux linking a choke with 800 turns is changed by 180 mWb in 0·9 s. Calculate the average EMF induced.

4 When the current in a 50-turn coil changes, an average EMF of 10 V is induced for 0·4 s. Calculate the total change of magnetic flux.

5 The magnetic flux in a coil collapses, reaching zero after 1·2 s, during which time an average EMF of 1 kV is induced in the 4000-turn coil. Calculate the initial magnetic flux.

6 The average EMF induced in a 300-turn coil by a magnetic flux change of 0·08 Wb is 12 V. Calculate the time taken for the flux change.

7 A coil has an average EMF of 1250 V induced in it when a linking-magnetic-flux change of 25 mWb takes place in 10 ms. Calculate the number of turns on the coil.

8 Calculate the mutual inductance between two coils if a current of 12 A in one coil causes a flux linkage of 25 mWb with the second, which has 480 turns.

9 The mutual inductance between two coils, each having 200 turns, is 25 mH. Calculate the magnetic flux linking one of the coils when the current in the other is $0 \cdot 4$ A.

10 When a current of 40 A is passed through coil A, a magnetic flux of 20 mWb links coil B. If the mutual inductance between the two coils is 3 mH, how many turns has coil B?

11 Two coils have a mutual inductance of $0 \cdot 1$ H, and one of the coils, with 700 turns, has a magnetic flux of 3 mWb linking with it when the other coil carries a current. Calculate the value of the current.

12 The mutual inductance between two coils is 40 mH, and a current change in one coil changes the magnetic flux linking the other by 220 mWb, the change taking 4 ms. Calculate the average EMF induced.

13 An average EMF of 110 V is induced in a coil when the current in a second coil changes over a period of 25 μs. If the change in magnetic flux in the first coil is 22 mWb, calculate the mutual inductance between the coils.

14 The mutual inductance between two coils is $0 \cdot 4$ H. When a supply is switched on to the first coil, it takes $0 \cdot 1$ s to reach its steady value, during which time the average EMF induced in the second coil is 3 V. Calculate the final steady magnetic flux linking the second coil.

15 The average EMF induced in one coil due to the current change in a second is 400 V, the magnetic-flux change in the second coil being 80 mWb. If the mutual inductance between the two coils is $0 \cdot 6$ H, calculate the time taken for the current change.

16 What is meant by the statement that a coil has inductance of 1 henry? What is the effect on the voltage induced of
(a) rate of change of current, and
(b) number of turns?
What effect has inductance in the field coil of a generator? Explain how the effect of opening a field coil can be limited.

17 (a) Define 'one henry of inductance' in terms of amperes, volts and seconds.
(b) What direct voltage increase is required to raise the current in a circuit uniformly from 10 A to 20 A in 2 seconds if the circuit inductance is $0 \cdot 1$ henry and the resistance negligible?

18 Give TWO definitions of the term 'self inductance' and state the unit in which it is measured.

19 Calculate the self inductance of a coil of 700 turns which provides a linking flux of 6 mWb when carrying a current of $3 \cdot 5$ A.

20 The current in an air-cored coil of self inductance 30 mH changes from 1 A to 3 A in 25 ms. Calculate the average EMF induced during this time.

21 What current will be needed in a 600-turn coil of self inductance $0 \cdot 2$ H in order to set up a linking flux of 50 mWb?

22 The current in a coil of self inductance 10 mH falls from $2 \cdot 5$ A to zero, during

which time an average EMF of 50 V is induced. Calculate the time taken for the current to change.

23 What linking magnetic flux will be set up by a 25-turn coil carrying a current of 40 mA if it has a self inductance of 0·15 H?

24 The time taken for the current in a coil to fall from 10 A to 3 A is 0·35 s, during which time an average EMF of 2 V is induced. Calculate the self inductance of the coil.

25 How many turns has a coil of self inductance 10 mH if it sets up a linking magnetic flux of 25 μWb when carrying a current of 0·1 A?

26 An initial current of 0·5 A in a coil of self inductance 80 mH is increased over a period of 0·5 s, during which time an average EMF of 40 mV is induced. Calculate the current in the coil at the end of the period.

27 Why is the EMF induced by a current change in an inductive circuit often called a 'back EMF'?

28 State Lenz's law, and explain how it affects any attempt to change the current in an inductive circuit.

29 (*a*) What is the unit of 'self inductance' and of 'mutual inductance'? Explain the meanings of the two terms. State TWO factors affecting the self inductance of a coil.
   (*b*) A 100-turn coil sets up a total magnetic flux of 0·25 Wb when carrying a current of 50 A. What is its self inductance?

30 Calculate the energy stored in the magnetic field of a 280 mH inductor carrying a current of 5 A.

31 An inductor has stored energy of 4 J in its magnetic field when carrying a current of 2 A. Calculate its self inductance.

32 How much current must be carried by a 200 mH inductor if it is to store energy of 10 J?

33 Sketch a graph to show how current varies with time (*a*) when the supply to a resistive coil is switched on, and (*b*) when it is switched off. For each case, indicate the effect on the curve of increasing self inductance, and of increasing circuit resistance.

34 When the switch S in the circuit of Figure 2.15*a* is opened, a spark appears across the contacts but this does not happen when the switch S in the circuit of Figure 2.15*b* is opened. Explain why.

35 Why is it often difficult to break the current to an inductive circuit? Indicate TWO methods which reduce the difficulty of such an operation.

## 2.12 Multiple-choice exercises

2M1  An electromotive force (EMF) is induced in a conductor when:
   (*a*) the voltage applied to it increases
   (*b*) it is moved in a space where there is no magnetic field
   (*c*) it is subject to a change in its linking magnetic field
   (*d*) a switch is closed in a circuit.

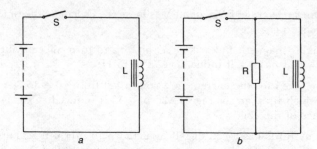

**Figure 2.15   Diagrams for Exercise 34**

2M2   The magnetic flux linking a coil of 120 turns falls to 20 mWb in 4 s. If the average EMF induced during the change is $2 \cdot 4$ V, the initial linking flux must have had a value of:
(*a*) 80 mWb        (*b*) $0 \cdot 24$ T            (*c*) 100 mWb      (*d*) 60 mWb.

2M3   The current in a coil of 1200 turns sets up a linking magnetic flux of $0 \cdot 15$ Wb. When the current is switched off, an average EMF of 90 V is induced in the coil while the magnetic flux falls to zero. How long has the flux change taken?
(*a*) $11 \cdot 25$ ms        (*b*) 2 s            (*c*) 270 min        (*d*) $0 \cdot 1125$ s.

2M4   Any induced EMF will always be subjected to an effect which opposes the change which produced it. This is known as:
(*a*) the rule of opposition      (*b*) Lenz's law
(*c*) Ohm's law                  (*d*) contrariness.

2M5   If a change of current in one circuit results in an EMF being induced in a separate circuit, the two circuits are said to have:
(*a*) a connecting link          (*b*) self inductance
(*c*) crossed connections        (*d*) mutual inductance.

2M6   The mutual inductance between two systems when a change of current at the rate of one ampere per second in the first induces an EMF of one volt in the second is:
(*a*) one volt                   (*b*) one weber
(*c*) one henry                  (*d*) one tesla.

2M7   If two coils are mounted on a common magnetic circuit, their mutual inductance compared with that if they were arranged close together in air will be:
(*a*) much greater               (*b*) zero
(*c*) infinite                   (*d*) very much smaller.

2M8   Two air-cored coils, each of 800 turns, are placed close together. If a current of 4 A in one coil produces a linking magnetic flux of $2 \cdot 5$ mWb in the other, the mutual inductance between the coils is:
(*a*) 8 H          (*b*) $0 \cdot 78$ mH        (*c*) 2 H          (*d*) $0 \cdot 5$ H.

2M9 A coil which has an EMF of one volt induced in it when it carries current which is changing at a rate of one ampere per second has a self inductance of:

(*a*) zero            (*b*) two henrys
(*c*) one henry       (*d*) one tesla.

2M10 If $\Phi$ is the linking magnetic flux, $N$ is the number of turns, and $I$ is the current carried, the self inductance of a coil is given by the formula:

(*a*) $L = \dfrac{\Phi \times I}{N}$            (*b*) $L = \dfrac{\Phi \times N}{I}$

(*c*) $L = \dfrac{N}{I \times \Phi}$            (*d*) $L = \dfrac{I}{N \times \Phi}$

2M11

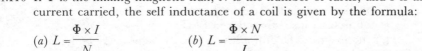

**Figure 2.16 Diagram for Exercise 2M11**

The symbol shown in Figure 2.16 is for:
(*a*) an iron-cored inductor    (*b*) a transformer
(*c*) a transistor              (*d*) an air-cored inductor.

2M12 A coil of 90 turns has a self inductance of 0·3 H. The current it must carry if it is to set up a linking magnetic flux of 18 mWb will be:
(*a*) 0·486 A      (*b*) 1500 A      (*c*) 5·4 A      (*d*) 0·185 A.

2M13 A coil with self inductance of 0·85 H carries a steady current of 1·2 A. If the current is increased to 3 A in a time of 0·15 s, the average EMF induced in the coil as the current changes will be:
(*a*) 1·56 V      (*b*) 10·2 V      (*c*) 17 V      (*d*) 98 mV.

2M14 When an EMF is induced in a conductor, its direction is always:
(*a*) such as to oppose the effect which produces it
(*b*) positive to negative
(*c*) that which assists the change of current inducing it
(*d*) from left to right.

2M15 The energy stored in the magnetic field of a coil of self inductance 0·15 H when it carries a steady current of 220 mA is:
(*a*) 7·26 mJ      (*b*) 3·6 J      (*c*) 3·63 mJ      (*d*) 16·5 mJ.

2M16 The energy stored in a coil of self inductance 1·8 H is 0·3 J. This means that the current carried by the coil is:
(*a*) 0·27 A      (*b*) 577 mA      (*c*) 167 mA      (*d*) 81 mA.

2M17 The time constant in seconds of an inductive circuit of self inductance $L$ henrys and resistance $R$ ohms is:

(*a*) $\dfrac{L}{R}$      (*b*) $\dfrac{R}{L}$      (*c*) $RL$      (*d*) $\dfrac{1}{LR}$

2M18 A series circuit is made up of an inductance and a 1 kΩ resistor in series. If the combination has a time constant of 2·5 ms, the inductance must have a value of:
(a) 2500 H        (b) 0·4 mH        (c) 2·5 H        (d) 400 mH.

2M19 When interrupting the current in an inductive circuit a problem may arise because:
(a) the switch is connected in the wrong circuit
(b) of arcing caused by induced EMF
(c) the magnetic field of a lifting electromagnet may be lost
(d) of induced current.

2M20 The diode D shown in the circuit diagram below is called:

(a) a zener diode              (b) a semiconductor diode
(c) a discharge diode          (d) a flywheel or free-wheel diode.

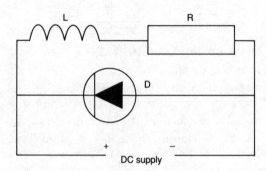

**Figure 2.17   Diagram for Exercise 2M20**

## Chapter 3

# Magnetic materials and iron losses

## 3.1 Introduction

Most electrical machines and instruments rely on magnetism for their operation, and in many cases ferromagnetic materials are used, since the magnetic flux produced is thereby increased. This chapter will consider some of the properties of these materials and will indicate that in many cases the application of magnetic materials brings with it problems which cannot completely be solved.

## 3.2 Electromagnetic units and the magnetisation curve

The unit of magnetic flux (symbol $\Phi$) is the weber (symbol Wb) which is defined in terms of the EMF induced by a change in the flux cutting or linking a coil. The magnetic flux density (symbol B) is a measure of the concentration of flux, and is measured in **teslas** (symbol T), which are also webers per square metre (Wb/m$^2$).

Thus
$$B = \frac{\Phi}{A}$$

where  $B$ = magnetic flux density (T)

   $\Phi$ = total magnetic flux (Wb)

   $A$ = area over which flux is spread (m$^2$)

When the magnetic flux is set up by an electromagnet, the effect setting up the flux is called the **magnetomotive force** (MMF) which is measured in ampere-turns (At). The amount of magnetic flux set up depends on the **magnetising force** (symbol H) which is measured in ampere-turns per metre

(At/m). Magnetising force strictly has the dimensions of amperes per metre (A/m), but the number of turns is deliberately included for clarity.

Thus
$$H = \frac{IN}{l}$$

where  $H$ = magnetising force (At/m)

$I$ = current in magnetising coil (A)

$N$ = number of turns on magnetising coil

$l$ = length of magnetic flux path (m)

When the magnetic flux is set up in a nonmagnetic material, such as air, brass, aluminium or any other material than those called 'ferromagnetics' and described in Section 3.3, there is a definite relationship between the magnetising force and the resulting magnetic flux density.

Thus
$$\frac{B}{H} = \mu_0$$

where  $B$ = magnetic flux density (T)

$H$ = magnetising force (At/m)

$\mu_0$ = permeability of free space, which is a constant having the value $4\pi/10^7$

A graph of $B$ against $H$ is called the magnetisation curve for the material concerned. Figure 3.1 shows the curve for any nonmagnetic material to be a straight line.

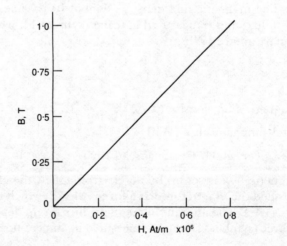

**Figure 3.1   Magnetisation curve for nonmagnetic materials**

Ferromagnetic materials behave differently from all other materials, providing a great deal more magnetic flux for a given magnetising force.

Thus
$$\frac{B}{H} = \mu_0\mu_r$$

where $\mu_r$ = relative permeability of the material concerned under certain conditions of magnetisation. For a ferromagnetic material, commonly called simply a magnetic material, there is not a fixed relationship between $B$ and $H$.

A magnetic material can be considered to be made up of a very great number of particles each of which is a permanent magnet. When the material is in the unmagnetised condition, these small magnets point in all directions so that their effects cancel out and no overall magnetism can be detected (see Figure 3.2a). As the material is subjected to an increasing magnetising force, the minute magnets begin to swing into line, so that, when the material is fully magnetised, all the north poles point in one direction (see Figure 3.2b). Although this account is oversimplified, it indicates the processes involved. After an initial curve, the graph tends to become linear as increasing magnetising force results in increasing flux densities. This is the period when the small magnets are swinging into line. A certain point is reached where the curve tends to flatten out, and 'saturation' has taken place. The small magnets are now in line, and no matter how much extra magnetising force is applied, very little improvement can be made in their directions, so very little extra flux density results.

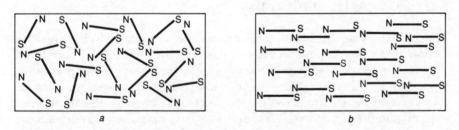

**Figure 3.2   Diagrams to illustrate simple theory of magnetism**

A typical magnetisation curve is shown in Figure 3.3. Note the different scale used for the horizontal axis from that in Figure 3.1, with much smaller values of magnetising force. The curve of Figure 3.1 drawn on the same scale as that of Figure 3.3 would be an almost horizontal line, indicating the much lower magnetic flux density produced by the same magnetising force. Figure 3.3 also shows a curve of $\mu_r$ drawn to a base of $H$; the value of $\mu_r$ must vary because the $B/H$ curve is not a straight line, and it falls rapidly after saturation occurs.

For most applications, it is necessary to produce as much magnetic flux as possible for a given value of magnetising force, and therefore the portion

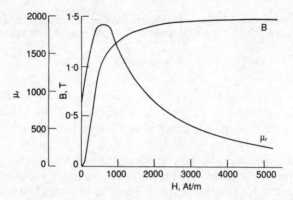

**Figure 3.3   Typical magnetisation curve, showing also the variation in relative permeability**

of the curve below saturation is commonly used. In many cases, however, the very large cross-sectional area needed to give the required magnetic flux at a flux density below saturation results in a heavy and expensive magnetic core, which cannot be tolerated, and saturation is allowed to occur. The designer must always compromise between the small, light, inexpensive core which saturates, and the large, heavy, costly core which works on the more linear part of the magnetisation curve.

## 3.3   Magnetic materials

Materials showing the ferromagnetic properties considered in the previous section are those composed wholly or partly of the metals iron, nickel and cobalt. A material made up of a mixture of metals (and sometimes other elements) is called an 'alloy', and most magnetic materials used in commercial practice are alloys of iron. Strangely enough, the addition of a small amount of a material which is not itself magnetic, can often have a considerable effect on the magnetic properties of the resulting alloy. For example, a little carbon added to iron will make a magnetic steel which is not only harder and tougher than the basic iron, but which allows almost twice as much magnetic flux to be set up for a given magnetising force. The magnetic behaviour of an alloy may be affected by the heat treatment it receives during manufacture or by the method of rolling it into sheets for laminations (see Section 3.6). Many magnetic irons have silicon added to them to reduce hysteresis losses (see Section 3.5).

Nickel is a more expensive material than iron and magnetic circuits containing nickel tend to saturate at lower flux densities than iron alloys. The flux set up for a small magnetising force is, however, comparatively high; many nickel-based materials tend to saturate suddenly, giving a sharp 'knee' to the magnetisation curve.

Although cobalt is a ferromagnetic material, it is seldom used except for permanent magnets. Figure 3.4 shows the magnetisation curves for a few of the more commonly used magnetic materials; Table 3.1 lists these and some other materials with their properties and uses. Many magnetic metals are known by trade names, such as 'Mumetal' and 'Stalloy'.

The range of materials suitable for various types of permanent magnets is very wide indeed, and is not considered here.

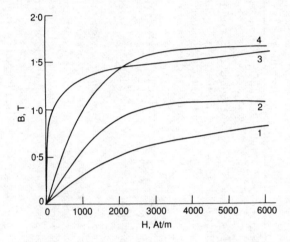

**Figure 3.4   Magnetisation curve for some magnetic materials**
1 Cast iron;   2 Radiometal;   3 Silicon–steel;   4 Cast steel

## 3.4   Magnetic shielding

It has already been shown that a magnetic circuit will allow very much more magnetic flux to be set up for a given magnetising force than a nonmagnetic circuit. In effect, the magnetic material provides an easier path for magnetic flux. Thus, a piece of magnetic material placed in a magnetic field will carry more of the field flux than its surroundings, distorting the field as shown in Figure 3.5. The magnetic flux distortion will be greatest when the relative permeability of the magnetic material is highest, and for weak magnetic fields the most suitable material for shielding will be one such as Mumetal (see Table 3.1).

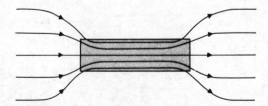

**Figure 3.5   Distortion of magnetic field by magnetic material**

**Table 3.1: Composition and properties of some magnetic materials**

| Material | Composition | Magnetic properties | Applications |
| --- | --- | --- | --- |
| Soft iron (or magnetic iron) | Iron (usually annealed by heat treatment) | Easily magnetised and demagnetised | Bell and telephone electromagnets. Relay cores, DC choke cores etc. |
| Silicon steel | Iron with carbon and up to 5% silicon | Low hysteresis losses and high relative permeability | Magnetic circuits for rotating machinery (up to about 3% silicon) and for transformers (up to 5% silicon) where subjected to alternating magnetic flux |
| Cast iron | Iron | High hysteresis loss, low relative permeability | Yokes for small rotating machines |
| Cast steel | Iron with carbon | Higher saturation than cast iron | Yokes for large rotating machines |
| Mumetal* } Permalloy C* } | 75–80% nickel, balance iron | Very high relative permeability and low hysteresis losses at low magnetising forces | Instrument magnetic circuits, current transformers, screens for magnetic shielding etc. |
| Radiometal* } Permalloy B* } | 50% nickel, 50% iron | Good relative permeability at medium magnetising force | Transformers for electronic communications equipment, quick-acting relays etc |
| Rhometal* } Permalloy D* } | 36% nickel, 64% iron | Poor relative permeability, but very low eddy current loss | High-frequency transformers. |
| Permendur* | 50% cobalt, 50% iron | Very high saturation level | Cathode-ray-tube deflection systems. Telephone-earpiece diaphragms |

*Indicates proprietary trade name.

This effect can be used to advantage where it becomes necessary to protect an instrument movement from the effects of the earth's magnetic field, or from other 'stray' fields owing to magnetic leakage from equipment such as rotating machines and transformers. The movement is enclosed within a box of nickel–iron, such as Mumetal, the interior of which is then free of the external magnetic flux as shown in Figure 3.6. Even a screen of thin plates of magnetic material will shield a delicate instrument movement from the effects of an external magnetic field, and will thus improve its accuracy.

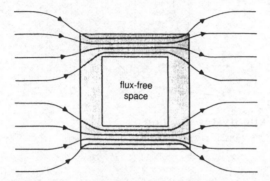

**Figure 3.6  To illustrate magnetic screening**

## 3.5  Hysteresis loss

We have already seen that, as the current in a magnetising coil wound on a core of magnetic material is increased, the graph of increasing magnetic flux density $B$ against increasing magnetising force $H$ is the magnetisation curve shown as section 0 to 1 of Figure 3.7. At point 1, $H$ has reached its highest value, corresponding to the maximum value of magnetising current, and $B$ shows signs of approaching saturation. If the magnetising current is now reduced to zero, the $B/H$ curve will not, in most cases, follow the curve traced out as the current increased. Instead, it will follow the path 1 to 2, the magnetic flux density corresponding to the length 0 to 2 being left in the material. This is called 'remanent flux density' (remaining flux). If the direction of the magnetising current is reversed, it will set up a magnetising force in the opposite direction, shown as $H-$. To begin with, this negative magnetising force will cause the magnetic flux density to fall to zero, so the graph follows the path 2 to 3. The magnetising force corresponding to length 0 to 3 is called the 'coercive force', or the magnetising force needed to provide for the removal of the remanent magnetic flux density.

If the reversed magnetising current continues to increase, it will set up magnetic flux in the opposite direction to that set up initially, the graph following the path 3 to 4. Reduction of the current to zero will result in

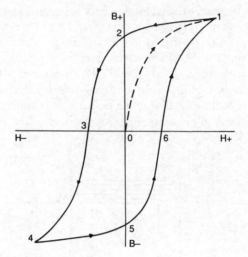

**Figure 3.7   The hysteresis loop**
See text for explanation

the graph following the line 4 to 5, 0 to 5 (equal to 0 to 2) being the remanent magnetism. A build up of current in the original direction will give the graph section 5 to 6 to 1, 0 to 6 (equal to 0 to 3) being the coercive force. The complete figure, which will be traced out once for each complete cycle of an alternating current in the magnetising coil, is called the hysteresis loop.

Since coercive force must be applied to overcome the remanent magnetism, work is done in completing the loop, the energy concerned appearing as heat in the magnetic material. This heat is known as hysteresis loss, the amount of loss depending on the value of coercive force necessary. If the coercive force is high, the loop will be a fat one and the loss will be high; if the coercive force is low, the resulting thin loop will indicate a low hysteresis loss (Figure 3.8). It will be seen that the hysteresis loss can be measured in terms of the area enclosed by the hysteresis loop.

In practice, the reduction of hysteresis loss can be achieved by using iron containing silicon. The more the silicon added, the smaller will be the hysteresis loss, but the harder and more brittle will the magnetic material become. With a content of about 6% silicon, the material is very difficult to stamp, machine and drill. Silicon contents of up to about 5% are added to the magnetic steel used for stationary machines like transformers which are not subjected to centrifugal forces, and up to about 3% for the rotating parts of machine magnetic circuits.

Hysteresis loss occurs when a coil wound on a magnetic circuit carries an alternating current, owing to the reversing magnetic flux set up by the reversing current. The higher the frequency of the alternating current, the greater will be the hysteresis loss. This loss also occurs in the rotating parts of DC machines, which pass alternate north and south magnetic poles,

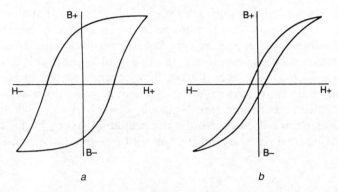

**Figure 3.8   Hysteresis loops for materials of high and low loss**
  *a* High-loss material
  *b* Low-loss material

and are thus subject to alternating magnetisation. In these cases, the hysteresis loss increases as the speed of rotation becomes greater.

## 3.6   Eddy-current loss

We have seen in Chapter 2 that any conductor which is subject to a changing magnetic flux will have an EMF induced in it. An alternating flux, such as that in the core of a transformer or the rotor of a DC machine, will be continuously changing and will thus induce an EMF in any conductor it cuts. The magnetic circuit itself will be an electrical conductor, and will have an alternating EMF induced in it by an alternating magnetic flux. This EMF will drive current against the resistance of the magnetic circuit; since the path of this current within the circuit is not easily defined, the current is referred to as an 'eddy current'.

Most magnetic circuits will behave as single turns of conductor, so the induced EMF will be low. However, the electrical resistance of a solid magnetic circuit will be very low, so the value of the eddy currents can be very high, giving rise to a considerable heat loss owing to the power dissipated in the core resistance ($P = I^2R$). The heat produced by eddy-current loss may not only raise the temperature of the core and possibly damage the insulation of the magnetising coil, but may also considerably reduce the efficiency of the machine concerned, because the heating power has to be provided in the form of additional input power.

As for hysteresis loss, eddy-current loss can never be completely removed. However, it can be reduced to an acceptable level by breaking up the magnetic circuit into small sections, each of which will have a lower EMF induced than the complete circuit, and will have higher resistance. Further reduction will be obtained by using a magnetic material having high resistivity. In practice, the usual method is to construct the magnetic circuit of thin sheets, called 'laminations', stacked together so that they appear to make a solid magnetic circuit but are, in fact, insulated from each other. Since the voltages concerned are very low, the insulation need not be very effective, and usually consists of a thin sheet of paper stuck to one side of each lamination, or a layer of varnish or of oxide on one or both sides.

The laminations are arranged so that the magnetic flux is set up along them, and not across them (see Figure 3.9). If the flux is set up across the laminations, it will have to cross the insulation layers, and since these are nonmagnetic the amount of flux set up for a given magnetising force

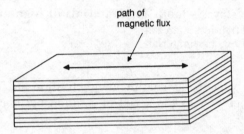

**Figure 3.9   Part of laminated magnetic circuit showing correct magnetic-flux direction**

will be much reduced. In many cases, it is easier to wind the magnetising coils separately, and to slide the completed coils onto the limbs of the magnetic circuit. To make this possible, the magnetic circuit must be built in parts, a joint occurring where two parts meet. To improve the magnetic fit of the parts, it is common to interleave laminations at the corners where joins are made as shown in Figure 3.10.

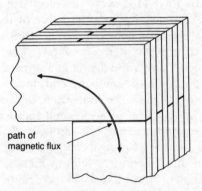

**Figure 3.10   Interleaved laminations at corner of magnetic circuit**

## 3.7  Iron losses

Both hysteresis loss and eddy-current loss give rise to heat in a magnetic circuit. The two losses are usually taken together and are called 'iron loss'.

Thus $$P_I = P_E + P_H$$

where  $P_I$ = iron loss (W)

$P_E$ = eddy-current loss (W)

$P_H$ = hysteresis loss (W)

Eddy-current loss is a loss due to current in a resistive system, which in this case is the magnetic circuit. Such a loss is an $I^2R$ loss, and depends on the square of the current. Current is driven by induced EMF which depends on the rate of change of magnetic flux, which depends in turn on the maximum value of flux, and of flux density, as well as on the frequency of the supply. Thus eddy-current loss depends on the square of the frequency and on the square of the maximum magnetic flux density.

$$P_E \propto f^2 B_{max}^2$$

where  $P_E$ = eddy-current loss

$f$ = supply frequency

$B_{max}$ = maximum magnetic flux density

An equation to enable the actual loss due to eddy currents to be calculated is extremely complicated, as is the equation for hysteresis loss. In practice, it is found that hysteresis loss depends directly on the supply frequency, and also on the maximum magnetic flux density. In the latter case, the relationship lies between the loss being directly proportional to $B$, and being proportional to the square of $B$.

$$P_H \propto f B_{max}^{1 \cdot 6}$$

where  $P_H$ = hysteresis loss

$f$ = supply frequency

$B_{max}$ = maximum magnetic flux density

Iron losses thus vary with both frequency and magnetic flux density. A power transformer is likely to be fed at constant frequency and at constant voltage, which means that magnetic flux density is almost constant. The iron losses in a transformer are therefore constant from no load to full load. With some modern magnetic materials used for transformer laminations these formulas tend to overestimate the losses.

The situation with rotating machines, particularly DC machines, is very much more complicated. The frequency of the alternating magnetic flux in the rotating parts depends on rotational speed and may be variable, as may be the magnetic flux density which depends on the variable field current.

## 3.8 Summary of formulas and proportionalities for Chapter 3

See text for definitions of symbols.

Magnetising force:

$$H = \frac{IN}{l} \qquad l = \frac{IN}{H} \qquad I = \frac{Hl}{N} \qquad N = \frac{Hl}{I}$$

Magnetic flux density:

$$B = \frac{\Phi}{A} \qquad \Phi = BA \qquad A = \frac{\Phi}{B}$$

$$\frac{B}{H} = \mu_0\mu_r \qquad B = \mu_0\mu_r H \qquad H = \frac{B}{\mu_0\mu_r}$$

Iron losses:

$$P_I = P_E + P_H \qquad P_E \propto f^2 B_{max}^2 \qquad P_H \propto f B_{max}^{1 \cdot 7}$$

## 3.9 Exercises

1 Sketch a magnetisation curve for
  (a) an air-cored coil,
  (b) a coil with a straight iron core (i.e. with a very large airgap), and
  (c) a coil wound on a closed iron magnetic circuit.

2 Sketch the magnetisation curves, labelling the axes, for coils with
  (a) an air core
  (b) an iron core which saturates fairly suddenly at a certain flux density, and
  (c) an iron core with an air gap.

3 Name, and indicate the approximate composition of, magnetic materials having the following properties:
  (a) easily magnetised and demagnetised,
  (b) low hysteresis loss and high permeability,
  (c) very low hysteresis loss and very high permeability at very low magnetising force, and
  (d) very low eddy-current loss.

4 What is meant by the term 'magnetic shielding'? Where may it be necessary, and how is it achieved?

5 Sketch a typical hysteresis loop for iron. Mark on the axes of the sketch
  (i) the remanent flux density
  (ii) the coercive force, and
  (iii) where saturation is being approached.

6 Using sketches of hysteresis loops to illustrate your answer, briefly state why silicon is added to some magnetic materials. What is the disadvantage of this process?

7 State the meaning of the term 'eddy current loss', and give two methods of reducing its value.

8 State briefly, with reasons, whether the steel for a transformer core should
(a) have a wide or a narrow hysteresis loop
(b) be made up of thick or thin plates or be solid
(c) have a low or high electrical resistance.

9 Make a labelled sketch of a square-cross-section core for a transformer, showing its detailed construction. Describe briefly its various components and the precautions to be taken during its assembly.

## 3.10 Multiple-choice exercises

3M1 The symbol for magnetic flux density and its unit are, respectively,
(a) tesla (T) and Weber (Wb)      (b) $B$ and tesla (T)
(c) $H$ and ampere-turns      (c) $A$ and Wb/m$^2$.

3M2 The formula relating magnetising force $H$ to magnetising current $I$, number of turns on the magnetising coil $N$ and the length of the magnetic flux path $l$ is:

(a) $H = \dfrac{I \times N}{l}$      (b) $\dfrac{B}{H} = \mu_0$

(c) $I = \dfrac{H \times N}{l}$      (d) $H = \dfrac{l}{N \times I}$

3M3 When an iron core is placed within an air-cored coil, the magnetic flux produced for a given current is increased. The number of times it increases is called the:
(a) multiplying factor      (b) relative permeability of the iron
(c) magnetic flux density      (d) magnetising force.

3M4 The magnetisation curve for a nonmagnetic material:
(a) can have any shape depending on the material concerned
(b) rises quickly at first but then flattens out
(c) is a straight line drawn through the origin of the curve
(d) shows the effect of magnetic saturation.

3M5 The permeability of free space is a constant with a value of:
(a) $8 \cdot 85 \times 10^{-12}$      (b) unity
(c) $0 \cdot 0000126$      (d) $4\pi \times 10^{-7}$

3M6 A ferromagnetic material will include at least one of the elements:
(a) iron, nickel and cobalt
(b) aluminium and copper
(c) hydrogen, oxygen and helium
(d) carbon and oxygen.

3M7　Magnetic shielding is the name given to:
   (a) switching off all heavy currents to prevent the presence of strong magnetic fields
   (b) keeping equipments sensitive to magnetic fields away from other equipment
   (c) surrounding magnetically sensitive equipment with high-permeability magnetic material
   (d) never using magnetically sensitive equipments.

3M8　The energy loss in a magnetic material due to remanent magnetism when it is subjected to alternating magnetisation is called:
   (a) iron loss　　　　　　　(b) remanence
   (c) lamination loss　　　　(d) hysteresis loss.

3M9　The hysteresis loop for one magnetic material is much thinner and encloses a smaller area than that for a second material drawn using the same scales. The hysteresis loss of the first material is:

   (a) much greater than
   (b) smaller than
   (c) the same as
   (d) not comparable with

   that of the second material.

3M10 Hysteresis loss can never be eliminated, but can be reduced by:
   (a) making the magnetic system with laminations
   (b) feeding the magnetic system with a direct current
   (c) using low-voltage systems
   (d) using a magnetic material containing a small percentage of silicon.

3M11 The energy loss in a magnetic material which is due to induced currents flowing within the magnetic material is called:
   (a) energy loss　　　　　　(b) eddy-current loss
   (c) power loss　　　　　　　(d) hysteresis loss.

3M12 Eddy-current loss can never be eliminated, but can be reduced by:
   (a) feeding the magnetic system with a direct current
   (b) using a magnetic material containing silicon
   (c) making the magnetic circuit with laminations
   (d) using low-hysteresis-loss magnetic materials.

3M13 The iron losses of a typical power transformer are constant because:
   (a) magnetic flux density and supply frequency are almost constant
   (b) the transformer will operate on low load
   (c) increasing eddy-current loss will be cancelled by reducing hysteresis loss
   (d) the reduction in frequency offsets increasing magnetic flux density.

## Chapter 4

# Alternating-current theory

## 4.1　Introduction

Provided that the effective, or root-mean-square (RMS) values of voltage and current are used, Ohm's law applies to both AC and DC circuits.

However, reactive systems (those containing inductors or capacitors) behave quite differently. To begin with, both inductors and capacitors have an effect which limits current when an alternating voltage is applied. These effects are called inductive reactance (for an inductor) or capacitive reactance (for a capacitor), and are found from the formulas

$$X_L = 2\pi f L \qquad X_C = \frac{1}{2\pi f C} = \frac{10^6}{2\pi f C'}$$

where $X_L$ = inductive reactance ($\Omega$)

　　　$X_C$ = capacitive reactance ($\Omega$)

　　　$f$ = supply frequency (Hz)

　　　$L$ = circuit inductance (H)

　　　$C$ = circuit capacitance (F)

　　　$C'$ = circuit capacitance ($\mu$F)

The second formula for capacitive reactance is given to help calculation when capacitance is given in microfarads. Using capacitive and inductive reactance alone,

$$I = \frac{V}{X_C} \quad \text{and} \quad I = \frac{V}{X_L}$$

where $I$ = circuit current (A)

　　　$V$ = supply voltage (V)

There is a phase difference between current and voltage for a reactive circuit.

In the pure inductive circuit, current LAGS voltage by 90°, and in the pure capacitive circuit current LEADS voltage by 90°. The phase relationships are shown in phasor form in Figure 4.1.

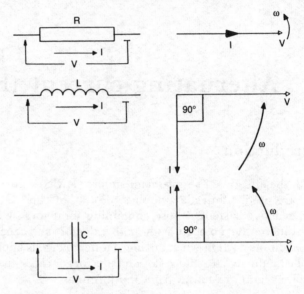

**Figure 4.1   Phase relationships between voltage and current for pure resistive, inductive and capacitive circuits.**

## 4.2   Resistive and inductive series circuit

An inductor consists of a length of wire wound into a coil. This wire will usually have significant resistance, so the concept of a pure inductor is not a practical one. In fact, resistance and inductance will be mingled, each turn of the coil having both properties. This effect can be represented by a pure inductor in series with a resistor of value equal to the resistance of the coil, the circuit arrangement being shown in Figure 4.2a. Taking the current, which is common to both resistor and inductor, as reference, the phasor diagram can be drawn as indicated in Figure 4.2b or c. These figures show alternative methods of drawing the same diagram.

Since voltage and current must be in phase for a resistor, the PD $V_R$ across the resistor must be in phase with the current $I$. Current lags by 90° in a pure inductive circuit, so the PD $V_L$ across the inductor leads the current $I$ by 90°. (NB: If current lags voltage then voltage leads current. Remember also that phasors are assumed to rotate anticlockwise at $\omega$ (omega) rad/s.) The supply voltage $V$ is the phasor sum of $V_R$ and $V_L$, and is found by completing the parallelogram in Figure 4.2b, or the triangle in Figure 4.2c. The angle between supply voltage and current is called the 'phase angle', and is given by the symbol $\phi$ (Greek, small 'phi'). In this

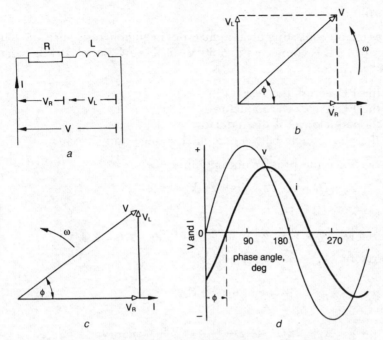

**Figure 4.2** **Circuit, phasor and wave diagrams for resistive and inductive AC series circuit**

case, current lags supply voltage by $\phi°$. This phase angle is shown on the phasor diagrams as well as on the wave diagram Figure 4.2*d*.

It can be seen from Figure 4.2*c* that the voltages $V$, $V_R$ and $V_L$ together form a right-angled triangle, to which Pythagorus's theorem applies. Thus

$$V^2 = V_R^2 + V_L^2$$

Simple trigonometry also applies, so that

$$\cos \phi = \frac{V_R}{V} \qquad \sin \phi = \frac{V_L}{V} \qquad \tan \phi = \frac{V_L}{V_R}$$

## Example 4.1

A resistor and an inductor of negligible resistance are connected in series to an AC supply. The PD across the resistor is found to be 24 V, while that across the inductor is measured at 32 V. Calculate the supply voltage and the phase angle between circuit voltage and current.

$$V^2 = V_R^2 + V_L^2 = 24^2 + 32^2 = 576 + 1024 = 1600 \text{ V}^2$$

Therefore

$$V = \sqrt{V^2} = \sqrt{1600} \text{ volts} = 40 \text{ V}$$

$$\tan \phi = \frac{V_L}{V_R} = \frac{32}{24} = 1 \cdot 333$$

From the tangent tables, $\phi = 53 \cdot 1°$

## Example 4.2

A series circuit consisting of an inductor of negligible resistance and a pure resistor of 12 $\Omega$ is connected to a 30 V, 50 Hz AC supply. If the current is 2 A, calculate:

(a) the PD across the resistor
(b) the PD across the inductor
(c) the inductance of the inductor
(d) the phase angle between applied voltage and current.

(a) Since the resistor obeys Ohm's law,

$$V_R = IR = 2 \times 12 \text{ volts} = 24 \text{ V}$$

(b) $V^2 = V_R^2 + V_L^2$

Therefore $V_L^2 = V^2 - V_R^2 = 30^2 - 24^2 = 900 - 576 = 324 \text{ V}^2$

Therefore $V_L = \sqrt{324}$ volts $= 18$ V

(c) $$X_L = \frac{V_L}{I} = \frac{18}{2} \text{ ohms} = 9 \ \Omega$$

$$X_L = 2\pi f L, \text{ so } L = \frac{X_L}{2\pi f} = \frac{9}{2\pi \times 50} \text{ henrys}$$

$$= 0 \cdot 0286 \text{ H or } 28 \cdot 6 \text{ mH}$$

(d) $$\tan \phi = \frac{V_L}{V_R} = \frac{18}{24} = 0 \cdot 75$$

From tangent tables, $\phi = 36 \cdot 9°$

For a resistor, the effect limiting current for a given voltage is resistance, and $R = V/I$. For a pure inductor, the effect is inductive reactance, and $X_L = V/I$. Thus, it can be seen that the ratio $V/I$ is the effect limiting the current in a circuit; for a circuit including resistance and inductive reactance this effect is due to a combination of both. In Figure 4.2a, the PD across the resistor obeys Ohm's law, so that $V_R = IR$.

Similarly, $V_L = IX_L$

But $V^2 = V_R^2 + V_L^2 = (IR)^2 + (IX_L)^2 = I^2R^2 + I^2X_L^2$

Therefore $\dfrac{V^2}{I^2} = R^2 + X_L^2$

and $\dfrac{V}{I} = \sqrt{(R^2 + X_L^2)}.$

For an AC circuit, the ratio $V/I$ is known as **'impedance'** (symbol $Z$) measured in ohms. The impedance of a circuit in ohms may be defined as the applied voltage in volts needed to drive a current of one ampere.

Therefore
$$Z = \frac{V}{I} = \sqrt{(R^2 + X_L^2)}$$

or
$$Z^2 = R^2 + X_L^2$$

We can see that this latter formula would apply to a right-angled triangle. Such a triangle, called the 'impedance triangle', is shown in Figure 4.3. The angle $\phi$ included between sides $R$ and $Z$ is the same as the phase angle between current and voltage for a given circuit. Applying simple trigonometry,

$$\cos\phi = \frac{R}{Z} \qquad \sin\phi = \frac{X_L}{Z} \qquad \tan\phi = \frac{X_L}{R}$$

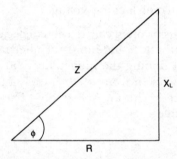

**Figure 4.3   Impedance triangle for resistive and inductive AC series circuit**

## Example 4.3

Calculate the impedance of the circuit in Example 4.2.

$$R = 12 \ \Omega \quad \text{and} \quad X_L = 9 \ \Omega$$

Therefore $\quad Z = \sqrt{(R^2 + X_L^2)} = \sqrt{(12^2 + 9^2)}$ ohms $= \sqrt{(144 + 81)}$ ohms

$$= \sqrt{225} \text{ ohms} = 15 \ \Omega,$$

We can check this by calculating the circuit current.

$$I = \frac{V}{Z} = \frac{30}{15} \text{ amperes} = 2 \text{ A}$$

## Example 4.4

A coil of inductance $1 \cdot 59$ mH and negligible resistance is connected in series with a 10 $\Omega$ resistor to a $2 \cdot 4$ V, 500 Hz supply. Calculate:

(a) the inductive reactance of the coil,
(b) the impedance of the circuit,
(c) the current in the circuit,

(*d*) the PD across each component, and

(*e*) the circuit phase angle.

(*a*) $X_L = 2\pi f L = 2\pi \times 500 \times 1\cdot59 \times 10^{-3}$ ohms $= 5\ \Omega$

(*b*) $Z = \sqrt{(R^2 + X_L^2)} = \sqrt{(10^2 + 5^2)}$ ohms $= 11\cdot2\ \Omega$

(*c*) $I = \dfrac{V}{Z} = \dfrac{2\cdot4}{11\cdot2}$ amperes $= 0\cdot215$ A

(*d*) $V_R = IR = 0\cdot215 \times 10$ volts $= 2\cdot15$ V

$V_L = IX_L = 0\cdot215 \times 5$ volts $= 1\cdot07$ V

(*e*) $\tan\phi = \dfrac{V_L}{V_R} = \dfrac{1\cdot07}{2\cdot15} = 0\cdot5$

$\phi = 26\cdot6°$, current lagging voltage.

An iron-cored inductance is often called a choke because its high inductive reactance has a 'choking' effect and limits the alternating current through it. A simple method of finding the inductance and reactance of a coil is to measure the current taken from a DC supply of known voltage, and then to repeat the process with an AC supply. The method is illustrated by the following example.

## Example 4.5

A choke takes a current of $2\cdot4$ A from a 12 V DC supply. When connected to a 195 V, 50 Hz AC supply the current is 15 A. Calculate the resistance, impedance, reactance and inductance of the coil. Neglect iron loss.

$$R = \frac{V_{DC}}{I_{DC}} = \frac{12}{2\cdot4} \text{ ohms} = 5\ \Omega$$

$$Z = \frac{V_{AC}}{I_{AC}} = \frac{195}{15} \text{ ohms} = 13\ \Omega$$

Since   $X_L^2 = Z^2 - R^2$ (from the impedance triangle — Figure 4.3)

$X_L = \sqrt{(Z^2 - R^2)} = \sqrt{(13^2 - 5^2)}$ ohms $= 12\ \Omega$

$X_L = 2\pi f L$ so $L = \dfrac{X_L}{2\pi f} = \dfrac{12}{2\pi \times 50}$ henrys $= 0\cdot0382$ H

## Example 4.6

A coil of inductance $0\cdot636$ H takes a current of 1 A from a 240 V, 50 Hz supply. What is the resistance of the coil?

$$Z = \frac{V}{I} = \frac{240}{1} \text{ ohms} = 240\ \Omega$$

$$X_L = 2\pi f L = 2\pi \times 50 \times 0\cdot636 \text{ ohms} = 200\ \Omega$$

From the impedance triangle (Figure 4.3),

$$R^2 = Z^2 - X_L^2$$

so $$R = \sqrt{(Z^2 - X_L^2)} = \sqrt{(240^2 - 200^2)} \text{ ohms} = 133 \ \Omega$$

## 4.3  Resistive and capacitive series circuit

Some types of capacitor, notably the electrolytic type, have considerable current leakage through their dielectrics. The effect of this leakage is similar to that of a resistor in series with the capacitor. In other cases, such as the mica- or paper-dielectric capacitor, the leakage is so small that it can be ignored altogether for practical purposes. Whatever type of capacitor is used, a resistor is often connected in series, so the concept of a resistive and capacitive series circuit is an important one.

The circuit arrangement is shown in Figure 4.4a. As in the inductive-resistive circuit, current is taken as the reference phasor (Figure 4.4b). The PD $V_R$ across the resistor will be in phase with this current, while the PD $V_C$ across the capacitor will lag the current by 90°. (NB: If current leads voltage, then voltage lags current.) The supply voltage $V$ is the phasor sum

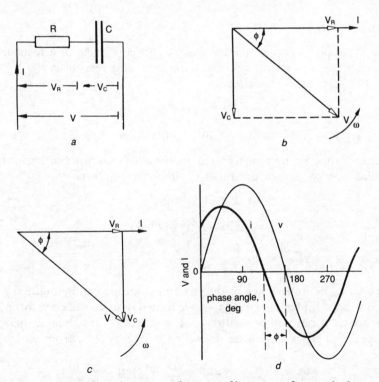

**Figure 4.4  Circuit, phasor and wave diagrams for resistive and capacitive AC series circuit**

of the component voltages $V_R$ and $V_C$, and is found by completing the parallelogram as in Figure 4.4b. An alternative method of drawing the phasor diagram is shown in Figure 4.4c. As for the resistive-inductive circuit, the angle between the voltage and current is the 'phase angle' and is denoted by the symbol $\phi$. Since current leads voltage in this case, the angle is called a 'leading phase angle'. Figure 4.4d shows the wave diagram for the circuit.

From Figure 4.4c the three voltages $V_R$, $V_C$ and $V$ form the sides of a right-angled triangle to which Pythagorus's theorem applies.

$$\text{Thus} \quad V^2 = V_R^2 + V_C^2$$

Simple trigonometry also applies, so that

$$\cos \phi = \frac{V_R}{V} \qquad \sin \phi = \frac{V_C}{V} \qquad \tan \phi = \frac{V_C}{V_R}$$

## Example 4.7

A capacitor and a resistor are connected in series, the PD across each being 12 V and 16 V, respectively. Calculate the supply voltage.

$$V = \sqrt{(V_R^2 + V_C^2)} = \sqrt{(12^2 + 16^2)} \text{ volts} = 20 \text{ V}$$

## Example 4.8

A 240 V supply is connected across a circuit consisting of a resistor and a capacitor connected in series. If the PD across the resistor is 200 V, what is the PD across the capacitor?

$$V^2 = V_R^2 + V_C^2 \quad \text{so} \quad V_C^2 = V^2 - V_R^2$$

Therefore $V_C = \sqrt{(V^2 - V_R^2)} = \sqrt{(240^2 - 200^2)} \text{ volts} = 133 \text{ V}$

The impedance of the resistive–capacitive series circuit can be found in a similar way to that of the resistive–inductive circuit.

$$V^2 = V_R^2 + V_C^2$$

$$(IZ)^2 = (IR)^2 + (IX_C)^2$$

Therefore $Z = \sqrt{(R^2 + X_C^2)} = \dfrac{V}{I}$

Once again, the impedance of the circuit (in ohms) can be defined as the applied voltage (in volts) needed to drive a current of one ampere through it.

The formula can also be written as $Z^2 = R^2 + X_C^2$. Once again, therefore, an impedance triangle can be drawn as in Figure 4.5, from which

$$\cos \phi = \frac{R}{Z} \qquad \sin \phi = \frac{X_C}{Z} \qquad \tan \phi = \frac{X_C}{R}$$

Notice that the impedance triangle for the $RC$ circuit is drawn the other

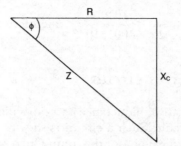

**Figure 4.5    Impedance triangle for resistive and capacitive AC circuit**

way up from that for the *RL* circuit. The reason for this will become apparent in Section 4.4.

## Example 4.9

A 15·9 nF capacitor and a 20 kΩ resistor are connected in series to a 5 V, 1 kHz supply. Calculate (*a*) circuit impedance, (*b*) circuit current, (*c*) the PD across each component, and (*d*) the circuit phase angle.

(*a*)    $X_C = \dfrac{10^9}{2\pi fC'} = \dfrac{10^9}{2\pi \times 1000 \times 15\cdot9}$ ohms = 10 kΩ

$Z = \sqrt{(R^2 + X_C^2)} = \sqrt{(10^2 + 20^2)}$ kilohms = 22·4 kΩ

(*b*)    $I = \dfrac{V}{Z} = \dfrac{5}{22\cdot4 \times 10^3}$ amperes = 0·224 mA

(*c*)    $V_R = IR = 0\cdot224 \times 10^{-3} \times 20 \times 10^3$ volts = 4·48 V

$V_C = IX_C = 0\cdot224 \times 10^{-3} \times 10 \times 10^3$ volts = 2·24 V

(*d*)    $\tan\phi = \dfrac{V_C}{V_R} = \dfrac{2\cdot24}{4\cdot48} = 0\cdot500$

From the tangent relationship, $\phi = 26\cdot6°$, current leading voltage.

## Example 4.10

A 10 Ω resistor and capacitor of unknown value are connected in series across a 115 V, 400 Hz supply. If a voltmeter connected across the resistor reads 80 V, what is the value of the capacitor?

$V_C = \sqrt{(V^2 - V_R^2)} = \sqrt{(115^2 - 80^2)}$ volts = 82·6 V

$I = \dfrac{V_R}{R} = \dfrac{80}{10}$ amperes = 8 A

$X_C = \dfrac{V_C}{I} = \dfrac{82\cdot6}{8}$ ohms = 10·3 Ω

$$C' = \frac{10^6}{2\pi f X_C} = \frac{10^6}{2\pi \times 400 \times 10 \cdot 3} \text{ microfarads} = 38 \cdot 6 \text{ } \mu\text{F}$$

## 4.4   General series circuit

A circuit having resistance, inductance and capacitance in series is known as a general series circuit. Such a circuit is shown in Figure 4.6*a*, whilst Figure 4.6*b* shows a typical phasor diagram. Current is taken as reference since the same current passes through all three components. The PD $V_R$ across the resistor is in phase with the current, while the PD $V_L$ across the inductor and the PD $V_C$ across the capacitor lead by 90° and lag by 90°, respectively. These two last-mentioned voltages are in direct opposition, so the effective value is the difference between them.

Thus $\qquad\qquad\qquad\qquad V^2 = V_R^2 + (V_L - V_C)^2$

and $\qquad\qquad\qquad\quad (IZ)^2 = (IR)^2 + (IX_L - IX_C)^2$

Therefore $\qquad\qquad\qquad Z = \sqrt{\{R^2 + (X_L - X_C)^2\}}$

From Figure 4.6*b*,

$$\cos\phi = \frac{V_R}{V} \qquad \sin\phi = \frac{V_L - V_C}{V} \qquad \tan\phi = \frac{V_L - V_C}{V_R}$$

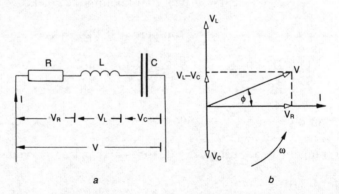

**Figure 4.6   Circuit and phasor diagrams for resistive, inductive and capacitive AC series circuit**

The impedance diagram for the circuit is shown in Figure 4.7; since inductive and capacitive reactance are drawn in different directions, their difference $(X_L - X_C)$ becomes the effective circuit reactance. From Figure 4.7,

$$\cos\phi = \frac{R}{Z} \qquad \sin\phi = \frac{X_L - X_C}{Z} \qquad \tan\phi = \frac{X_L - X_C}{R}$$

The phasor diagram of Figure 4.6 and the impedance diagram Figure 4.7 are drawn for a condition where $X_L$ is greater than $X_C$, so that $V_L$ exceeds $V_C$. In this case, inductive reactance is greater, and circuit current lags supply voltage. Had $X_C$ been greater than $X_L$, with $V_C$ exceeding $V_L$, the circuit current would lead the supply voltage (see Example 4.12).

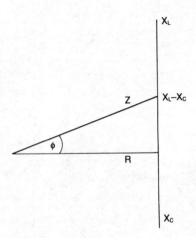

**Figure 4.7   Impedance diagram for resistive, inductive and capacitive AC series circuit**

## Example 4.11

A choke of inductance $0 \cdot 318$ H and resistance 30 $\Omega$ is connected in series with a 53 µF capacitance across a 24 V, 50 Hz supply. Calculate the circuit impedance, current and phase angle. Draw scale phasor and impedance diagrams.

$$X_L = 2\pi f L = 2\pi \times 50 \times 0 \cdot 318 \text{ ohms} = 100 \ \Omega$$

$$X_C = \frac{10^6}{2\pi fC'} = \frac{10^6}{2\pi \times 50 \times 53} \text{ ohms} = 60 \ \Omega$$

$$Z = \sqrt{\{R^2 + (X_L - X_C)^2\}} = \sqrt{\{30^2 + (100 - 60)^2\}} \text{ ohms} = 50 \ \Omega$$

$$I = \frac{V}{Z} = \frac{24}{50} \text{ amperes} = 0 \cdot 48 \text{ A}$$

$$V_L = IX_L = 0 \cdot 48 \times 100 \text{ volts} = 48 \text{ V}$$

$$V_C = IX_C = 0 \cdot 48 \times 60 \text{ volts} = 28 \cdot 8 \text{ V}$$

$$V_R = IR = 0 \cdot 48 \times 30 \text{ volts} = 14 \cdot 4 \text{ V}$$

$$\tan \phi = \frac{V_L - V_C}{V_R} = \frac{48 - 28 \cdot 8}{14 \cdot 4} = \frac{19 \cdot 2}{14 \cdot 4} = 1 \cdot 33$$

From tangent tables, $\phi = 53 \cdot 1°$ lagging

Phasor and impedance diagrams are as shown in Figure 4.8.

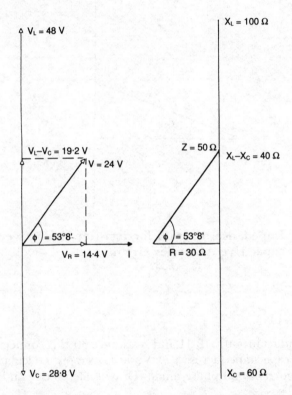

**Figure 4.8   Phasor and impedance diagrams for Example 4.11**

## Example 4.12

The circuit of Example 4.11 remains unchanged, but while the supply voltage is maintained constant at 24 V, the frequency is reduced to 25 Hz. Calculate the circuit impedance, the circuit current and the circuit phase angle. Draw to scale phasor and impedance diagrams.

$$X_L = 2\pi f L = 2\pi \times 25 \times 0 \cdot 318 \text{ ohms} = 50 \ \Omega$$

$$X_C = \frac{10^6}{2\pi f C'} = \frac{10^6}{2\pi \times 25 \times 53} \text{ ohms} = 120 \ \Omega$$

NB: Note that halving the frequency has halved the inductive reactance and doubled the capacitive reactance.

$$Z = \sqrt{\{R^2 + (X_L - X_C)^2\}} = \sqrt{\{30^2 + (50 - 120)^2\}} \text{ ohms}$$

$$= \sqrt{\{30^2 + (-70)^2\}} \text{ ohms} = 76\cdot2 \ \Omega$$

$$I = \frac{V}{Z} = \frac{24}{76\cdot2} \text{ amperes} = 0\cdot315 \text{ A}$$

$$\cos \phi = \frac{R}{Z} = \frac{30}{76\cdot2} = 0\cdot3937$$

From the cosine tables, $\phi = 66\cdot8°$ leading.

NB: Since $X_C$ is greater than $X_L$, current leads voltage. The scale phasor and impedance diagrams are shown in Figure 4.9. Voltage values are not worked here, but can be calculated as in Example 4.11. The values are shown on the phasor diagram.

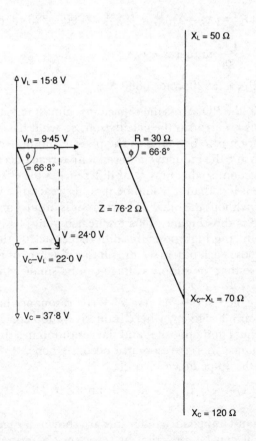

**Figure 4.9   Phasor and impedance diagrams for Example 4.12**

## 4.5   Series resonance

Look again at the last two examples. In Example 4.11, with a supply voltage of 24 V, the PD across the inductive component of the choke was 48 V, while that across the capacitor was $28 \cdot 8$ V. In Example 4.12, with the same supply voltage, the PD across the capacitor was $37 \cdot 8$ V. At first sight this appears impossible; now consider the following example.

### Example 4.13

A $14 \cdot 5$ µF capacitor and a $0 \cdot 637$ H inductor of series resistance 15 $\Omega$ are connected in series across a 100 V, 50 Hz supply. Calculate the circuit impedance, the current and the PD across the capacitor.

$$X_L = 2\pi fL = 2\pi \times 50 \times 0 \cdot 637 \text{ ohms} = 200 \ \Omega$$

$$X_C = \frac{10^6}{2\pi fC'} = \frac{10^6}{2\pi \times 50 \times 14 \cdot 5} \text{ ohms} = 220 \ \Omega$$

$$Z = \sqrt{\{R^2 + (X_L - X_C)^2\}} = \sqrt{\{15^2 + (200 - 220)^2\}} \text{ ohms} = 25 \ \Omega$$

$$I = \frac{V}{Z} = \frac{100}{25} \text{ amperes} = 4 \text{ A}$$

$$V_C = IX_C = 4 \times 220 \text{ volts} = 880 \text{ V}$$

In Example 4.13, the PD across the capacitor is almost nine times the supply voltage. There is no error in the calculation, so what has happened? The clue to the situation is to be found in the phasor and impedance diagrams of Figure 4.9. While the inductive and capacitive reactances are both high, they act in opposition, so that only their difference is the effective reactance. The impedance is less than it would be if either reactance were connected with the resistor without the other. The current is limited only by this lower impedance, and is thus higher. This greater current passes through both reactances, producing high potential differences across them, but as the two voltages oppose each other (see the phasor diagram of Figure 4.9) they tend to cancel, so that the supply voltage can be smaller than both or one of them.

This effect is called 'series resonance'. True resonance has not occurred in any of the examples given, where inductive and capacitive reactances have not been equal and opposite, and the condition has simply been close to that of resonance. Series resonance occurs when $X_L = X_C$.

Considering the impedance formula,

$$Z = \sqrt{\{R^2 + (X_L - X_C)^2\}}, \ X_L - X_C = 0, \text{ and } Z = \sqrt{(R^2 + 0^2)} = \sqrt{R^2} = R.$$

Circuit, phasor and impedance diagrams are shown in Figure 4.10. From these diagrams it can be seen that at resonance a series circuit has the following properties:

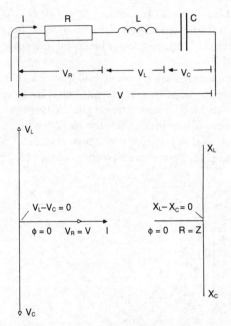

**Figure 4.10   Circuit, phasor and impedance diagrams for a series-resonant circuit**

(i) The current and supply voltage are in phase so that $\phi = 0°$.

(ii) The circuit impedance $Z$ will be equal to the circuit resistance $R$, and will thus be at its minimum possible value due to the cancellation of reactances.

(iii) The current will be at a maximum value due to minimum impedance. The series-resonant effect, with inductive and capacitive reactances equal and opposite, may be brought about in a number of ways:

(a) Change in inductance, giving a proportional change in inductive reactance (note that $X_L = 2\pi fL$, so $X_L \propto L$ if $f$ is constant).

(b) Change in capacitance, giving an inversely proportional change in capacitive reactance

(note that $X_C = \dfrac{1}{2\pi fC}$   so   $X_C \propto \dfrac{1}{C}$ if $f$ is constant).

(c) Change in frequency. If $L$ and $C$ are constant, $X_L \propto f$ and $X_C \propto 1/f$, so an increase in frequency will increase inductive reactance and decrease capacitive reactance. For example, look back to Examples 4.11 and 4.12. In both cases, a circuit had a self inductance of $0\cdot318$ H and a capacitance of 53 µF. In Example 4.11, with a frequency of 50 Hz, the reactances were $X_L = 100$ Ω and $X_C = 60$ Ω. In Example 4.12, with a frequency of 25 Hz, $X_L = 50$ Ω, while $X_C = 120$ Ω. At some frequency between these two values (in fact at 38·8 Hz) the inductive and capacitive reactances would be equal and series resonance would occur.

Figure 4.11 shows a number of values all plotted to a common base of frequency. The circuit resistance $R$ is unaffected by frequency change and remains constant. Inductive reactance $X_L$ increases with frequency, while capacitive reactance $X_C$ reduces. At the point where these two curves intersect, inductive and capacitive reactances become equal but opposite, and cancel, leaving the circuit resistance alone to limit current. The

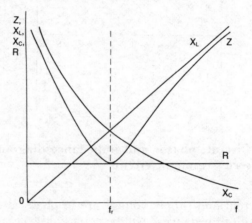

**Figure 4.11   Variation of resistance, reactance and impedance with frequency in a general series circuit**

impedance falls to a minimum, equal to the resistance at this frequency, which is called the **resonant frequency**, and given the symbol $f_r$. As frequency increases, inductive reactance exceeds capacitive reactance, and their difference joins with resistance to increase the impedance again. Figure 4.12 shows how the current in the circuit will vary as frequency changes, reaching a peak at the resonant frequency $f_r$. Because impedance is minimum and current maximum, series resonance is sometimes called '**acceptor resonance**' and occurs in 'acceptor circuits'.

Series resonance has useful applications in filter circuits, where one frequency is allowed to pass through while others are blocked, and in ripple control, where street lights or other equipment may be switched remotely by superimposing on the supply mains a higher frequency at which the operating circuits will resonate. In many other cases, however, its high current and high component voltages present dangers to be avoided.

A formula from which the resonant frequency may be determined is

$$f_r = \frac{1}{2\pi\sqrt{(LC)}} = \frac{10^3}{2\pi\sqrt{(LC')}}$$

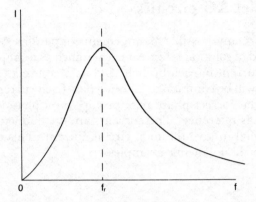

**Figure 4.12** **Variation of current with frequency in a general series circuit**

where        $f_r$ = resonant frequency of series circuit (Hz)

   $L$ = circuit inductance (H)

   $C$ = circuit capacitance (F)

   $C'$ = circuit capacitance ($\mu$F)

## Example 4.14

A 10 $\Omega$ resistor, a pure inductor of $0\cdot398$ H and a $6\cdot37$ $\mu$F capacitor are connected to a 100 V variable-frequency supply. The frequency is altered until the current reaches a maximum value, which occurs at 100 Hz. Calculate the value of the current, and the PDs across the inductor and the capacitor.

   Since the current is maximum, resonance occurs and impedance is equal to resistance.

Thus        $I = \dfrac{V}{Z} = \dfrac{V}{R} = \dfrac{100}{10}$ amperes = 10 A

$$X_L = 2\pi fL = 2 \times 3\cdot142 \times 100 \times 0\cdot398 \text{ ohms} = 250 \text{ } \Omega$$

$$V_L = IX_L = 10 \times 250 \text{ volts} = 2500 \text{ V or } 2\cdot5 \text{ kV}$$

Since this is a condition of resonance, capacitive and inductive reactances should be equal, as should the PDs across the inductor and the capacitor.

$$X_C = \frac{10^6}{2\pi fC'} = \frac{10^6}{2 \times 3\cdot142 \times 100 \times 6\cdot37} \text{ ohms} = 250 \text{ } \Omega$$

$$V_C = IX_C = 10 \times 250 \text{ volts} = 2500 \text{ V or } 2\cdot5 \text{ kV}$$

## 4.6  Parallel AC circuits

There are very many possible arrangements of parallel AC circuits. The simplest method of solution is to treat each branch as a simple series circuit. Calculate the current in each branch, and its phase relative to the supply voltage, which will be common to all branches. Each current is then drawn to scale, with the correct phase relationship, on a phasor diagram, with supply voltage as reference. The currents are then added using a simple geometric method to give the total circuit current. This type of solution is illustrated in the following examples.

### Example 4.15

A 60 Ω resistor, a pure 0·382 H inductor and a 66·3 µF capacitor are connected in parallel to a 240 V 50 Hz supply. Calculate the current taken from the supply and its phase angle relative to the supply voltage.

First, calculate the current in each branch. The circuit diagram is shown in Figure 4.13*a*.

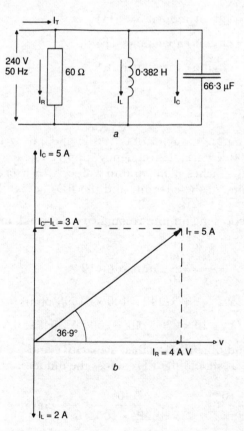

**Figure 4.13    Circuit and phasor diagrams for Example 4.15**

$$I_R = \frac{V}{R} = \frac{240}{60} \text{ amperes} = 4 \text{ A in phase with the voltage}$$

$$X_L = 2\pi f L = 2 \times 3 \cdot 142 \times 50 \times 0 \cdot 382 \text{ ohms} = 120 \ \Omega$$

$$I_L = \frac{V}{X_L} = \frac{240}{120} \text{ amperes} = 2 \text{ A lagging voltage by } 90°,$$

$$X_C = \frac{10^6}{2\pi f C'} = \frac{10^6}{2 \times 3 \cdot 142 \times 50 \times 66 \cdot 3} \text{ ohms} = 48 \ \Omega$$

$$I_C = \frac{V}{X_C} = \frac{240}{48} \text{ amperes} = 5 \text{ A leading voltage by } 90°$$

These three currents are drawn to scale on a phasor diagram as shown in Figure 4.13$b$. Since $I_L$ is directly opposing $I_C$, their difference $I_C - I_L$ is 5 A – 2 A = 3 A as shown. This current is added to $I_R$ by completing the parallelogram, and the phasor $I_T$ representing the current from the supply is drawn in. By measurement,

$$I_T = 5 \text{ A, and } \phi = 36 \cdot 9° \text{ leading voltage}$$

Thus the total current is 5 A, leading supply voltage by $36 \cdot 9°$.

## Example 4.16

A circuit, shown in Figure 4.14$a$, consists of a choke of resistance 15 $\Omega$ and inductance $7 \cdot 96$ mH, connected in parallel with a $7 \cdot 96$ μF capacitor, across a 125 V, 400 Hz supply. Calculate the current in each branch, and the current from the supply with its phase relative to supply voltage.

For the resistive–inductive series branch,

$$X_L = 2\pi f L = 2 \times 3 \cdot 142 \times 400 \times \frac{7 \cdot 96}{1000} \text{ ohms} = 20 \ \Omega$$

$$Z_1 = \sqrt{(R^2 + X_L^2)} = \sqrt{(15^2 + 20^2)} \text{ ohms} = 25 \ \Omega$$

$$I_1 = \frac{V}{Z_1} = \frac{125}{25} \text{ amperes} = 5 \text{ A}$$

$$\cos \phi = \frac{R}{Z} = \frac{15}{25} = 0 \cdot 6 \text{ lagging, since the circuit is inductive}$$

From the cosine table, $\phi = 53 \cdot 1°$ lagging

The current $I_1$ can now be drawn to scale on the phasor diagram (Figure 4.14$b$).

NB: the fractions of angle cannot be drawn with accuracy and may be ignored.

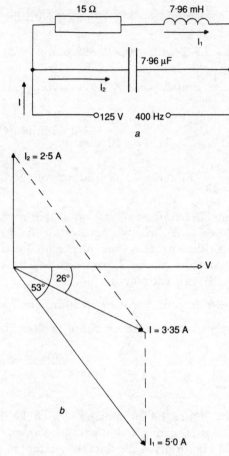

**Figure 4.14   Circuit and phasor diagrams for Example 4.16**

Turning to the capacitive branch,

$$X_C = \frac{10^6}{2\pi f C'} = \frac{10^6}{2 \times 3 \cdot 142 \times 400 \times 7 \cdot 96} \text{ ohms} = 50 \text{ } \Omega$$

$$I_2 = \frac{V}{Z_2} = \frac{V}{X_C} = \frac{125}{50} \text{ amperes} = 2 \cdot 5 \text{ A}$$

As this branch is purely capacitive, this current will lead supply voltage by 90°, and can be drawn to scale on Figure 4.14*b* as shown.

The phasor sum of $I_1$ and $I_2$ is found by completing the parallelogram, and by scale measurement,

$$I = 3 \cdot 35 \text{ A and } \phi = 26° \text{ lagging.}$$

Thus supply current is $3 \cdot 35$ A lagging supply voltage by 26°.

## 4.7   Parallel resonance

Figure 4.14*b* shows a scale phasor diagram for a parallel circuit, the supply current lagging the supply voltage by 26°. If the supply frequency were increased, the capacitive reactance would become smaller, and the current $I_2$ would become larger, although its phase relative to the constant supply voltage would not be affected. The increased frequency would increase the inductive reactance, and hence the impedance, of branch 1, which would reduce the value of the current $I_1$ and increase its angle of lag relative to supply voltage. If $I_2$ increases and $I_1$ reduces, both with increasing frequency, a point will be reached where the supply current $I$, which is the phasor sum of $I_1$ and $I_2$, is in phase with the supply voltage. Circuit and phasor diagrams for such a condition are shown in Figure 4.15.

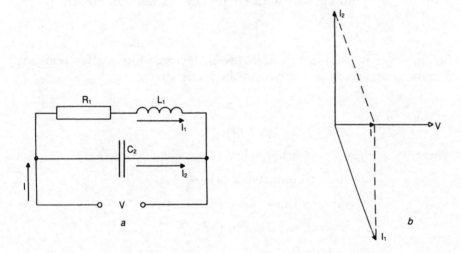

**Figure 4.15   Circuit and phasor diagrams for parallel-resonant circuit**

It will be noticed from Figure 4.15*b* that supply current will be small, since it is the phasor sum of the branch currents which, although large, are acting in opposition. This is the condition of **parallel resonance**, where current oscillates round the parallel circuit at the resonant frequency, the only current taken from the supply being that needed to supply the power loss in the resistance of the inductive branch. Curves showing the variation of branch impedances, total impedance and current with changing frequency are shown in Figure 4.16. These show that the supply current is minimum at resonance, consistent with the maximum value of impedance. For this reason, parallel resonance is often called **rejector resonance**, and occurs in a **rejector circuit**. Provided that the resistance of the inductor is small

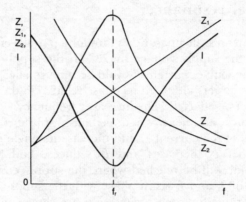

**Figure 4.16    Variation of branch impedances, total impedance and current with frequency for parallel circuit**

(it can never be zero in practical cases), the same formula for resonant frequency applies as in the case of the series circuit.

$$f_r = \frac{1}{2\pi\sqrt{(LC)}} = \frac{10^3}{2\pi\sqrt{(LC')}}$$

where                    $f_r$ = resonant frequency of parallel circuit (Hz)

$L$ = branch inductance (H)

$C$ = branch capacitance (F)

$C'$ = branch capacitance ($\mu$F)

This formula for the resonant frequency of parallel circuits must be treated with caution, because in many cases the resistance associated with the inductor will be large enough to make its use suspect. For example, if the simple formula shown is applied to the circuit values of Example 4.16, the calculated value for the resonant frequency is 632 Hz. If a more complex and correct formula is applied (it is not given here) the true resonant frequency for the circuit is found to be 557 Hz, which is some 14% less than that found by the simple formula.

The high impedance associated with parallel circuits is most often used in the tuning circuit of a radio or television receiver, where maximum amplification of the signal received occurs when circuit impedance is maximum. The device thus amplifies the station broadcasting on the frequency at which the parallel circuit resonates, but not those at other frequencies. The station received can be changed by using a variable capacitor in the parallel circuit (called a 'tuned circuit' in this case), and thus altering the frequency at which resonance occurs.

## 4.8   Three-phase AC systems

The basic principles applying to three-phase supplies were considered in Section 12.4 of 'Electrical Craft Principles', Volume 1, which indicated that such a system consists of three supplies of equal voltage, but with a phase displacement of 120° relative to each other. The supply is carried by three conductors, called 'lines', which are coloured red, yellow and blue, respectively, for identification. The currents in these conductors are known as the **line currents** ($I_L$) and the potential differences between them are known as the **line voltages** ($V_L$). A fourth conductor, called the 'neutral', and coloured black, is often connected to the star point of some systems, and is also connected to the general mass of earth. The potential difference between any one of the three lines and the neutral conductor or earth is called '**phase voltage**' ($V_P$). Loads may be connected to a three-phase supply in either of two ways, but in both cases the current in each load is called a '**phase current**' ($I_P$).

### Star-connected loads

This method of connection is shown in Figure 4.17. The values $V_{RY}$, $V_{YB}$ and $V_{BR}$ are line voltages, and usually have equal values. The values $V_{RN}$, $V_{YN}$ and $V_{BN}$ are phase voltages, which are also usually equal to each other. The diagram shows that for this method of connection the line currents $I_R$, $I_Y$ and $I_B$ are equal to the phase currents $I_1$, $I_2$ and $I_3$, respectively. A balanced system, which is one having identical loads, will have equal line currents. For a balanced load there will be no neutral current. Similarly, the phase currents will be equal to each other in magnitude.

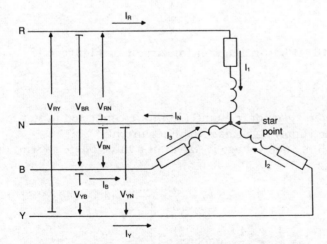

**Figure 4.17   Circuit diagram for balanced load star-connected to three-phase supply**

A complete phasor diagram for the circuit shown in Figure 4.17 is given in Figure 4.18. By measurement, it can be seen that the line voltages (for example, $V_{RY}$) are $\sqrt{3}$ times larger than the phase voltages (such as $V_{RN}$).

For the star-connected system,

$$I_L = I_P \qquad V_L = V_P\sqrt{3}$$

where   $I_L$ = line current (A)

$I_P$ = phase current (A)

$V_L$ = line voltage (V)

$V_P$ = phase voltage (V).

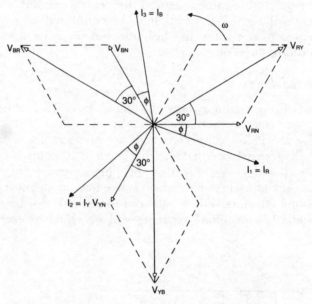

**Figure 4.18   Phasor diagram for circuit of Figure 4.17**

## Example 4.17

Three heaters, each of resistance 20 $\Omega$, are connected in star to a 415 V three-phase supply. Calculate the line current.

Note: If only one voltage is stated for a three-phase system, this is the line voltage.

$$V_L = V_P\sqrt{3} \quad \text{so} \quad V_P = \frac{V_L}{\sqrt{3}} = \frac{415}{1\cdot73} \text{ volts} = 240 \text{ V}$$

$$I_P = \frac{V_P}{R} = \frac{240}{20} \text{ amperes} = 12 \text{ A}$$

$$I_L = I_P = 12 \text{ A}.$$

## Example 4.18

A star-connected load consists of three identical inductors, each of self inductance 76·4 mH and resistance 10 Ω. If the line current is 10 A, calculate the line voltage if the supply frequency is 50 Hz.

$$X_L = 2\pi f L = 2 \times 3 \cdot 14 \times 50 \times 0 \cdot 0764 \text{ ohms} = 24 \ \Omega.$$

$$Z = \sqrt{(R^2 + X_L^2)} = \sqrt{(10^2 + 24^2)} \text{ ohms} = 26 \ \Omega$$

$$I_L = I_P = \frac{V_P}{Z}$$

so

$$V_P = I_L Z = 10 \times 26 \text{ volts} = 260 \text{ V}$$

$$V_L = (\sqrt{3})V_P = 1 \cdot 73 \times 260 \text{ volts} = 450 \text{ V}.$$

## *Delta-connected loads*

This method of connection is shown in Figure 4.19. It can be seen that each load is connected across two lines, so the line voltages ($V_{RY}$, $V_{YB}$ and $V_{BR}$) are equal to the voltages across the loads ($V_1$, $V_2$ and $V_3$) which are called phase voltages.

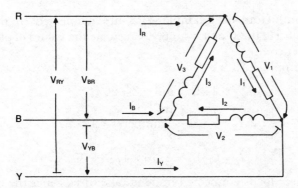

**Figure 4.19   Circuit diagram for balanced load delta-connected to three-phase supply**

The currents in the loads are driven by line voltages, and these are the phase currents $I_1$, $I_2$ and $I_3$. Each line current is the phasor sum of two phase currents as shown in the phasor diagram of Figure 4.20. The delta connection (sometimes referred to as a mesh connection) has no star point, and cannot have a fourth (neutral) wire connected to it.

For the delta-connected system,

$$V_L = V_P \qquad I_L = (\sqrt{3})I_P$$

where the symbols have the same meanings as those defined for the star-connected system.

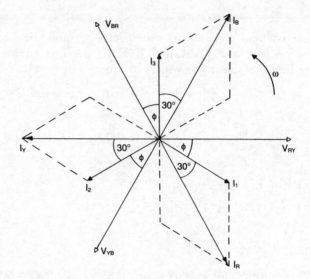

**Figure 4.20   Phasor diagram for the circuit of Figure 4.19**

## Example 4.19

The three inductors of Example 4.18 are connected in delta to the 450/260 V, 50 Hz three-phase supply. Calculate the values of phase current and of line current.

As in Example 4.18

$$X_L = 24 \ \Omega \quad \text{and} \quad Z = 26 \ \Omega$$

$$I_P = \frac{V_L}{Z} = \frac{450}{26} \text{ amperes} = 17 \cdot 3 \text{ A}$$

$$I_L = (\sqrt{3})I_P = 1 \cdot 73 \times 17 \cdot 3 = 30 \text{ A}$$

Note that three loads connected in delta take three times the line current from the supply that they would take if connected in star.

## Example 4.20

A power-factor-correction device consists of three capacitors delta-connected to a 415 V, 50 Hz three-phase supply, the line current to the unit being 24 A. Calculate the capacitance of each of the three capacitors.

$$I_P = \frac{I_L}{\sqrt{3}} = \frac{24}{1 \cdot 73} \text{ amperes} = 13 \cdot 9 \text{ A}$$

$$X_C = \frac{V_L}{I_P} = \frac{415}{13 \cdot 9} \text{ ohms} = 30 \ \Omega$$

$$X_C = \frac{10^6}{2\pi f C'}, \quad \text{so} \quad C' = \frac{10^6}{2\pi f X_C} = \frac{10^6}{2 \times 3 \cdot 14 \times 50 \times 30} \text{ microfarads}$$

$$= 106 \ \mu\text{F}.$$

## 4.9   Summary of formulas for Chapter 4

See text for definitions of symbols.
Inductive reactance:

$$X_L = 2\pi f L \qquad f = \frac{X_L}{2\pi L} \qquad L = \frac{X_L}{2\pi f}$$

Capacitive reactance:

$$X_C = \frac{1}{2\pi f C} = \frac{10^6}{2\pi f C'}$$

$$f = \frac{1}{2\pi C X_C} \quad \text{or} \quad \frac{10^6}{2\pi C' X_C}$$

$$C = \frac{1}{2\pi f X_C} \qquad C' = \frac{10^6}{2\pi f X_C}$$

For purely reactive circuits:

$$I = \frac{V}{X_C} \qquad X_C = \frac{V}{I} \qquad V = I X_C$$

$$I = \frac{V}{X_L} \qquad X_L = \frac{V}{I} \qquad V = I X_L$$

For the *RL* series circuit:

$$Z = \frac{V}{I} = \sqrt{(R^2 + X_L^2)} \qquad R = \sqrt{(Z^2 - X_L^2)} \qquad X_L = \sqrt{(Z^2 - R^2)}$$

$$\cos\phi = \frac{R}{Z} \qquad \sin\phi = \frac{X_L}{Z} \qquad \tan\phi = \frac{X_L}{R}$$

For the *RC* series circuit:

$$Z = \frac{V}{I} = \sqrt{(R^2 + X_C^2)} \qquad R = \sqrt{(Z^2 - X_C^2)} \qquad X_C = \sqrt{(Z^2 - R^2)}$$

$$\cos\phi = \frac{R}{Z} \qquad \sin\phi = \frac{X_C}{Z} \qquad \tan\phi = \frac{X_C}{R}$$

For the *RLC* (general) series circuit:

$$Z = \frac{V}{I} = \sqrt{\{R^2 + (X_L - X_C)^2\}} \qquad R = \sqrt{\{Z^2 - (X_L - X_C)^2\}}$$

$$X_L - X_C = \sqrt{(Z^2 - R^2)} \qquad X_L = X_C \pm \sqrt{(Z^2 - R^2)}$$

$$X_C = X_L \pm \sqrt{(Z^2 - R^2)}$$

$$\cos \phi = \frac{R}{Z} \qquad \sin \phi = \frac{X_L - X_C}{Z} \qquad \tan \phi = \frac{X_L - X_C}{R}$$

(Note: current LAGS voltage where $X_L$ is greater than $X_C$, and LEADS voltage when $X_L$ is smaller than $X_C$).

At resonance in the *RLC* circuit:

$$Z = R$$

$$f_r = \frac{1}{2\pi\sqrt{(LC)}} = \frac{10^3}{2\pi\sqrt{(LC')}}$$

For balanced three-phase AC systems:
  Star connected:

$$V_L = (\sqrt{3})V_P \qquad V_P = \frac{V_L}{\sqrt{3}}$$

$$I_L = I_P$$

  Delta-connected:

$$I_L = (\sqrt{3})I_P \qquad I_P = \frac{I_L}{\sqrt{3}}$$

$$V_L = V_P$$

## 4.10   Exercises

1 A resistor and a pure inductor are connected in series to an AC supply. A voltmeter connected across the resistor reads 100 V, while a voltmeter connected across the inductor reads 218 V. Calculate the supply voltage.

2 A resistor and a pure inductor are connected in series to a 100 V, 400 Hz supply. A high-resistance voltmeter connected across the resistor reads 50 V. What will it read if connected across the inductor?

3 The PD across a pure inductor is 20 V when it is connected in series with a resistor to a 25 V AC supply. Calculate the PD across the resistor.

4 If the supply of Exercise 1 is at 50 Hz and the current in the circuit is 5 A, calculate the circuit resistance, circuit reactance, circuit impedance, circuit inductance and circuit phase angle.

5 A circuit consisting of a resistor in series with a pure inductor is connected to a 115 V, 60 Hz supply, when the circuit current is $2 \cdot 3$ A. If the resistor value is 40 $\Omega$, calculate the resistor PD, the inductor PD, the inductance of the inductor and the circuit phase angle.

6 Calculate the impedance of the circuit of Exercise 5.

7 If the resistor of Exercise 2 has a value of 5 Ω, calculate the circuit impedance, the circuit reactance, the circuit inductance and the circuit phase angle.

8 A choke takes a current of 12 A when connected to a 240 V, 50 Hz supply, and 2 A from a 12 V DC supply. Calculate the resistance and the inductance of the coil.

9 A battery-operated ohm-meter gives a reading of 10 Ω when connected across a choke. When the same coil is connected to a 115 V, 400 Hz supply, the current is 5 A. What is the self inductance of the coil?

10 (*a*) When a certain inductive coil is connected to a DC supply at 240 V, the current in the coil is 16 A. When the same coil is connected to an AC supply at 240 V, 50 Hz, the current is 12·27 A. Calculate (i) the resistance, (ii) the impedance, (iii) the reactance, (iv) the inductance, of the coil.
  (*b*) If the supply frequency were to be altered to 60 Hz at 240 V, would the current in the coil be greater or less than the value given above? Give reasons for your answer.

11 What is the function of a choke in an alternating-current circuit?
  When a DC supply at 240 V is applied to the ends of a certain choke, the current in the coil is 20 A. If an AC supply at 240 V, 50 Hz is applied to the coil, the current in the coil is 12·15 A. Calculate the impedance, reactance, inductance and resistance of the coil.
  What would be the general effect on the current if the frequency of the AC supply were increased?

12 A single-phase alternating-current circuit consists of a resistance of 20 Ω and an inductive coil of inductance 0·1 H and of negligible resistance connected in series. A voltage of 240 V at 50 Hz is applied to the ends of the circuit. Calculate:
  (*a*) the current in the circuit, and
  (*b*) the potential differences across each element in the circuit.

13 A 10 Ω resistor connected in series with an inductive coil of negligible resistance to a 50 V AC supply carries a current lagging the supply voltage by 30°. By means of a scaled phasor diagram calculate the PD across the coil.

14 A coil of wire takes a current of 10 A when connected to a DC supply of 100 V. When connected to an AC supply of 100 V the current taken is 5 A. Explain the difference and calculate the reactance of the coil.

15 A capacitor and a resistor are connected in series to a 50 Hz supply. A high-resistance voltmeter connected across the resistor reads 132·7 V, while when connected across the capacitor it reads 200 V. What is the supply voltage?

16 A resistor and a capacitor are connected in series to a 100 V, 1 kHz supply. The PD across the resistor is 75 V. What is the PD across the capacitor?

17 A 240 V, 50 Hz supply is connected to a circuit consisting of a resistor in series with a capacitor. A voltmeter connected across the capacitor reads 200 V. What will it read when connected across the resistor?

18 If the current in the circuit given in Exercise 16 is 100 mA, calculate the resistance, reactance, impedance, capacitance and phase angle of the circuit.

19 A 2 μF capacitor and a 100 Ω resistor are connected in series across a 415 V,

50 Hz supply. Calculate (i) the circuit impedance, (ii) the current, (iii) the PD across the resistor, (iv) the PD across the capacitor, and (v) the circuit phase angle.

20 A circuit consisting of a 20 Ω resistor connected in series with a capacitor is connected to a 115 V, 60 Hz supply, when the current is 4·6 A. Calculate the capacitive reactance and capacitance of the circuit, and the phase angle between supply voltage and current.

21 A pure inductor of 0·03 H, a 10 Ω resistor and a 150 μF capacitor are connected in series across a 240 V, 50 Hz supply. Calculate (*a*) the supply current, (*b*) the phase angle between the current and the supply voltage, and (*c*) the potential difference across each component.

22 The circuit of Exercise 19 is broken and a choke of resistance 100 Ω and an inductance of 4 H is connected in series with the resistor and the capacitor. Calculate
   (i) the circuit impedance,
   (ii) the current,
   (iii) the PD across the separate resistor,
   (iv) the PD across the capacitor, and
   (v) the circuit phase angle.

23 (*a*) A 20 Ω resistor is connected in series with an inductor and a capacitor to a 100 V variable-frequency supply. At a given frequency, the inductive and capacitive reactances of the circuit are 20 Ω and 60 Ω, respectively. Calculate the circuit current, the PD across the capacitor, and the phase angle between voltage and current.
   (*b*) Calculate the current, capacitor PD and the phase angle of the same circuit if the frequency of the applied voltage is doubled.

24 Name three properties of a circuit in which series resonance occurs.

25 Name three changes which, when made to a series *LCR* circuit or to its supply, will result in a condition of series resonance.

26 Using a common base of frequency, sketch on the same sheet the variation of circuit resistance, inductive reactance, capacitive reactance, impedance and current as frequency is varied to make a series *LCR* circuit pass through a condition of resonance.

27 A 0·15 H coil, a 12 Ω resistor and a 12 μF capacitor are connected in series to a 24 V variable-frequency supply. The frequency is adjusted until the current reaches a maximum, when it is shown by the calibrated control to be 119 Hz. Calculate the value of the maximum current, and the PD across each component when this current is passing. Draw a scale phasor diagram for the system.

28 A choke of self inductance 0·5 H but of unknown resistance is connected in series with a capacitor to a 240 V, 50 Hz supply, when the current in the circuit is 10 A and is in phase with the voltage. Calculate the resistance of the choke and the PD across the capacitor.

29 A 1 μF capacitor and a choke of inductance 152 mH and resistance 80 Ω are connected in series across a 12 V, 500 Hz supply. Calculate (i) the circuit

impedance, (ii) the current, (iii) the circuit phase angle, and (iv) the PD across the capacitor.

30 What is meant by the following terms used in connection with alternating current: resistance; impedance; reactance? A PD of 240 V, at a frequency of 50 Hz, is applied to the ends of a circuit containing a resistance of 5 Ω, an inductance of 0·02 H and a capacitance of 150 μF, all in series. Calculate the current in the circuit.

31 Explain the meaning of the following terms used in connection with alternating current:
inductance, capacitance, reactance and impedance.
   A circuit comprises a resistance of 6 Ω, an inductance of 0·05 H, and a capacitance of 320 μF, all connected in series. An alternating supply at 240 V 50 Hz, is applied to the ends of the circuit. Calculate the current taken.

32 A 10 μF capacitor is connected in series with a 1·58 H inductor and a 20 Ω resistor across a 30 V, 40 Hz supply. Calculate the circuit current, its phase relative to supply voltage, and the PD across each component. Draw a phasor diagram for the circuit in this condition.

33 A resistor of 50 Ω, a pure inductance of 0·15 H and a 100 μF capacitor are connected in parallel to a 240 V, 50 Hz supply. By means of a scale phasor drawing of the current in each branch, calculate the current from the supply and its phase relative to applied voltage.

34 In Figure 4.21, the apparatus and the supply voltage in the two circuits are identical. Write down the current values at 25 Hz.

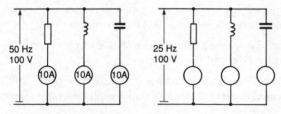

**Figure 4.21   Diagrams for Exercise 34**

35 The currents in three single-phase circuits in parallel are
   (*a*) 10 A at unity power factor
   (*b*) 20 A at a power factor of 0·8 lagging
   (*c*) 30 A at a power factor of 0·6 leading.
   Find the resultant current from a phasor diagram drawn to a suitable scale.

36 (*a*) An alternating current of 10 A flows through a series circuit comprising a 1 Ω resistor and a coil of 3 Ω resistance, 3 Ω inductive reactance. Draw a phasor diagram to show current and voltage relationships.
      Determine the overall voltage by measurement or calculation.
   (*b*) An alternating voltage of 25 V is applied to two circuits in parallel. One circuit has a resistance of 5 Ω; the other circuit has resistance 3 Ω, and inductive reactance 4 Ω in series. Draw a phasor diagram to show current

and voltage relationships.
Determine by measurement or calculation the total current.

37  A choke of inductance 0·08 H and resistance 12 Ω is connected in parallel with a 120 μF capacitor to a 240 V, 50 Hz supply. Determine the current from the supply and its phase angle by means of a scale phasor diagram.

38  Explain the meaning of the term 'parallel resonance'. Sketch a circuit in which parallel resonance may occur, and indicate where it could be used.

39  Draw a circuit diagram of a resistive choke in parallel with a capacitor across an AC supply. Sketch a phasor diagram for this circuit when it is in a condition of resonance.

40  Using a common base of frequency, plot curves to show the variation of impedance and current in a parallel circuit passing through the resonant condition.

41  Three identical 24 Ω resistors are connected in star to a 415 V three-phase supply. Calculate the line current.

42  Three identical 159 μF capacitors are star-connected to a 400 V three-phase 50 Hz supply system. What will be the value of the line current?

43  Three identical 50 Ω resistors are connected in delta to a 400 V three-phase supply. Calculate the line current.

44  Each load of a delta-connected system comprises a choke of inductance 0·1 H in series with resistance 10 Ω. If the supply voltage of the three-phase system is 415 V at 50 Hz, calculate the line current.

45  The line current to a balanced star-connected three-phase load from a 415 V, 50 Hz supply is 20 A, the current lagging the phase voltage by 30°. Calculate the resistance and self inductance of each of the three loads.

46  The loads of Exercise 45 are delta-connected to the same supply. Calculate the values of phase and line currents.

47  A power-factor correction capacitor consists of three 200 μF capacitors, delta-connected to the 415 V, 50 Hz three-phase supply. Calculate the phase and line currents.

48  Calculate the line current to the chokes of Exercise 44 if they are reconnected in star to the same supply.

## 4.11   Multiple-choice exercises

4M1   The formula used to calculate the inductive reactance of an inductor is:

(a) $X_L = \dfrac{fL}{2\pi}$          (b) $X_L = 2\pi fL$

(c) $L = 2\pi fX_L$          (d) $X_L = \dfrac{1}{2\pi fL}$

4M2  In a pure capacitor connected to an alternating supply:
(*a*) current lags voltage by 90°
(*b*) voltage and current are in phase
(*c*) current leads voltage by 45°
(*d*) current leads voltage by 90°.

4M3  If the frequency of the alternating supply to a pure capacitor reduces, its capacitive reactance will:
(*a*) reduce          (*b*) be unchanged
(*c*) increase        (*d*) become zero.

4M   The phase angle of a resistive–inductive series circuit can be found from the expression:

(*a*) $\cos \phi = \dfrac{R}{Z}$          (*b*) $\sin \phi = \dfrac{Z}{X_L}$

(*c*) $\phi = 2\pi fL$          (*d*) $\tan \phi = \dfrac{V_R}{V_L}$

4M5  A pure capacitor and a resistor are connected in series to an alternating supply. The potential differences across the capacitor and across the resistor are measured at 100 V and 140 V, respectively, so the supply voltage is:
(*a*) 240 V          (*b*) 98 V          (*c*) 172 V          (*d*) 40 V.

4M6  The phase angle between the voltage and the current in exercise 4M5 will be:
(*a*) 35·5°          (*b*) 54·6°          (*c*) 90°          (*d*) 39·1°

4M7  If a coil of self inductance 390 mH and resistance 1·9 kΩ is connected to a 24 V supply with a frequency of 1 kHz, the current will be:
(*a*) 129 A          (*b*) 12·6 mA          (*c*) 77·4 μA          (*d*) 7·74 mA.

4M8  The phase relationship of current to supply voltage for the circuit of Exercise 4M7 will be:
(*a*) in phase              (*b*) 52·2° lagging
(*c*) 37·8° leading        (*d*) 90° lagging.

4M9  An inductive circuit connected to a 2 V DC supply draws a current of 45 mA. When connected to a 240 V 50 Hz supply, the current is 2·67 A. The self inductance of the circuit is:
(*a*) 78 Ω          (*b*) 0·249 H          (*c*) 141 mH          (*d*) 24·9 mH.

4M10 If a coil with a self inductance of 24 mH takes a current of 3·83 mA from a 15 V 24 kHz supply, its series resistance is:
(*a*) 6·27 Ω          (*b*) 3·92 kΩ          (*c*) 3·62 kΩ          (*d*) 1·5 kΩ.

4M11 If a pure inductor is connected to an AC supply, the phase relationship of the current to the voltage is:
(*a*) 90° lagging          (*b*) in phase
(*c*) 45° lagging          (*d*) 90° leading.

4M12 A circuit consisting of a capacitor connected in series with a resistor is connected to an AC supply. If the potential differences across the capacitor and resistor are measured as 56 V and 45 V, respectively, the supply voltage is:
(*a*) 101 V          (*b*) 33·3 V          (*c*) 71·8 V          (*d*) 11 V

4M13  When a 56 nF capacitor and a $2 \cdot 2$ kΩ resistor are connected in series to a $2 \cdot 4$ kHz supply, the circuit phase angle will be:
(a) $88 \cdot 5°$ leading          (b) $28 \cdot 3°$ leading
(c) $61 \cdot 7°$ lagging          (d) $61 \cdot 7°$ leading.

4M14  If the supply to the circuit of Exercise 4M13 has a PD of 24 V, the current in the circuit will be:
(a) $10 \cdot 9$ mA      (b) $0 \cdot 429$ A      (c) $0 \cdot 917$ mA      (d) $9 \cdot 6$ mA.

4M15  The formula for the impedance of a series circuit containing resistance, inductance and capacitance is:
(a) $Z = \sqrt{\{R^2 + (X_L^2 - X_C^2)\}}$
(b) $Z = R + X_L - X_C$
(c) $Z = 2\pi f L - \dfrac{1}{2\pi f C}$
(d) $X_C = 2\pi f L$

4M16  If the frequency of the supply applied to a resistive and capacitive series circuit is increased, the circuit phase angle will:
(a) increase          (b) remain the same
(c) reduce            (d) become zero.

4M17  When a 5 V, 50 Hz supply is connected to a coil of resistance 80 Ω and self inductance 191 mH which is in series with a $26 \cdot 5$ μF capacitor, the circuit current will be:
(a) $62 \cdot 5$ mA      (b) $4 \cdot 17$ mA      (c) 50 mA      (d) $26 \cdot 2$ mA.

4M18  The potential difference across the capacitor in the circuit of Exercise 4M17 will be:
(a) 6 V          (b) 3 V          (c) $2 \cdot 09$ V          (d) 5 V.

4M19  Resonance occurs in a series circuit containing resistance, inductance and capacitance when:
(a) circuit resistance is zero
(b) inductive and capacitive reactances are equal
(c) circuit phase angle is $90°$
(d) $R = 2\pi L C$.

4M20  The formula for the frequency at which resonance occurs in a series $RLC$ circuit is:
(a) $R = 2\pi L C$          (b) $f_r = \dfrac{2\pi}{LC}$

(c) $f_r = \dfrac{1}{2\pi L C}$          (d) $f_r = \dfrac{1}{2\pi \sqrt{(LC)}}$

4M21  One of the following statements is UNTRUE for a series resonant circuit. The untrue statement is:
(a) supply voltage is $45°$ out of phase with current
(b) circuit impedance is equal to circuit resistance
(c) supply voltage and current are in phase
(d) circuit current will be a maximum at resonant frequency.

4M22 Every circuit containing inductance and capacitance has a frequency at which it will resonate. The foregoing statement is:
(*a*) false                      (*b*) true only if resistance is high
(*c*) true                      (*d*) true only if capacitance is fixed.

4M23 A $2 \cdot 2$ μF capacitor and a coil are connected in series. If the combination takes a maximum current from an AC supply when its frequency is 200 Hz, the self inductance of the coil must be:
(*a*) $2 \cdot 2$ H       (*b*) $0 \cdot 288$ H       (*c*) $57 \cdot 6$ mH       (*d*) 144 mH.

4M24 A 2 μF capacitor, a $1 \cdot 6$ H pure inductor and a 2 kΩ resistor are all connected in parallel to a 25 V, 100 Hz supply. The supply current will be:
(*a*) $24 \cdot 8$ mA       (*b*) $14 \cdot 1$ mA       (*c*) $6 \cdot 5$ mA       (*d*) $12 \cdot 5$ mA.

4M25 In a parallel circuit at resonance:
(*a*) the current in the capacitive branch will be at a maximum
(*b*) the overall circuit phase angle will be 90°
(*c*) circuit impedance will be equal to circuit resistance
(*d*) current will be minimum when it is in phase with the supply voltage.

4M26 In a star-connected three-phase system the relationship between phase and line voltage values is:
(*a*) $V_L = V_P \sqrt{3}$
(*b*) $V_L = V_P$
(*c*) $V_L \sqrt{3} = V_P$
(*d*) $V_L = \dfrac{V_P}{2}$

4M27 If three identical loads were connected in star and are reconnected in delta to the same three-phase supply, the line current will:
(*a*) remain the same
(*b*) lag by 30° instead of by 60°
(*c*) increase by a factor of three
(*d*) decrease by a factor of three.

4M28 The phase angles between the line currents of a balanced three-phase system are:
(*a*) 90°            (*b*) 0°            (*c*) 60°            (*d*) 120°.

4M29 If three identical resistors, each of resistance 45 Ω are star-connected to a three-phase supply with a line voltage of 240 V, the line current will be:
(*a*) $5 \cdot 33$ A       (*b*) $3 \cdot 08$ A       (*c*) $1 \cdot 78$ A       (*d*) $0 \cdot 89$ A.

4M30 Three identical coils, each of self inductance $1 \cdot 4$ H, are connected in delta to a 55 V, 100 Hz three-phase supply. If the line current is 104 mA, the resistance of each coil must be:
(*a*) 916 Ω       (*b*) 440 Ω       (*c*) 255 Ω       (*d*) 880 Ω.

*Chapter 5*

# Power in alternating-current circuits

## 5.1  Introduction

Power is the rate of doing work, or of expending energy. The electrical unit of power is the **watt**, which represents a rate of expending energy of one joule each second. If a resistor of $R$ ohms has a direct voltage of $V$ volts applied to it, so that a direct current of $I$ amperes flows, the power dissipated, $P$ watts, will be given by

$$P = VI = I^2R = \frac{V^2}{R}$$

Now, for an AC circuit, $P$ is the average power, while $V$ and $I$ are RMS voltage and RMS current, respectively. This power is dissipated in the resistor as heat. Power is dissipated in much the same way when an alternating voltage is applied to the resistor; indeed, as we have seen, the effective (RMS) value of an alternating current is defined in terms of its heating effect. Since both voltage and current values are continuously changing in the AC system, power will also fluctuate and the rate of dissipating energy is the **instantaneous power** which is given by

$$p = vi$$

where   $p$ = instantaneous power dissipated (W)
$v$ = instantaneous voltage (V)
$i$ = instantaneous current (A)

## 5.2  Power in the resistive AC circuit

For a resistive AC circuit, current and voltage are in phase, and the power at any instant can be found by multiplying the voltage and current at that instant. This gives the power wave shown in Figure 5.1, from which the power can be seen to consist of a series of pulses. When current is negative, so is voltage, the product of the two negative values giving a positive power pulse. For a purely resistive load fed from an AC supply,

$$P = VI = I^2R = \frac{V^2}{R}$$

where   $P$ = average power (W)
  $V$ = RMS supply voltage (V)
  $I$ = RMS supply current (A)

Many electrical loads, such as electric fires, irons, kettles, water heaters, filament lamps etc., can be considered as resistive.

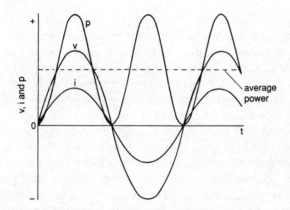

**Figure 5.1   Voltage, current and power waves for resistive AC circuit**

## Example 5.1

A 3 kW immersion heater is connected to a 240 V AC supply. Calculate the current.

$$I = \frac{P}{V} = \frac{3000}{240} \text{ amperes} = 12 \cdot 5 \text{ A}$$

## 5.3   Power in the capacitive AC circuit

Let us consider a loss-free capacitor connected to an AC supply. Although a current flows to the ideal capacitor, it does not become warm, and this suggests that no average power is dissipated. The wave diagram for the capacitor (Figure 5.2) shows the current leading the voltage by 90°, and to sketch the instantaneous power curve we must remember that $p = vi$. In the first quartercycle of voltage, $v$ and $i$ are both positive, so their product,

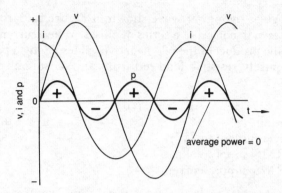

**Figure 5.2    Voltage, current power waves for capacitive AC circuit**

the power wave, is also positive. In the second quartercycle of voltage, $v$ is positive but $i$ is negative, so the power wave goes negative. In the third quartercycle of voltage, both $v$ and $i$ are negative, so the power wave is positive, while in the fourth quartercycle of voltage, negative voltage and positive current multiply to give negative power. The power wave is thus a series of identical positive and negative pulses whose average value over any halfcycle of voltage or current is zero.

During its first and third quartercycles, the voltage is increasing, and the supply provides energy to charge the capacitor. During the second and fourth quartercycles of voltage, the reducing PD across the capacitor allows it to discharge, returning its energy to the supply. The positive pulses represent energy supplied to the capacitor, while the negative pulses represent energy supplied by the capacitor as it discharges. The interchange of energy dissipates no average power in a pure capacitor, so no heating occurs.

Since we have voltage and current, but no average power, the expression $P = VI$ is no longer true. The product of voltage and current in this case is called '**reactive power**' and is measured in reactive voltamperes (VAr). The current to a capacitor which does not contain resistance does not dissipate energy, and is called '**reactive current**'.

## Example 5.2

A 10 µF capacitor is connected to a 240 V, 50 Hz supply. Calculate the reactive current and the reactive voltamperes.

$$X_C = \frac{10^6}{2\pi f C'} = \frac{10^6}{2\pi \times 50 \times 10} \text{ ohms} = 318 \ \Omega$$

$$I = \frac{V}{X_C} = \frac{240}{318} \text{ amperes} = 0 \cdot 755 \text{ A}$$

reactive voltamperes = $240 \times 0 \cdot 755$ voltamperes reactive = 181 VAr

## 5.4   Power in the inductive AC circuit

A similar state of affairs to that in the capacitive AC circuit applies to a pure inductance connected to an AC supply. This time the current lags the voltage by 90°, but a power curve sketched from $p = vi$ again consists of equal positive and negative pulses of power with an average value, over any halfcycle of voltage or current, of zero as shown in Figure 5.3. This time the energy is stored in the magnetic field of the inductor for a quartercycle of supply voltage, being fed back into the supply as the magnetic field collapses during the next quartercycle. An explanation of how energy is stored in a magnetic field has been given in Section 2.6.

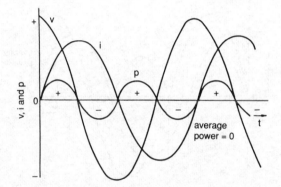

**Figure 5.3   Voltage, current and power waves for purely inductive AC circuit**

As explained in Chapter 10 of 'Electrical Craft Principles' Volume 1, a pure inductor is unlikely to be found in practice, as such a device would have to be completely without resistance. Since a practical inductor consists of a coil of wire, which will have resistance, the concept of a pure inductor is usually a theoretical one. If such an inductor could be made, it would carry very high currents without dissipating any power at all. If a conductor made of a special material such as niobium–tin is cooled to a temperature approaching the absolute zero (zero K or – 273°C), it loses its resistance and is known as a superconductor. Some very powerful electromagnets for use in research employ superconductors, as do a few experimental motors, but the cost of achieving and maintaining very low temperatures is extremely high.

Examples similar to Example 5.2 could be worked for pure inductors, but the practical absence of inductors without resistance would mean that such work would have little practical application. Losses in a resistive inductor are considered in Section 5.6.

## 5.5   Power in resistive and capacitive AC circuits

In a circuit consisting of resistance and capacitive reactance in series, the voltage and current will have a relative phase angle between 0° and 90°, depending on the ratio of resistance to reactance (see Chapter 4). Figure 5.4 shows a wave diagram for a resistive and capacitive circuit, with the current leading the voltage by $\phi$°. Although there is still some energy returned to the supply, as shown by the negative pulses, the energy drawn from the supply, and represented by the positive pulses, is greater. Thus, the net energy drawn from the supply will be dissipated as heat in the resistive part of the circuit.

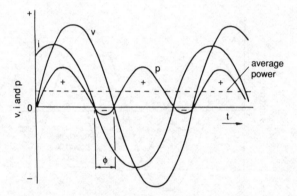

**Figure 5.4   Voltage, current and power waves for resistive and capacitive AC series circuit**

The ratio of resistance to reactance in the circuit must have some bearing on the power dissipated, because power is expended in a resistive circuit, but not in a reactive circuit. Figure 5.5 shows the phasor diagram to correspond with the wave diagram of Figure 5.4, the current leading the voltage by $\phi$°.

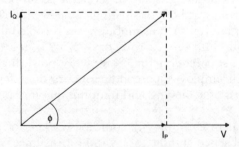

**Figure 5.5   Phasor diagram for resistive and capacitive AC circuit, showing components of current**

The current phasor can be split up into two component currents. These currents will not actually flow in the circuit, but, since their phasor sum is equal to the actual current, it is convenient to assume that they do so. $I_p$ is the 'inphase' or 'active' component of current, being in phase with the voltage while the 'quadrature' or 'reactive' component $I_Q$ leads the voltage by 90°. In a purely resistive circuit, where voltage and current are in phase, the power dissipated can be found by multiplying together the RMS voltage and the RMS current. It follows that, in the resistive and reactive circuit, power dissipated can be found by multiplying together the voltage and the component of current in phase with it.

$$P = VI_p$$

but
$$\cos \phi = \frac{I_p}{I}$$

so
$$I_p = I \cos \phi \quad \text{and} \quad P = VI \cos \phi$$

Cos $\phi$ is known as the 'power factor' of the circuit (see Section 5.8).

Since this result has been derived from the phasor diagram it will apply only to sinusoidal waveforms and in such cases average power = RMS voltage × RMS current × the cosine of the phase angle between voltage and current.

## Example 5.3

A circuit connected to a 240 V AC supply consists of a resistance of $28 \cdot 8$ Ω in series with a capacitor of reactance $38 \cdot 4$ Ω. Calculate (i) the circuit current, (ii) the circuit phase angle, and (iii) the power dissipated.

(i)
$$Z = \sqrt{(R^2 + X_C^2)} = \sqrt{(28 \cdot 8^2 + 38 \cdot 4^2)} \text{ ohms} = 48 \text{ Ω}$$

$$I = \frac{V}{Z} = \frac{240}{48} \text{ amperes} = 5 \text{ A}$$

(ii) It is often useful to be able to calculate the circuit phase angle from the resistance, reactance and impedance of the circuit. From Section 4.2,

$$\cos \phi = \frac{V_R}{V} = \frac{IR}{IZ} \quad \text{so} \quad \cos \phi = \frac{R}{Z}$$

and
$$\sin \phi = \frac{V_C}{V} = \frac{IX_C}{IZ} \quad \text{so} \quad \sin \phi = \frac{X_C}{Z}$$

and
$$\tan \phi = \frac{V_C}{V_R} = \frac{IX_C}{IR} \quad \text{so} \quad \tan \phi = \frac{X_C}{R}$$

From the first expression,
$$\cos \phi = \frac{R}{Z} = \frac{28 \cdot 8}{48} = 0 \cdot 6$$

From cosine tables, $\phi = 53 \cdot 1°$

(iii) $P = VI \cos \phi = 240 \times 5 \times 0 \cdot 6$ watts = 720 W

This power is dissipated in the resistive part of the circuit, and no power is lost in the capacitor. It follows that expressions for power using only resistive components will be true for RMS values of AC systems as well as for average values of DC supplies.

Thus
$$P = I^2 R = \frac{V_R^2}{R}$$

where $V_R$ is the PD across the resistor.

To check these expressions for the case of Example 5.3,

$$P = I^2 R = 5^2 \times 28 \cdot 8 \text{ watts} = 720 \text{ W}$$

$$V_R = IR = 5 \times 28 \cdot 8 \text{ volts} = 144 \text{ V}$$

$$P = \frac{V_R^2}{R} = \frac{144^2}{28 \cdot 8} \text{ watts} = 720 \text{ W}$$

## Example 5.4

A 10 Ω resistor and a capacitor are connected in series to a 120 V, 60 Hz supply. If the power lost in the circuit is 360 W, calculate the capacitance.

$$P = I^2 R \quad \text{so} \quad I^2 = \frac{P}{R} \quad \text{and} \quad I = \sqrt{\frac{P}{R}}$$

$$I = \sqrt{\frac{P}{R}} = \sqrt{\frac{360}{10}} \text{ amperes} = \sqrt{36} \text{ amperes} = 6 \text{ A}$$

$$Z = \frac{V}{I} = \frac{120}{6} \text{ ohms} = 20 \text{ } \Omega$$

$$X_C = \sqrt{(Z^2 - R^2)} = \sqrt{(20^2 - 10^2)} \text{ ohms} = 17 \cdot 3 \text{ } \Omega$$

$$X_C = \frac{10^6}{2\pi f C'} \quad \text{so} \quad C' = \frac{10^6}{2\pi f X_C}$$

$$C' = \frac{10^6}{2\pi f X_C} = \frac{10^6}{2\pi \times 60 \times 17 \cdot 3} \text{ microfarads} = 153 \text{ } \mu\text{F}$$

## 5.6   Power in resistive and inductive AC circuits

When resistance and inductive reactance are in series, current lags supply voltage by an angle of $\phi°$, which will vary from almost $0°$ to nearly $90°$. The wave diagram is shown in Figure 5.6, with instantaneous values of voltage and current multiplied as previously explained to give the resulting power wave. Once again, energy is both taken from the supply and returned to it, that taken from the supply exceeding the energy returned. The average power $P$ is shown on the wave diagram.

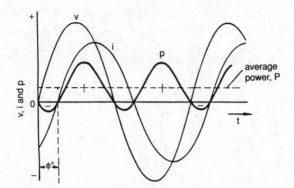

**Figure 5.6** **Voltage, current and power waves for resistive and inductive AC circuit**

Figure 5.7 shows the phasor diagram corresponding to the wave diagram of Figure 5.6. As explained in the previous section, average power is equal to the supply voltage multiplied by the component of current in phase with it.

Thus
$$P = VI_p = VI \cos \phi$$

Once again, this expression for power is true only if the current and voltage are sine waves, but this is not an unreasonable assumption for most electrical power systems.

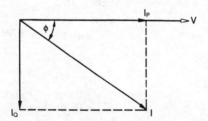

**Figure 5.7** **Phasor diagram for resistive and inductive AC circuit, showing components of current**

## Example 5.5

A 4 $\Omega$ resistor and a pure inductive reactance of 3 $\Omega$ are connected in series to a 200 V AC supply. Calculate (i) the current (ii) the circuit phase angle, and (iii) the power dissipated.

(i)
$$Z = \sqrt{(R^2 + X^2)} = \sqrt{(4^2 + 3^2)} \text{ ohms} = 5 \ \Omega$$

Therefore
$$I = \frac{V}{Z} = \frac{200}{5} \text{ amperes} = 40 \text{ A}$$

It is useful to be able to find the circuit phase angle from the series circuit resistance, reactance and impedance. From Section 4.2,

$$\cos \phi = \frac{V_R}{V} = \frac{IR}{IZ} \quad \text{so} \quad \cos \phi = \frac{R}{Z}$$

and

$$\sin \phi = \frac{V_L}{V} = \frac{IX_L}{IZ} \quad \text{so} \quad \sin \phi = \frac{X_L}{Z}$$

and

$$\tan \phi = \frac{V_L}{V_R} = \frac{IX_L}{IR} \quad \text{so} \quad \tan \phi = \frac{X_L}{R}$$

(ii) Therefore

$$\cos \phi = \frac{R}{Z} = \frac{4}{5} = 0\cdot 8$$

From the cosine tables, $\phi = 36\cdot9°$ lagging.

(iii) $P = VI \cos \phi = 200 \times 40 \times 0\cdot8$ watts $= 6400$ W

All of this power is dissipated in the resistor, and none in the reactor. Thus, the power dissipated could be found from:

$$P = I^2R = \frac{V_R^2}{R}$$

where $V_R$ is the PD across the resistor.

To check these expressions for the last example,

$$P = I^2R = 40 \times 40 \times 4 \text{ watts} = 6400 \text{ W}$$

and

$$V_R = IR = 40 \times 4 \text{ volts} = 160 \text{ V}$$

$$P = \frac{V_R^2}{R} = \frac{160 \times 160}{4} \text{ watts} = 6400 \text{ W}$$

## Example 5.6

A choke connected to a 130 V, 50 Hz supply has a resistance of 5 $\Omega$ and dissipates 500 W. Calculate its inductance.

$$P = I^2R \quad \text{so} \quad I = \sqrt{\frac{P}{R}} = \sqrt{\frac{500}{5}} \text{ amperes} = 10 \text{ A}$$

$$Z = \frac{V}{I} = \frac{130}{10} \text{ ohms} = 13 \text{ }\Omega$$

$$X_L = \sqrt{(Z^2 - R^2)} = \sqrt{(13^2 - 5^2)} \text{ ohms} = 12 \text{ }\Omega$$

$$L = \frac{X_L}{2\pi f} = \frac{12}{2\pi \times 50} \text{ henrys} = 0\cdot0382 \text{ H}$$

Alternating-current electric motors can be likened to resistive–inductive series circuits, but the expression $Z = V/I$ cannot directly be used to calculate the impedance of the machine. This is because the effective voltage applied to the windings is not usually the supply voltage, but the difference between the supply voltage and the EMF induced in the windings of the machine. This effect will be more fully discussed in Chapter 10. However, the power input and output of a motor can be calculated, as shown in Example 5.7.

## Example 5.7

The current input to a single-phase motor on full load is 7·55 A from a 240 V AC supply, the current lagging the voltage by 20°. If the machine is 85% efficient, calculate the input and output powers.

Input power $P_{in} = V I \cos \phi = 240 \times 7·55 \times \cos 20°$ watts.
From the cosine relationship,

$$\cos 20° = 0·9397$$

Correct to three significant figures,

$$\cos 20° = 0·940$$

$$P_{in} = 240 \times 7·55 \times 0·940 \text{ watts} = 1700 \text{ W} = 1·7 \text{ kW}$$

Output power $\qquad P_{out} = P_{in} \times \text{efficiency} = 1·7 \times \dfrac{85}{100}$ kilowatts

$$= 1·45 \text{ kW}$$

## 5.7   Power in general series circuits

The two preceding sections have dealt with $RC$ and $RL$ series circuits, respectively. Since no power is dissipated in the inductive or capacitive components of AC circuits, the $RCL$ (general) series circuit follows the same rules. The circuit is first solved following the pattern laid down in Section 4.4, and the power can then be calculated by any one of the three methods:

$$P = VI \cos \phi \qquad P = I^2 R \qquad P = \frac{V_R^2}{R}$$

where     $P$ = power dissipated (W)
$V$ = supply voltage (V)
$I$ = circuit current (A)
$\phi$ = circuit phase angle
$R$ = circuit resistance ($\Omega$)
$V_R$ = PD across the resistive component (V)

## Example 5.8

Calculate the power dissipated in the general series circuit of Example 4.11. From the working of Example 4.11, the circuit current is 0·48 A and the resistance is 30 $\Omega$.

$$P = I^2 R = 0 \cdot 48^2 \times 30 \text{ watts} = 6 \cdot 91 \text{ W}$$

The two alternative methods are also shown.

For Example 4.11, $\phi = 53 \cdot 1°$, so $\cos \phi = 0 \cdot 6$

$$P = VI \cos \phi = 24 \times 0 \cdot 48 \times 0 \cdot 6 \text{ watts} = 6 \cdot 91 \text{ W}$$

From Example 4.11, $V_R = 14 \cdot 4$ V.

$$P = \frac{V_R^2}{R} = \frac{14 \cdot 4^2}{30} \text{ watts} = 6 \cdot 91 \text{ W}$$

## Example 5.9

Calculate the power dissipated in the resonant circuit of Example 4.14. From Example 4.14, circuit resistance is 10 $\Omega$ and the current is 10 A.

$$P = I^2 R = 10^2 \times 10 \text{ watts} = 1000 \text{ W or } 1 \text{ kW}$$

Two alternative methods are:

$$P = VI \cos \phi, \text{ where } \phi = 0° \text{ since the circuit is at resonance.}$$
$$\text{Thus } \cos \phi = \cos 0° = 1$$

$$P = 100 \times 10 \times 1 \text{ watts} = 1000 \text{ W or } 1 \text{ kW}$$

or $\qquad V_R = IR = 10 \times 10 \text{ volts} = 100 \text{ V}$

$$P = \frac{V_R^2}{R} = \frac{100^2}{10} \text{ watts} = 1000 \text{ W or } 1 \text{ kW}$$

Note that since the current reaches a maximum at the condition of series resonance, the power is also maximum. If the frequency is changed, or component values are altered, the circuit will cease to resonate, and both current and power dissipated will decrease.

## 5.8   Power factor

We saw in Sections 5.3 and 5.4 that it is possible for current to flow in a circuit and to dissipate no power. In most practical cases this will not happen, but where the phase angle between current and voltage is large the 'in-phase' or 'active' component of current will be smaller than the quadrature or 'reactive' component. In AC circuits, the product of voltage and current need not result in the power dissipated in watts; this product gives voltamperes, sometimes called '**apparent power**'. The term apparent power is not a good one, as it suggests that voltamperes and watts are alike. In fact, whether they are similar depends on the phase angle between current and voltage, or on the '**power factor**' of the circuit.

Power factor (often abbreviated to PF) is defined as

$$\frac{\text{true power}}{\text{apparent power}} \text{ or } \frac{\text{watts}}{\text{voltamperes}}$$

This definition is a true one under all circumstances. If the voltage and current are sinusoidal, true power $= VI \cos \phi$, and since apparent power $= VI$,

$$\text{power factor} = \frac{\text{true power}}{\text{apparent power}} = \frac{VI \cos \phi}{VI} = \cos \phi$$

Since $\cos \phi$ can be derived from a phasor diagram for voltages (as in Figure 4.2) or an impedance diagram (as in Figure 4.3), for sinusoidal systems,

$$\text{PF} = \frac{P}{VI} = \cos \phi = \frac{V_R}{V} = \frac{R}{Z}$$

## Example 5.10

Calculate the power factor of the circuit in Example 5.6.

$$\text{power factor} = \cos \phi = \frac{R}{Z} = \frac{5}{13} = 0 \cdot 385$$

In a predominantly inductive series circuit, where current lags voltage, the power factor is called a '**lagging power factor**'. The solution to Example 5.10 could thus be more correctly given as a power factor of $0 \cdot 385$ lagging. Similarly, in a predominantly capacitive series circuit, where current leads voltage, the power factor is called a '**leading power factor**'.

The power factor can vary between definite limits, being 1 (unity) for purely resistive circuits, where the phase angle is $0°$ and $P = VI$; or 0 (zero) for purely reactive (inductive or capacitive) circuits, where the phase angle is $90°$ and $P = 0$.

## Example 5.11

An AC single-phase motor takes 5 A at $0 \cdot 7$ power factor lagging when connected to a 240 V, 50 Hz supply. Calculate the power input to the motor. If the motor efficiency is $70\%$, calculate the output.

$$P = VI \cos \phi = 240 \times 5 \times 0 \cdot 7 \text{ watts} = 840 \text{ W}$$

$$\text{Output power} = \text{input power} \times \text{efficiency} = 840 \times 70/100 \text{ watts} = 588 \text{ W}$$

## Example 5.12

A 200 V AC circuit comprises a 40 $\Omega$ resistor in series with a capacitor of reactance 30 $\Omega$. Calculate the current and the power factor.

$$Z = \sqrt{(R^2 + X_C^2)} = \sqrt{(40^2 + 30^2)} \text{ ohms} = 50 \text{ }\Omega$$

$$I = \frac{V}{Z} = \frac{200}{50} \text{ amperes} = 4 \text{ A}$$

$$\text{PF} = \frac{R}{Z} = \frac{40}{50} = 0 \cdot 8 \text{ leading}$$

Alternatively, power factor could have been calculated from values of true and apparent power.

$$VI = 200 \times 4 \text{ voltamperes} = 800 \text{ VA}$$

$$P = VI \cos \phi = VI \frac{R}{Z} = 800 \times \frac{40}{50} \text{ watts} = 640 \text{ W}$$

$$PF = \frac{P}{VI} = \frac{640}{800} = 0 \cdot 8 \text{ leading}$$

Chapter 11 will show that an instrument can be made to measure the average power in an AC circuit. Such an instrument is called a wattmeter, and can be used in conjunction with a voltmeter and an ammeter to measure power factor. The instruments are connected as shown in Figure 5.8, the readings being used to calculate power factor as illustrated by the following example.

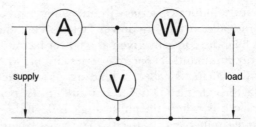

**Figure 5.8   Connection of instruments to allow calculation of power factor**

## Example 5.13

Instruments connected as in Figure 5.8 to a single-phase AC motor give the following readings:

   wattmeter, 1800 W, voltmeter, 240 V, ammeter 10 A.

Calculate the operating power factor of the motor.

$$PF = \frac{P}{VI} = \frac{1800}{240 \times 10} = 0 \cdot 75$$

The motor will present a resistive and inductive load, so the power factor will be lagging.

   power factor = $0 \cdot 75$ lagging.

An instrument directly measuring power factor, called a power-factor meter, is also available. These instruments are not common, and accuracy is often suspect. The method of connection is shown in Figure 5.9.

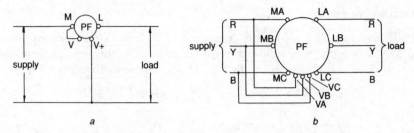

**Figure 5.9** **Connection for single-phase and three-phase power-factor meters**
(*a*) Single-phase
(*b*) Three-phase

## 5.9 Components of power

We have already seen in Sections 5.5 and 5.6 that the current in an AC circuit may be considered to have 'in-phase' and 'quadrature' components (see Figures 5.5 and 5.7). The voltamperes, or apparent power, of a circuit is similarly broken down into components. Figure 5.10 shows a power triangle for a resistive and inductive circuit. Since watts/voltamperes = $\cos \phi$, the true or active power makes an angle of $\phi$ with the apparent power, the angle being also the phase angle for the circuit concerned.

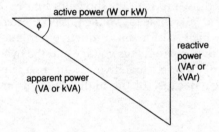

**Figure 5.10** **Power diagram for resistive and inductive AC circuit**

True power is the inphase component of apparent power; thus

$$P = VI \cos \phi$$

$$= \text{apparent power} \times \text{power factor}$$

Reactive power is the quadrature component of apparent power; thus

$$\text{VAr} = VI\sin \phi$$

Since these three power relationships form the sides of a right-angled triangle,

$$(\text{VA})^2 = (\text{W})^2 + (\text{VAr})^2$$

## Example 5.14

A 10 Ω resistor and a capacitive reactance of 20 Ω are connected in series to a 240 V supply. Calculate the apparent power, the true power, the reactive power and the power factor. Draw a power diagram to scale.

$$Z = \sqrt{(R^2 + X_C^2)} = \sqrt{(10^2 + 20^2)} \text{ ohms} = 22 \cdot 4 \ \Omega$$

$$I = \frac{V}{Z} = \frac{240}{22 \cdot 4} \text{ amperes} = 10 \cdot 7 \text{ A}$$

apparent power = $VI$ = 240 × 10·7 voltamperes = 2570 VA

or 2·57 kVA

$$\text{Power factor} = \frac{R}{Z} = \frac{10}{22 \cdot 4} = 0 \cdot 446 \text{ leading}$$

true power = apparent power × PF = 2570 × 0·446 watts = 1150 W

or 1·150 W

or

true power = $I^2 R$ = 10·7² × 10 watts = 1150 W

reactive power = $\sqrt{\{(VA)^2 - W^2\}}$

$= \sqrt{(2570^2 - 1150^2)}$ voltamperes reactive

$= 2 \cdot 30$ kVAr (leading)

The power diagram, drawn to scale, is shown in Figure 5.11. Because the circuit is capacitive and current leads voltage, it is inverted when compared with Figure 5.10 which is for an inductive circuit. If a circuit is predominantly inductive, current lags voltage and the overall kVAr or VAr is described as '**lagging**'. For a capacitive circuit, reactive voltamperes are described as '**leading**'. Because of the high values of power used, power

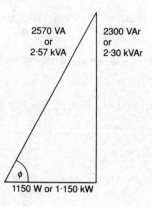

**Figure 5.11   Power diagram for Example 5.14**

diagrams are often labelled in terms of kilowatts (kW), kilovoltamperes (kVA) and kilovoltamperes reactive (kVAr).

Loads at differing power factors can be added using a power diagram to show the resultant voltamperes and power factor. The method is illustrated in Example 5.15.

## Example 5.15

A single-phase load consists of:

  (i) 12 kW of lighting and heating at unity power factor,
 (ii) 8 kW of motors at $0 \cdot 8$ power factor lagging, and
(iii) 10 kVA of motors at $0 \cdot 7$ power factor lagging.

Calculate (*a*) the total kW, (*b*) the total kVAr, (*c*) the total kVA, (*d*) the overall power factor, and (*e*) the total supply current at 240 V.

All loads are drawn on the power diagram in values of voltamperes or kVA (see Figure 5.12).

Load (i): At unity power factor, kW = kVA, so the load of 12 kW is drawn as a horizontal line to a suitable scale.

Load (ii): $\text{PF} = \dfrac{\text{kW}}{\text{kVA}}$    so    $\text{kVA} = \dfrac{\text{kW}}{\text{PF}} = \dfrac{8}{0 \cdot 8}$ kilovoltamperes = 10 kVA

The angle of lag has a cosine of $0 \cdot 8$, and from the cosine table this is $36 \cdot 9°$. A line equal to 10 units is drawn to represent the 10 kVA, making an angle of 37° with the horizontal.

The first two loads are then added by completing the parallelogram, to give the resultant A shown as a broken line in Figure 5.12.

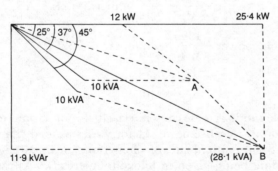

**Figure 5.12    Power diagram for Example 5.15**

Load (iii) is given directly in kVA, the angle being that which has a cosine of $0 \cdot 7$. From the cosine table this is $45 \cdot 6°$, so a line of length 10 units is drawn at this angle to the horizontal. This load is then added to the resultant A, and gives a total kVA of B, measured off as $28 \cdot 1$ kVA. The inphase (horizontal) component of this load is $25 \cdot 4$ kW, and represents the power. The quadrature (vertical) component is $11 \cdot 9$ kVAr, and represents the reactive kilovoltamperes. The angle made by the load is $25°$, and the cosine of this angle is the power factor, which is $0 \cdot 91$ lagging.

$$I = \frac{\text{kVA} \times 10^3}{\text{V}} = \frac{28100}{240} \text{ amperes} = 117 \text{ A.}$$

## 5.10    Rating of alternating-current cables and plant

The following example illustrates how the current carried by cables and supplied to a motor depends on the power factor as well as the rating.

### Example 5.16

A single-phase $3 \cdot 73$ kW motor is 85% efficient at full load and is fed from a 240 V supply. Calculate its full-load current if it operates at a power factor of (i) unity (ii) $0 \cdot 85$ lag, and (iii) $0 \cdot 6$ lag.

$$\text{input power} = 3 \cdot 73 \times 1000 \times \frac{100}{85} \text{ watts} = 4390 \text{ W}$$

(i)    $$\text{apparent power} = \frac{4390}{1} \text{ voltamperes} = 4390 \text{ VA}$$

Therefore    $$I = \frac{\text{VA}}{\text{V}} = \frac{4390}{240} \text{ amperes} = 18 \cdot 3 \text{ A}$$

(ii)    $$\text{apparent power} = \frac{4390}{0 \cdot 85} \text{ voltamperes} = 5170 \text{ VA}$$

Therefore    $$I = \frac{\text{VA}}{\text{V}} = \frac{5170}{240} \text{ amperes} = 21 \cdot 5 \text{ A}$$

(iii)    $$\text{apparent power} = \frac{4390}{0 \cdot 6} \text{ voltamperes} = 7320 \text{ VA}$$

Therefore    $$I = \frac{\text{VA}}{\text{V}} = \frac{7320}{240} \text{ amperes} = 30 \cdot 5 \text{ A}$$

If a correct cable rating is to be chosen, clearly it is important to take power factor into account. Because power factor varies as load changes, motors and other electrical equipment are often rated in apparent power, the maximum loading being given in kilovoltamperes (kVA) rather than in kilowatts (kW).

A power-factor meter is sometimes used, but such instruments are not readily available (see Figure 5.9). Figure 5.8 shows how an RMS-reading ammeter, an RMS-reading voltmeter and a wattmeter may be connected to calculate power factor. From the readings,

$$\text{power factor} = \frac{\text{wattmeter reading}}{\text{voltmeter reading} \times \text{ammeter reading}}$$

Although the power factor of a motor varies with load, it is usually higher at full load than at lower loadings. Full-load power factor can be safely used to calculate the current rating of cables, since the rise in current owing to reducing power factor is offset by the fall in load current.

## 5.11   Disadvantages of a low power factor

Example 5.16 shows clearly the way in which the current to a load is affected by power factor. If a motor takes a current of 30 A when operating at unity power factor, it will take 60 A if its power factor is $0 \cdot 5$. All the disadvantages of a low power factor are due to the fact that a given load takes more current at a low power factor than it does at a high power factor. The most important disadvantages of operating a load at a low power factor are:

   (i)  Larger cables, switchgear and transformers may be necessary both within an installation, and in the supply mains feeding it.

  (ii)  Low-power-factor working causes operating difficulties on high-voltage transmission lines.

 (iii)  Because of the effects of items (i) and (ii), electricity companies usually penalise the consumer whose load is at a poor power factor by charging more for the electrical energy used.

 (iv)  Larger cables may be needed within an installation to carry the extra current at low power factor. Alternatively, extra load can be connected to a cable if the power factor of the existing load it carries is improved.

  (v)  Higher currents give rise to higher copper losses in cables and transformers.

 (vi)  Higher currents give larger voltage drop in cables, and a change in load gives a larger change in voltage drop if the power factor is low. This is called 'poor voltage regulation'.

## 5.12   Power-factor correction

As indicated in the previous section, there are very great disadvantages in operating a system at a low power factor, no matter whether it is lagging or leading.

Since the majority of low power factors are lagging, owing to the effects of inductive loads such as motors and transformers, improvement may be effected by connection of capacitors in parallel with the load. Figure 5.13 shows circuit and phasor diagrams for a typical inductive load; the current $I_L$ lagging the voltage by $\phi_1$. Figure 5.14 shows the same circuit with the addition of parallel-connected capacitors, which carry the leading reactive current $I_C$. The resultant current $I$ will be the phasor sum of $I_L$ and $I_C$, the phase angle being reduced to $\phi_2$. This results in an improvement in power factor.

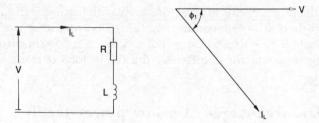

**Figure 5.13   Circuit and phasor diagrams for an inductive circuit before power-factor correction**

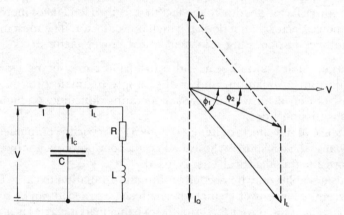

**Figure 5.14   Diagrams as Figure 5.13 with power-factor correction applied**

The power factor will be corrected to unity when the capacitor current is equal and opposite to the quadrature or reactive component $I_Q$ of the uncorrected current $I$. If too much capacitance is used, $I_C$ will exceed $I_Q$ and overcorrection, with a leading power factor, will result. This will give rise to the ill effects of a lagging power factor, and must be avoided.

Note that the current with the improved power factor only flows in the circuit up to the point of connection of the power-factor-correction capacitor. For this reason it is better to correct each load with its own locally connected capacitor. To save expense, one capacitor at the mains position may serve a whole installation, but for large installations it must be variable to prevent overcorrection if circuit inductance is reduced by switching off motors. With such a system, no reduction can be made in the size of circuit cables, although the overall improvement in power factor may result in a tariff saving. In fact, for many industrial consumers it is foolish not to spend money on power-factor-correction equipment. The reduction in the tariff will usually repay the capital outlay in less than two years, and after this the reduction is a direct saving (see Figure 5.18).

Example 5.17 illustrates the saving in load current due to power-factor correction.

## Example 5.17

At full load and at $0 \cdot 75$ power factor lagging, the efficiency of a 2 kW, 240 V single-phase motor is 85%. Calculate the supply current at full load:
- (*a*) with the lagging power factor above,
- (*b*) when power-factor correction is applied to give unity power factor, and
- (*c*) when the power factor is overcorrected so that the current leads the voltage by 20°.

$$\text{input power} = \frac{\text{output power}}{\text{efficiency}} = \frac{2000}{0 \cdot 85} \text{ watts} = 2350 \text{ W}$$

This input power remains the same regardless of power factor.

(*a*) $I = \dfrac{\text{voltamperes}}{\text{volts}} = \dfrac{\text{input power}}{\text{power factor} \times \text{volts}} = \dfrac{2350}{0 \cdot 75 \times 240} \text{ amperes} = 13 \cdot 1 \text{ A}$

(*b*) At unity power factor, voltamperes = input power.

$$I = \frac{\text{input power}}{\text{volts}} = \frac{2350}{240} = 9 \cdot 79 \text{ A}$$

(*c*) With a leading phase angle of 20°, the power factor is cos 20° leading, or $0 \cdot 940$ leading.

$$I = \frac{\text{input power}}{\text{power factor} \times \text{volts}} = \frac{2350}{0 \cdot 94 \times 240} \text{ amperes} = 10 \cdot 4 \text{ A}$$

The current to be carried by a capacitor for correction to a certain value, and the value of the capacitor itself, may be calculated as shown in Example 5.18.

## Example 5.18

A 240 V, 5 kW, 50 Hz single-phase motor working at full load with an efficiency of 85% has a power factor of $0 \cdot 7$ lagging.

- (*a*) Calculate the current taken by the motor on full load.
- (*b*) If a capacitor is connected across the motor terminals to raise the overall power factor to unity, calculate:
  - (i) the current to be carried by the capacitor, and
  - (ii) the capitance of the capacitor in microfarads.
- (*c*) Repeat the calculations of part (*b*) for correction to a power factor of $0 \cdot 9$ lagging.

(*a*) The output of the motor is 5 kW

$$\text{input power} = \frac{\text{output power}}{\text{efficiency}} = 5 \times \frac{100}{85} \text{ kilowatts} = 5 \cdot 88 \text{ kW}$$

$$P = VI \cos \phi \quad \text{so} \quad I = \frac{P}{V \cos \phi} = \frac{5 \cdot 88 \times 10^3}{240 \times 0 \cdot 7} \text{ amperes} = 35 \text{ A}$$

(b) (i) A phasor diagram is drawn to scale for the solution of capacitor current. A voltage phasor of any length is drawn horizontally, and then a current phasor representing 35 A is drawn at a lagging angle of 45½° (cos 45·6° = 0·7) as shown by line AB in Figure 5.15. The capacitor current needed for correction to unity is $I_{CA}$ (AC) which is equal to the opposite side of the parallelogram BD. By measurement, BD has a scale length of 25 A, which is the current carried by the capacitor.

(ii)
$$X_C = \frac{V}{I_{CA}} = \frac{10^6}{2\pi f C'}$$

Therefore $C' = \dfrac{I_{CA} \times 10^6}{2\pi f \times V} = \dfrac{25 \times 10^6}{2\pi \times 50 \times 240}$ microfarads = 331 μF

(c) (i) If the power factor is raised to 0·9, the new load current will lag the voltage by 26° (cos 25·8° = 0·9) and is drawn in at this angle as line AE on Figure 5.15. The new capacitor current $I_{CB}$ (AF) is thus equal to the opposite side of the parallelogram BE. From scale measurement $I_{CB} = 13$ A.

(ii) As in part (b) (ii),

$$C' = \frac{I_{CB} \times 10^6}{2\pi f \times V} = \frac{13 \times 10^6}{2\pi \times 50 \times 240} \text{ microfarads} = 172 \text{ μF}$$

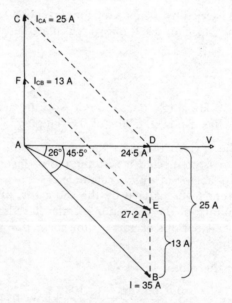

**Figure 5.15   Phasor diagram for Example 5.18**

NB: In this case, almost twice as much capacitance is needed to correct the power factor to unity as for correction to $0 \cdot 9$ lagging. The unity-power-factor current of $24 \cdot 5$ A is only $2 \cdot 7$ A smaller than the current at $0 \cdot 9$ lagging of $27 \cdot 2$ A. (These figures can be verified by scaling from Figure 5.15.)

In practice, it is not economic to correct the power factor to unity, since when the phase angle becomes small the inphase component of current is virtually the same as the actual current. The level to which power factor is corrected depends on the tariff applied and on the cost of capacitors, but is usually between $0 \cdot 97$ and $0 \cdot 99$ lagging.

In most industrial situations, power-factor-correction calculations are carried out using power diagrams rather than current-phasor diagrams, and the rating of correcting capacitors is given in kilovoltamperes reactive rather than in microfarads.

## Example 5.19

A factory has a maximum demand of 220 kVA at $0 \cdot 75$ power factor lagging. Calculate the kVAr rating of a capacitor bank to improve the power factor to (*a*) unity, and (*b*) $0 \cdot 95$ lagging.

The power diagram is drawn to scale in Figure 5.16. The apparent power of 220 kVA is drawn at an angle of $41 \frac{1}{2}°$ (cos $41 \cdot 4° = 0 \cdot 75$) below the horizontal, line AB.

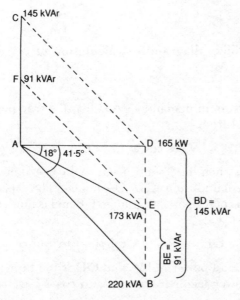

**Figure 5.16    Power diagram for Example 5.19**

(*a*) The reactive kVA of the capacitor to correct to unity is given by AC, equal to the opposite side of the parallelogram, BD, which scales at 145 kVAr.

(*b*) The new kVA line at 0·95 power factor is drawn as AE, 18° lagging kW (cos 18·2° = 0·95). The reactive kVA of the capacitor for this correction is AF, equal to the opposite side of the parallelogram, BE, which scales at 91 kVAr.

This indicates that to correct power factor from 0·75 to 0·95 requires 91 kVAr, but to correct it to unity requires 145 kVAr, showing the very large increase in capacitor-bank size for a small increase in power-factor correction.

The method of scale drawing used in Example 5.19 is perfectly valid, but rather time consuming. The kVAr rating of the required capacitor bank may be calculated more accurately and more quickly. Consider the power diagram shown in Figure 5.17.

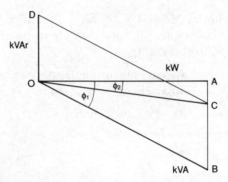

**Figure 5.17   Power diagram for calculation of correcting capacitor kVAr**

Given the maximum demand kVA (OB), the true power in kW (OA) can be calculated from

$$kW = kVA \times \cos \phi_1, \text{ i.e. } OA = OB \cos \phi_1$$

kVAr (AB) can then be found from AB = OA tan $\phi_1$; kVAr after correction can be found from AC = OA tan $\phi_2$. The capacitor kVAr (OD) is the opposite side of the parallelogram to CB and is thus equal to it. Hence, OD = CB = AB − AC.

Thus                    $$OD = CB = OA \tan\phi_1 - OA \tan \phi_2$$

Since OA is the load power in kW and OD is the rating of the capacitor to correct the power factor from cos $\phi_1$ to cos $\phi_2$, we could say that

$$kVAr = kW \tan \phi_1 - kW \tan \phi_2 = kW(\tan \phi_1 - \tan \phi_2)$$

## Example 5.20

An industrial building with a maximum demand of 420 kVA at $0 \cdot 82$ power factor lagging is to have its power factor improved to $0 \cdot 97$ lagging by connection of a central power-factor-correction capacitor. Calculate the rating of the necessary capacitor.

The true-power maximum demand is kVA cos $\phi_1 = 420 \times 0 \cdot 82$

$$= 344 \cdot 4 \text{ kW}$$

cos $\phi_1 = 0 \cdot 82$ so $\phi_1 = 34 \cdot 9°$ and tan $\phi_1 = 0 \cdot 698$

cos $\phi_2 = 0 \cdot 97$ so $\phi_2 = 14 \cdot 1°$ and tan $\phi_2 = 0 \cdot 251$

Correcting kVAr = kW(tan $\phi_1$ – tan$\phi_2$) = $344 \cdot 4(0 \cdot 698 - 0 \cdot 251)$

$$= 344 \cdot 4 \times 0 \cdot 447 = 154 \text{ kVAr}$$

The example suggests that a single capacitor bank at the incoming mains position (known as **bulk correction**) is the standard method for power-factor correction. In practice this is often not the case. The given maximum demand is unlikely to apply at all times, and when load is low there is the possibility that a single capacitor bank will cause overcorrection of power factor. A low leading power factor can result in excessive tariff costs as can a low lagging figure.

Power-factor-sensing relays can be obtained to switch the capacitor banks so as to ensure that the selected correction is applied at all times. Alternatively, individual loads may have correction capacitors fitted to them (known as **load correction**) so that the capacitors are switched on and off with the load. While this is an almost ideal solution to the problem, it is expensive, since the cost of buying and fitting a large number of small units is far higher than the provision of a single central correction bank.

In some cases, a mixture of bulk and load correction is applied, with larger loads being individually power-factor corrected and a central unit being used to deal with the smaller loads (see Chapter 13).

## *Power-factor correction with synchronous motor*

A synchronous motor running at low load and with its rotor overexcited will operate at a leading power factor, and can thus be used for power-factor correction. The costs of buying and maintaining a motor solely for this purpose are high, so the method is not widely used, although it does have application where such a machine will be required anyway, such as to drive a large fan. The advantage of the system is that the leading power factor of the synchronous motor can be continuously adjusted by variation of the rotor excitation, and thus the load power factor can be kept at the optimum level to give the lowest tariff.

## 5.13 Power in three-phase systems

Basic three-phase systems were considered in Chapter 12 of 'Electrical Craft Principles' Volume 1 and in Section 4.8 of this book. Both basic methods for connection of three-phase supplies involve the use of three separately-connected loads. It follows that the total power dissipated in a three-phase load is the sum of the powers dissipated in its three phases or, for a balanced load, three times the power dissipated in each phase.

For a balanced star-connected load,

$$P = 3V_P I_P \cos \phi$$

But $\qquad I_P = I_L \quad \text{and} \quad V_P = \dfrac{V_L}{\sqrt{3}}$

Therefore $\qquad P = 3 \dfrac{V_L}{\sqrt{3}} I_L \cos \phi = (\sqrt{3}) V_L I_L \cos \phi$

For a balanced delta-connected load,

$$P = 3V_P I_P \cos \phi$$

But $\qquad V_P = V_L \quad \text{and} \quad I_P = \dfrac{I_L}{\sqrt{3}}$

Therefore $\qquad P = 3 V_L \dfrac{I_L}{\sqrt{3}} \cos \phi = (\sqrt{3}) V_L I_L \cos \phi$

Thus, for any balanced three-phase load, total power dissipated is equal to the square root of three multiplied by line voltage multiplied by line current, multiplied by phase power factor.

## Example 5.21

Three identical 10 $\Omega$ impedances, each with a power factor of $0 \cdot 8$, are connected to a 415 V three-phase supply (*a*) in star, and (*b*) in delta. Calculate the power dissipated in each case.

(*a*) $V_P = \dfrac{415}{\sqrt{3}}$ volts = 240 V

$\qquad I_L = I_P = \dfrac{V}{Z} = \dfrac{240}{10}$ amperes = 24 A

$\qquad P = (\sqrt{3}) V_L I_L \cos \phi = \sqrt{3} \times 415 \times 24 \times 0 \cdot 8$ watts = $13 \cdot 8$ kW

(*b*) $I_P = \dfrac{V_P}{Z} = \dfrac{V_L}{Z} = \dfrac{415}{10}$ amperes = $41 \cdot 5$ A

$\qquad I_L = (\sqrt{3}) I_P = (\sqrt{3}) \times 41 \cdot 5$ amperes = $71 \cdot 8$ A

$\qquad P = (\sqrt{3}) V_L I_L \cos \phi = (\sqrt{3}) \times 415 \times 71 \cdot 8 \times 0 \cdot 8$ watts = $41 \cdot 4$ kW

**Figure 5.18    Four automatically controlled 100 kVAr, 415 V power-factor correction capacitors**

For an unbalanced load, the power for each of the three separate sections must be calculated separately and added to give the total power.

## Measurement of power in three-phase systems

For a balanced three-phase load, the total power dissipated is three times the power in any one phase. The power in one of the phases may be measured by a wattmeter (see Section 11.6).

If a neutral is available (four-wire system) the wattmeter may be connected as shown in Figure 5.19, and will read the power in the phase concerned directly. Thus

total power = three times wattmeter reading

For a three-wire system, the wattmeter cannot be connected with its voltage coil between lines because of the extra 30° phase difference between line and phase voltage (see Figure 4.20). If two resistors of value equal to the resistance of the wattmeter voltage circuit are connected as shown in Figure 5.20, an artificial star point will be formed, and

total power = three times wattmeter reading

If the load is four-wire and unbalanced, it is necessary to use three

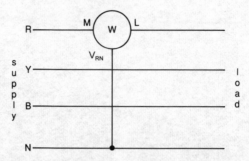

**Figure 5.19   One-wattmeter connection (four-wire balanced load)**
$P = 3 \times$ wattmeter reading

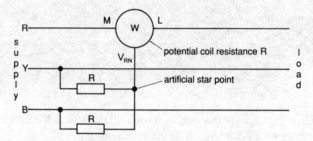

**Figure 5.20   One-wattmeter connection (three-wire balanced load)**
$P = 3 \times$ wattmeter reading

wattmeters connected as in Figure 5.21. Each wattmeter will read the power of the line in which it is connected, and

total power = sum of three wattmeter readings

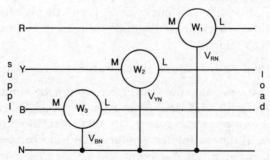

**Figure 5.21   Three-wattmeter connection (four-wire unbalanced load)**
$P = W_1 + W_2 + W_3$

The two-wattmeter method will apply to a three-wire unbalanced load. It is connected as shown in Figure 5.22, and

total power = sum of two wattmeter readings

Note that the voltage coils are connected between lines, which have a voltage 30° out of phase with phase voltage (see Figure 4.19). For this reason, if the angle between phase voltage and current exceeds 60°, the angle between line voltage and current for one of the meters will exceed 90° and that instrument will attempt to read backwards. Power can then be measured by reversing the voltage connections on the wattmeter to make it read forwards, and subtracting its reading from that of the other instrument.

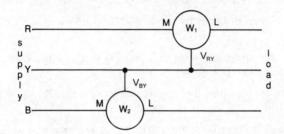

**Figure 5.22   Two-wattmeter connection (three-wire unbalanced load)**
$P = W_1 + W_2$

## 5.14   Summary of formulas for Chapter 5

See text for definitions of symbols.

Instantaneous power:

$$p = vi \qquad v = \frac{p}{i} \qquad i = \frac{p}{v}$$

For series AC circuits:

$$P = VI \cos \phi \qquad V = \frac{P}{I \cos \phi} \qquad I = \frac{P}{V \cos \phi} \qquad \cos \phi = \frac{P}{VI}$$

$$P = I^2 R \qquad R = \frac{P}{I^2} \qquad I = \sqrt{\frac{P}{R}}$$

$$P = \frac{V_R^2}{R} \qquad R = \frac{V_R^2}{P} \qquad V_R = \sqrt{PR}$$

Power factor:

$$\mathrm{PF} = \frac{P}{VI} = \cos \phi = \frac{V_R}{V} = \frac{R}{Z}$$

The last three equalities are true only for sinusoidal waveforms. For balanced three-phase systems:

$$P = (\sqrt{3})V_L I_L \cos \phi \qquad\qquad V_L = \frac{P}{(\sqrt{3})I_L \cos \phi}$$

$$I_L = \frac{P}{(\sqrt{3})V_L \cos \phi} \qquad\qquad \cos \phi = \frac{P}{(\sqrt{3})V_L I_L}$$

## 5.15   Exercises

1 Calculate the current taken by a 100 W, 240 V tungsten filament lamp.

2 Calculate the current taken by a 1 kW, 240 V fire, and the resistance of the element.

3 A resistive load connected to a 100 V, 60 Hz supply has a resistance of 12 $\Omega$. Calculate the power consumed by the load.

4 Calculate the current and the reactive voltamperes when a 30 $\mu$F capacitor is connected to a 115 V, 60 Hz supply.

5 When a capacitor is connected to a 415 V, 50 Hz supply, the reactive power is 2·71 kVAr. Calculate the capacitance of the capacitor.

6 A 10 $\Omega$ resistor and a capacitor with a reactance of 10 $\Omega$ are connected in series across a 200 V AC supply. Calculate the circuit impedance, the current, the phase angle and the power dissipated.

7 A 3·18 $\mu$F capacitor and a 30 $\Omega$ resistor are connected in series to a 100 V, 1 kHz supply. Calculate the power dissipated in the resistor.

8 A capacitor and a 50 $\Omega$ resistor are connected in series to an alternating-current supply. The voltage across the capacitor is 200 V RMS and across the resistor is 150 V RMS. Determine:
   (*a*) the RMS value of the supply voltage,
   (*b*) the peak value of the voltage across the capacitor, assuming a sinusoidal waveform,
   (*c*) the power used in the resistor.

9 A 25 $\Omega$ resistor and a capacitor are connected in series to a 240 V, 50 Hz supply, when the power dissipated in the resistor is 625 W. Calculate the capacitance of the capacitor.

10 A resistive–capacitive circuit connected to a 100 V, 60 Hz supply takes a current of 10 A and dissipates a power of 600 W. Calculate the values of the resistor and of the capacitor.

11 A 0·5 H choke of resistance 200 $\Omega$ is connected to a 50 V, 50 Hz supply. Calculate the current and the power dissipated.

12 A choke of resistance 4 $\Omega$ and self inductance 3·18 mH is connected to a 100 V, 400 Hz supply. Calculate the current, its phase relative to voltage, and the power dissipated.

13 A 10 Ω resistor and a 31·8 mH inductance are connected to a 240 V, 50 Hz supply. Calculate the power and the energy dissipated if the circuit is connected for two hours.

14 A choke of inductance 0·1 H is connected to a 115 V, 400 Hz supply and takes a current of 0·4 A. Calculate the power dissipated.

15 A choke of resistance 10 Ω dissipates 1 kW when connected to a 240 V, 50 Hz supply. Calculate its inductance.

16 A series resistive–inductive circuit has an impedance of 100 Ω and dissipates 0·4 kW when connected to a 240 V AC supply. Calculate the circuit resistance and reactance.

17 A single-phase motor takes 16 A at 240 V. Assuming the current lags by 30°, calculate the power input to the motor. What is the output power of the motor if its efficiency is 0·8?

18 A single-phase 240 V motor gives an output of 1 kW. If its efficiency is 82% and it operates at a power factor of 0·8, calculate the current taken from the supply.

19 Calculate the power dissipated in the circuit given in Chapter 4, Exercise 21.

20 Calculate the power in the circuit of Chapter 4, Exercise 23 at both the frequency of part (*a*) and that of part (*b*).

21 Calculate the power in the resonant series circuit of Chapter 4, Exercise 27.

22 Calculate the power factor of the circuit of Exercise 6.

23 Calculate the power factor of the circuit of Exercise 7.

24 Calculate the power factor of the circuit of Exercise 8.

25 Calculate the power factor of the circuit of Exercise 11.

26 Calculate the power factor of the circuit of Exercise 12.

27 Calculate the power factor of the motor of Exercise 17.

28 Calculate the power factor of the circuit of Exercise 20 for both frequencies.

29 (*a*) Define the meaning of the term 'power factor' as applied to an AC circuit.
   (*b*) An AC single-phase motor takes 20 A at 0·8 power factor lagging when connected to a 240 V, 50 Hz supply. What is the power taken by the motor?

30 A circuit supplied at 240 V, 50 Hz consists of a resistor of 15 Ω in series with an inductor of 20 Ω reactance (and negligible resistance). Calculate the current and the power factor.

31 Define the term power factor, and illustrate by means of a phasor diagram.
   A coil of insulated wire of resistance 9 Ω and inductance 0·02 H, is connected to a single-phase AC supply at 240 V, 50 Hz. Calculate (*a*) the current in the soil, (*b*) the power factor, and (*c*) the power in the circuit.

32 A current of 1 A is produced by 100 V AC in each of the following separately: (*a*) a resistor, (*b*) a capacitor, and (*c*) a coil of 60 Ω resistance and 80 Ω reactance. Find the power in watts and the reactive voltamperes in each case.

33 A single-phase motor develops 11·2 kW. The input to the motor is recorded

by instruments with readings as follows: 240 V, 75 A, 13 120 W. Calculate the efficiency of the motor and its power factor. Draw a diagram of connections of the instruments. Account for the energy lost in the motor.

34 You are required to record the input to a single-phase AC motor in kW and kVA. Make a connection diagram showing the instruments you would use. A $22 \cdot 4$ kW single-phase motor delivers full-load output at $0 \cdot 75$ power factor. If the input is $33 \cdot 5$ kVA, calculate the efficiency of the motor.

35 A single-phase load consists of heating and lighting (10 kW at unity power factor) and induction motors (8 kW at a power factor of $0 \cdot 8$ lagging). Calculate

    (*a*) the total kW,                    (*b*) the total kVAr,
    (*c*) the total kVA,                  (*d*) the overall power factor,
    (*e*) the total line current at 240 V.

36 A small workshop has 4 kW of lighting and 8 kW of heating, both at unity power factor, as well as 10 kVA of motors at $0 \cdot 8$ PF lagging. Calculate the overall kVA, kW, kVAr and power factor of the workshop.

37 A pumping station has two main motors, each with a full-load output of 100 kW at an efficiency of 90% and a power factor of $0 \cdot 9$ lagging. The station also has a motor with a full-load output of 48 kW at an efficiency of 85% and a power factor of $0 \cdot 8$ lag. Calculate the demand of the station when on full load in kVA, kW and kVAr, as well as the overall power factor.

38 Why are transformers and motors rated in kVA and not in kW?

39 (*a*) What is meant by power factor?
    (*b*) The installation in a factory has the following installed load: lighting 50 kW, heating 30 kW, power $44 \cdot 8$ kW.
    Assuming that the lighting and heating loads are noninductive, and the power has an overall efficiency of 87% at a power factor of $0 \cdot 7$ lagging, calculate:
    (i) the total loading in kW, and
    (ii) the kVA demand at full load.

40 (*a*) Define the term power factor and explain how a low power factor affects the size of cable required to carry a given AC load.
    (*b*) A 240 V, single-phase induction motor delivers $16 \cdot 4$ kW at full load. The efficiency of the motor at this load is 86% and the power factor $0 \cdot 75$ lagging. Calculate the following:
    (i) the current to the motor,
    (ii) the kW output, and
    (iii) the kVA input.

41 List and explain the disadvantages of operating an electrical load at a low power factor.

42 Explain clearly the meaning of power factor, and illustrate by means of a phasor diagram.
At full load, and at $0 \cdot 766$ power factor lagging, the efficiency of a $22 \cdot 4$ kW, 240 V single-phase motor is 87%. Calculate the value of the supply current at full load:
(*a*) with lagging power factor as above,

(*b*) when power factor correction is applied to the terminals of the motor so that voltage and current are in phase, and

(*c*) when the power factor is overcorrected so that the current leads the voltage by 40°.

43 (*a*) Explain the meaning of power factor, with the aid of a phasor diagram.

(*b*) A 240 V, 3·03 kW, single-phase motor works at full load at a power factor of 0·7 lagging, with an efficiency of 85%.

Assuming that the efficiency remains constant in each case, find the value of the supply current at full load

(i) with lagging power factor as above,

(ii) with power factor overcorrected so that the current leads the voltage by 30°, and

(iii) with the power factor so corrected that the current and voltage are in phase.

44 Why is a low lagging-power-factor load undesirable on a power-supply system? A single-phase motor takes 5 A at 0·8 power factor lagging. Show by phasor diagrams how the power factor can be improved by shunting the motor by a capacitor. What current should the capacitor pass to improve the power factor to unity?

45 A 240 V, 7·46 kW, 50 Hz single-phase motor working at full load with an efficiency of 87% has a power factor of 0·75 lagging.

(*a*) Calculate the current supplied to the motor at full load.

(*b*) If a capacitor is connected across the motor terminals, to raise the overall power factor to unity, find

(i) the current to be carried by the capacitor, and

(ii) the value of the capacitor in microfarads.

46 Calculate the capacitor current and the capacitance of the capacitor needed to improve the power factor of the motor of Exercise 44 to 0·95 lagging. Assume a supply at 240 V, 50 Hz.

47 If the power factor of the motor of Exercise 45 is to be improved to 0·97 lagging, calculate the current carried by the capacitor and its value.

48 A factory has a maximum demand of 300 kW at a power factor of 0·8 lagging. Calculate the kVAr rating of capacitors to correct the power factor to (*a*) unity and (*b*) 0·97 lagging.

49 A mine has maximum demand of 550 kVA at 0·6 power factor lagging. Calculate the kVAr rating of capacitors to correct the power factor to (*a*) unity and (*b*) 0·9 lagging.

50 A small workshop has a maximum load of 50 kW at 0·75 power factor. Calculate the rating of a capacitor bank to improve the power factor to (*a*) unity, and (*b*) 0·95 lagging.

51 Three chokes, each of inductance 0·1 H and resistance 10 Ω, are star-connected to a 415 V, 50 Hz supply. Calculate the line current and the total power dissipated.

52 Calculate the line current and total power dissipated if the same three chokes of Exercise 51 are delta-connected to the same supply.

## 5.16   Multiple-choice exercises

5M1   A formula for the power dissipated in an AC circuit which has a power factor of unity is:

(a) $P = V^2 R$        (b) $P = I^2 R$        (c) $P = V/R$        (d) $P = R/V^2$

5M2   A 7 kW shower unit connected to a 240 V single-phase supply takes a current of:

(a) $34 \cdot 3$ A        (b) $2 \cdot 92$ A        (c) 13 A        (d) $29 \cdot 2$ A.

5M3   The power dissipated by a 10 μF capacitor connected across a 100 V, 500 Hz supply will be:

(a) $3 \cdot 14$ A        (b) 10W        (c) zero        (d) 318 W.

5M4   The reactive voltamperes taken by the capacitor of Exercise 5M3 is:

(a) zero        (b) 314 VAr        (c) $3 \cdot 14$ VA        (d) 100 VA.

5M5   The capacitor of Exercise 5M3 has a 20 Ω resistor connected in series with it across the same supply. The power dissipated will be:

(a) 142 W        (b) $37 \cdot 6$ W        (c) 266 W        (d) 1 kW.

5M6   A 1 kΩ resistor and a capacitor are connected in series to a 240 V, 50 Hz supply. If the power dissipated is $18 \cdot 6$ W the capacitor value will be:

(a) 22 μF        (b) $57 \cdot 6$ pF        (c) 10 μF        (d) $2 \cdot 2$ μF.

5M7   A choke connected to a 115 V, 60 Hz supply has a resistance of 8 Ω and dissipates a power of 400 W. The inductance of the choke is:

(a) $37 \cdot 6$ mH        (b) $3 \cdot 48$ H        (c) $45 \cdot 1$ mH        (d) $1 \cdot 94$ H.

5M8   A series circuit has a self inductance of 56 mH, capacitance of $1 \cdot 8$ μF and resistance of 19 Ω. Calculate the power dissipated in the circuit when connected to a 50 V, 500 Hz supply:

(a) $1 \cdot 53$ W        (b) $2 \cdot 63$ W        (c) 132 W        (d) $13 \cdot 2$ W.

5M9   Power factor can be defined as:

(a) the power dissipated in a reactive circuit

(b) $\dfrac{\text{true power}}{\text{apparent power}}$

(c) $\dfrac{\text{voltamperes}}{\text{watts}}$

(d) the quadrature component of the power dissipated.

5M10   The power factor of the circuit given in Exercise 5M6 is:

(a) $0 \cdot 57$ leading        (b) $1 \cdot 75$ leading
(c) $0 \cdot 43$ lagging        (d) $0 \cdot 57$ lagging.

5M11   A method of calculating power factor is to use:

(a) $\sin \phi = \dfrac{Z}{R}$        (b) $P = I^2 R$

(c) $\cos \phi = \dfrac{R}{Z}$        (d) $\tan \phi = \dfrac{R}{X}$

5M12 The relationship between true power ($P$), apparent power (VA) and reactive
voltamperes (VAr) is:
(a) $P = \text{VA} + \text{VAr}$        (b) $\text{VA} = \text{VAr}/P$
(c) $P^2 = (\text{VA})^2 + (\text{VAr})^2$    (d) $(\text{VA})^2 = P^2 + (\text{VAr})^2$

5M13 Correct connection of an ammeter, a voltmeter and a wattmeter in a single-
phase circuit is:

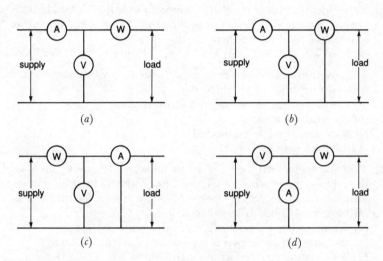

Figure 5.23    **Diagrams for Exercise 5M13**

5M14 If a 25 $\Omega$ resistor and a capacitor of reactance 35 $\Omega$ are connected in series
to a 100 V supply, the reactive voltamperes will be:
(a) 189 VAr      (b) $2 \cdot 33$ A      (c) 135 W      (d) 43 VAr.

5M15 The single example below which is NOT a disadvantage of operating plant
at a low power factor is:
(a) larger cables may be needed to feed the load
(b) higher voltage drops will result in lower plant voltages
(c) there will be lower copper losses in cables and transformers
(d) the cost of electrical energy will be higher for larger consumers.

5M16 A typical industrial load has a lagging power factor which is corrected by:
(a) reducing supply voltage by using a transformer
(b) connecting a lightly loaded induction motor to the system
(c) using larger cables to supply the loads
(d) parallel connection of capacitors.

5M17 The kVAr rating of a capacitor bank to improve the power factor of a load
with a power of kW from $\cos \phi_1$ lagging to $\cos \phi_2$ lagging is:
(a) $\text{kVAr} = \text{kW}(\tan \phi_1 + \tan \phi_2)$
(b) $\text{kVAr} = \text{kW}(\cos \phi_1 - \cos \phi_2)$
(c) $\text{kVAr} = \text{kVA}(\tan \phi_1 - \tan \phi_2)$
(d) $\text{kVAr} = \text{kW}(\tan \phi_1 - \tan \phi_2)$

5M18  An industrial load with a maximum demand of 420 kVA at a power factor of 0·8 lagging will have its power factor improved to 0·96 lagging by the parallel connection of capacitors with a rating of:
(a) 243 kVAr    (b) 154 kVAr    (c) 350 μF    (d) 96 kVAr.

5M19  The power dissipated in a three-phase system can be calculated from:
(a) $P = V_L I_L \cos \phi$    (b) $P = (\sqrt{3}) V_L I_L \cos \phi$
(c) $P = V I \cos \phi$    (d) $P = 3 V_L I_L \cos \phi$.

5M20  Three identical loads are delta-connected to a 400 V, 60 Hz three-phase supply. If each load has an impedance of 25 Ω at a lagging power factor of 0·9, the total power dissipated is:
(a) 5·76 kW    (b) 17·3 kW    (c) 10 kW    (d) 27 kW.

5M21  The power in a three-wire three-phase system may be measured accurately with two wattmeters provided that:
(a) the power in the neutral is also measured
(b) the load is balanced
(c) they are correctly connected
(d) the load is unbalanced.

5M22  Two wattmeters are being used to measure the power dissipated in a three-phase three-wire unbalanced load when it is observed that one wattmeter tries to deflect backwards. The reason for this is that:
(a) the power factor is less than 0·5
(b) the reversing wattmeter is wrongly connected
(c) the phase angle between voltage and current is exactly 30°
(d) both wattmeters are wrongly connected.

5M23  A single wattmeter can be used to measure the power dissipated in a balanced three-phase load provided that its reading is trebled and provided it is connected as follows:

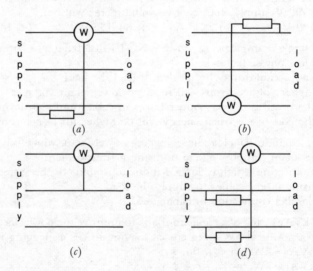

**Figure 5.24   Diagrams for Exercise 5M23**

## Chapter 6

# Further electronics

## 6.1 Introduction

The subject of electronics is a vast and specialist field, and we cannot hope here to do more than cover the basics. It will be assumed that the reader is conversant with Chapter 15 of Volume 1. Attention is also drawn to the work in Chapter 11 of this book covering the principles and applications of the cathode-ray oscilloscope and other measuring instruments.

No attempt is made here to consider the syllabus content of C & G 236-4-31 'Supplementary studies in electronics'. It should be appreciated that most electronic equipments look very complex at first sight, as do the diagrams of their circuits. However, all equipments and circuits are made up of a number of interconnected components. When the operation of these parts is understood, then the functioning of the complete apparatus or circuit will become clear.

## 6.2 Transducers

A **transducer** is a device which converts a physical quantity (such as force, vibration, sound, light, movement etc.) into an electrical quantity of corresponding value. Common examples are photocells, microphones, television cameras, strain gauges and so on. It is unusual for the transducer itself to have a sufficient output to operate the device it feeds, and amplification is usually necessary. For example, a microphone must have its signal increased by an amplifier before the signal becomes large enough to operate a loudspeaker. The arrangement is shown in Figure 6.1.

There are very many types of transducer, some of which are considered here.

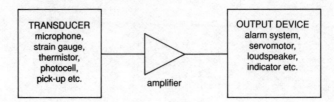

**Figure 6.1   Block diagram for a basic transducer system**

## Light-dependent resistors (LDRs)

Some semiconductor materials, such as selenium, cadmium sulphide, indium antimonide, selenide and lead sulphide, alter their resistance when light falls on them. The change occurs because the energy in the electromagnetic radiation (light) dislodges electrons from their atoms; they thus become available for conduction, and the resistance of the conductor reduces. A typical arrangement is shown in Figure 6.2 where interleaved conducting strips are deposited onto a glass slip and covered with a layer of semiconductor material typically a few thousandths of a millimetre thick.

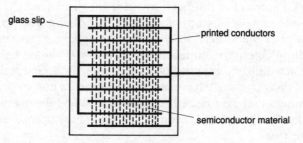

**Figure 6.2   Arrangement of typical LDR**

## Photodiode

The photodiode is a rectifier like other diodes, but is arranged so that light can fall onto its junction. The forward current then passing depends on the forward voltage as well as on the light level at the junction. The device can be used as the basis for equipments which respond to light.

## Phototransistors

A normal bipolar-junction transistor operates by amplification of its base current. A phototransistor has a hole in its case to allow incident light to fall on the collector–base junction; this generates hole–electron pairs which have the same effect as base current in a normal transistor. The phototransistor is often connected in circuit with other transistors to control the required load, as shown in Figure 6.3. For example, the presence of sunlight can operate a motor to result in the operation of a window blind.

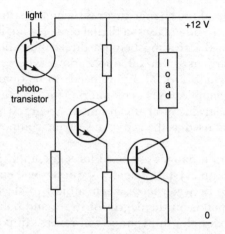

**Figure 6.3　Phototransistor in simple control circuit**

## *Strain gauges*

The resistance of a conductor is proportional to its length and inversely proportional to its cross-sectional area. If a wire is stretched, its length increases and its cross-sectional area reduces, so its resistance increases. Conversely, a wire which is compressed reduces in resistance. This is the basic theory of the strain gauge.

In practice a grid of fine wires is arranged as shown in Figure 6.4*a* and tightly cemented to a backing of strong paper. The gauge is then cemented to the surface whose strain is to be determined; any extension of the surface will result in an increase in strain-gauge resistance, or a compression of the surface in resistance reduction. The change in resistance concerned is usually very small, so a bridge circuit is used (see Section 11.15) to detect and to measure the change, often arranged as shown in Figure 6.4*b*. In some cases the bridge is balanced by adjustment of the variable arm,

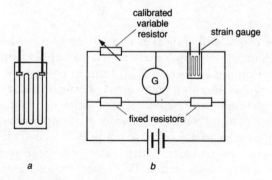

**Figure 6.4　The strain gauge**
*a* Arrangement of typical strain gauge
*b* Bridge circuit for measuring gauge resistance

allowing calculation of the change in strain-gauge resistance and hence of the strain involved. In other cases, a digital or deflecting instrument is used as a galvanometer, the reading being a direct indication of strain.

Strain gauges are used for all applications where a change in the dimensions of a structure or of a part needs to be quantified. They are used in such diverse applications as monitoring the structure of large bridges and measuring the strains in artificial limbs. Accuracy is so good that the strain gauge is often used as the active measuring component in scales and weighbridges.

Piezo-electric strain gauges are used for some applications. They rely on the fact that certain crystals, usually of quartz, will exhibit a very small potential difference between faces when subject to electrical strain. The level of voltage depends on the degree of strain and polarity on the sense of the strain; for example, the voltage will reverse in polarity if the force on a crystal changes from compression to tension. This type of strain gauge has the advantage that it requires no power source, although one will be needed for the amplifier which is certain to be used in conjunction with it.

## Thermocouples

When two junctions are made between different metals, a potential difference will appear if the temperatures of the two junctions are different. This is known as the **Seebeck effect**. If two ends of the metals are connected to a millivoltmeter (the cold junction) the reading will depend on the difference in temperature between the voltmeter and the other connection (the hot junction). Since the instrument can be remote from the hot junction, this provides an excellent method for measuring the temperature of objects, and is widely used in applications such as cooking, food storage, hot water systems and so on.

The thermocouple may also be used as a temperature-sensing device in many kinds of control system. For example, it could be used in an alarm to give indication of overtemperature in an industrial process.

## Thermistors

Almost all resistors will experience a change in resistance as their temperature changes, because nearly all have a temperature coefficient of resistance (see Volume 1, Section 2.5). In most cases we wish to keep the resistance change as small as possible. A thermistor is a resistor with a deliberately high temperature coefficient of resistance so that its resistance will change significantly with a change in its temperature.

Most thermistors are of the negative-temperature-coefficient type (NTC) so that their resistance falls as the temperature rises. Positive-temperature-coefficient (PTC) types are also available, and the circuit symbols for the two types are shown in Figure 6.5. Thermistors are used in many measuring and control applications where temperature is to be assessed, and in electronic circuits where their change in resistance

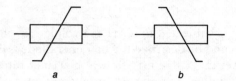

**Figure 6.5   Circuit symbols for thermistors**
  *a* Positive-temperature-coefficient (PTC) type
  *b* Negative-temperature-coefficient (NTC) type

with temperature is used to prevent effects such as thermal runaway, where increasing current in a circuit leads to a rise in temperature, still higher current and temperature, and so on so that the circuit runs out of control.

## Variable capacitors

The capacitance of a capacitor is proportional to its plate area and inversely proportional to the thickness of its dielectric. Thus there are two types of variable-capacitor transducer.

The first type is one with shaped air-spaced plates. One set of plates will rotate relative to the other so that the effective plate area depends on the position of the moving plates. An example is shown in Figure 1.14, although the photograph shows a laboratory-type variable capacitor rather than the smaller type used as a transducer to respond to small rotary movements.

A second type uses a dielectric which is compressible, so that a small movement of one plate effectively reduces the spacing between plates and increases capacitance. This type is used where a very small linear movement must be detected.

## Variable resistors

The resistor whose resistance depends on its temperature (the thermistor) has been considered. Resistance can also change if a slider is moved across a resistance wire as shown in Figures 2.3*a* and 2.3*b* of Volume 1. Thus resistance transducers responding to linear or to rotary movement can be produced.

## 6.3   Devices

There are so many different electronic devices that to catalogue all of them would be a near-impossible task. Some of those most likely to be encountered by the electrical craftsman are considered in this section.

## Diodes and transistors

These devices have been considered in sufficient detail on Chapter 15 of Volume 1, and this material will not be repeated.

## Zener diodes

If a normal diode is subjected to a reverse voltage in excess of its rating, it will break down and be destroyed. The Zener diode is designed to break down and allow reverse current to flow at a certain rated voltage, but to recover when that voltage is removed, provided that the power dissipated within it does not exceed its power rating. Zener diodes are manufactured in a range of preferred values (e.g. 4·7 V, 6·8 V, 18 V, 39 V, 47 V etc.) and two or more devices may be connected in series to make up a required voltage rating. The circuit symbol for the Zener diode, which is also known as the voltage-reference diode, is shown in Figure 6.6*a*.

The Zener diode can be used in many circuits as a voltage stabiliser. A simple application is shown in Figure 6.6*b*. The voltage across the load will remain constant at the breakdown voltage of the Zener diode even if the supply voltage varies. An increase in supply voltage will increase the current carried by the diode, and hence increase the current in the stabilising resistor R. The increased voltage drop in R will keep the load voltage constant.

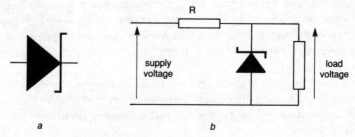

**Figure 6.6   The Zener diode**

    *a* Circuit symbol for a Zener diode

    *b* Simple Zener voltage stabilising circuit

## Diacs

The diac is a two-directional trigger diode which is used in the firing circuits of thyristors and triacs. It breaks down like a Zener diode when its rated voltage is applied and passes a sudden pulse of voltage instead of, for example, a steadily rising ramp voltage. Applying such a pulse to the gate of a thyristor or triac will result in firing at a more precise instant. Figure 6.7 shows the circuit symbol for the diac and typical wave diagrams for input and output voltage.

## Thyristors

The thyristor is a semiconductor-controlled rectifier which is used extensively in power control circuits such as lamp dimmers, motor-speed control and temperature control. Basically direct-current devices, a pair of thyristors

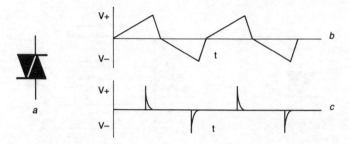

**Figure 6.7    The diac**
  *a* Circuit symbol for diac
  *b* Typical diac input-voltage wave
  *c* Corresponding output wave for Figure 6.7*b* input

can be arranged to control an alternating supply with one handling positive, and the other negative, halfcycles of current. The thyristor will not usually conduct until it is triggered (fired) by a voltage pulse at its gate connection; once conducting, it remains so until the current carried falls to zero, when it returns to the nonconducting state.

The circuit symbol for a thyristor is shown in Figure 6.8*a*. Control can be exercised in two ways. '**Burst control**' is a system where the thyristor is switched on for a number of complete halfcycles of the supply, and then left off for another batch of halfcycles. Current control is exercised by adjusting the time 'on' compared with the time 'off' and is used in applications such as temperature control by switching heaters on and off where the output will not be affected by the current being switched for complete halfcycles. In the '**phase control**' method, the exact time in the halfcycle of supply voltage at which the thyristor switches on is controlled by the instant at which the gate is pulsed. In this way, average current depends on the firing instant, becoming less as firing takes place later in the halfcycle. A simple phase controlled thyristor circuit is shown in Figure 6.8*b* and appropriate waveforms for early and late firing in the halfcycle (small and larger firing angles) in Figure 6.8*c* and 6.8*d*. The method is used in applications such as lighting control, where lamp flicker would be a problem if burst control were applied.

## *Reactive load control*

The thyristor is triggered into the conducting mode by feeding its gate and will then switch itself off only when the current through it falls to zero (but see reference to the GTO device later in this section). If the current is still flowing when forward voltage appears, conduction will again take place regardless of the signal at the gate. The effect can occur when the thyristor feeds reactive circuits, such as a direct-current motor which has considerable self inductance. This results in the induction of an EMF as the current falls, which tries to keep it flowing. A typical wave diagram in such a case is shown in Figure 6.9.

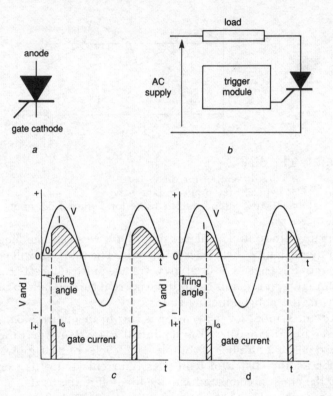

**Figure 6.8   The thyristor**

*a* Circuit symbol for a thyristor
*b* Block diagram for thyristor operation
*c* Wave diagrams for a thyristor fired early in the halfcycle
*d* Wave diagrams for a thyristor fired late in the halfcycle

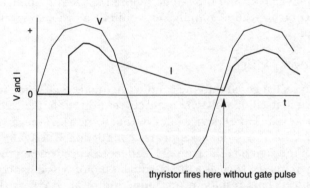

**Figure 6.9   To illustrate how a thyristor with an inductive load may misfire**

The solution to the problem is to provide a low-resistance path for the decaying current to enable it to fall to zero before the commencement of the next positive halfcycle of voltage. This is accomplished by connecting a diode, called a **flywheel diode**, as shown in Figure 6.10. There is no effect on the operation of the positive halfcycle because the diode is reverse biased and has high resistance. The decaying current finds a low-resistance path through the forward-biased diode.

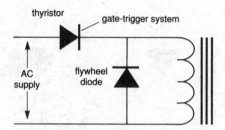

**Figure 6.10   Flywheel diode used to prevent misfiring**

## *Triacs*

The triac behaves in the same way as two thyristors connected back to back so that one controls positive halfcycles of an AC system, while the other controls negative halfcycles. The circuit symbol is shown in Figure 6.11*a* while Figure 6.11*b* shows a typical waveform. One triac can take the place of two thyristors, so there is an obvious advantage in cost terms. However, while the thyristor has the whole of the negative halfcycle to turn off before foward voltage is applied, the triac must turn off much more quickly. This limits the use of triacs to applications at power frequencies such as 50 Hz.

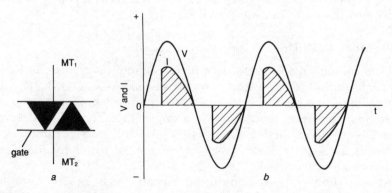

**Figure 6.11   The triac**
  *a* Triac circuit symbol
  *b* Wave diagrams for a triac

## GTO *devices*

Both the thyristor and the triac can be switched on by a trigger signal to the gate, but must wait until the end of the halfcycle before they will switch off. A gate-turn-off (GTO) device can be a thyristor or a triac, but is capable of being turned off as well as on by a signal to its gate. The circuit symbols are shown in Figure 6.12.

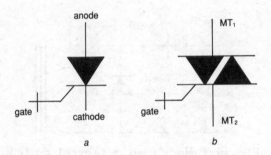

**Figure 6.12   Circuit symbols for GTO devices**
    *a* GTO (gate-turn-off) thyristor
    *b* GTO (gate-turn-off) triac

## 6.4   Packaging

### *Discrete components*

This is the name given to individual components in their normal condition, such as resistors, diodes, transistors, thyristors and so on (see Figure 6.13). So far we have considered no alternative to this construction, but a very widely used method, the integrated circuit, will be considered. Strictly speaking, circuit diagrams should show discrete components circled as shown, for example, in Figure 6.3, but this rule is widely ignored. Where there is no circle, as for example in Figure 6.12, the component is assumed strictly to be part of an integrated circuit.

### *Integrated circuits*

All the components of these circuits, commonly referred to as 'ICs', are made on top of, or in, a single chip of silicon. As well as the advantage of reliability, this method is cheap and takes up very little space, thousands of components being made on or in a single chip. The cost of developing such circuits is very high but, since they are made in large numbers, each one sells for a price which is far below that of the individual discrete components which would otherwise be needed.

Many thousands of common circuits are now available in integrated-circuit form. Figure 6.14 shows one of the more usual forms taken by integrated circuits, which are most often mounted on printed-circuit boards (see below).

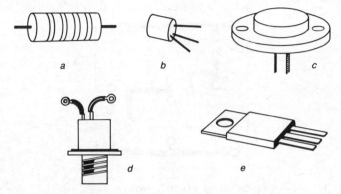

**Figure 6.13   Some discrete components**
    *a* Resistor
    *b* Bipolar-junction transistor
    *c* Power transistor
    *d* Thyristor
    *e* Triac.

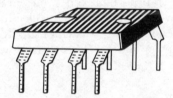

**Figure 6.14   One type of integrated circuit**

## *Printed-circuit boards*

To connect discrete components together with wired and soldered connections is extremely expensive and potentially there are numerous possibilities for failure. The printed circuit-board (PCB) overcomes such problems. A thin insulating board coated with a conducting layer, usually of copper, has the required circuit drawn on it using a pen containing special ink. The board is then etched to remove all the copper except that protected by the ink, which is then removed to leave the required circuit in the form of copper conducting strips on the surface of the board.

The connections of discrete components and/or integrated circuits are pushed through holes in the board and are soldered to the copper track. In this way simple or complex electronic circuits can be produced in a durable and economical form. Figure 6.15 shows the layout of a typical PCB. Where the circuit is particularly complex or is required to take up less space, both sides of the board may have printed circuits and components or integrated circuits on them.

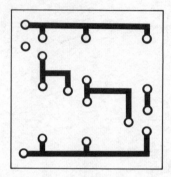

**Figure 6.15   Simple printed circuit board**
This board would be suitable for the circuit shown in Figure 6.3

## 6.5   DC power supplies

Direct-current power supplies are needed for many purposes. Most electronic circuits need such supplies for their operation; if they are low-power-consumption transportable types, they are likely to use batteries as their DC source. However, there are many electronic and other circuits and equipments which require a direct-current supply and are best fed from a rectified public AC power source. This section will deal with methods used in such cases.

### Halfwave rectification

A single diode as shown in Figure 6.16a will only allow current to flow in one direction, and rectifies the current flow as shown in Figure 6.16b. The resulting current is a series of isolated pulses and will be unsuitable for the operation of electronic circuits, although it will suffice for applications

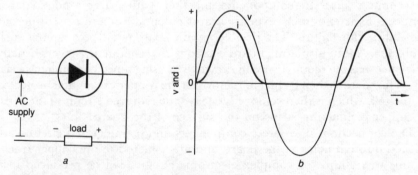

**Figure 6.16   Half-wave rectification**
*a* Single-phase halfwave rectifier circuit
*b* Voltage and current waveforms for single-phase halfwave rectification

such as battery charging. A transformer is often used at the input to ensure that the output direct voltage is at the required level.

Smoothing will be needed for many applications, and this is considered later.

## *Fullwave rectification*

An output which is more easily smoothed into a steady direct current than that from a single rectifier can be obtained by using two units as shown in Figure 6.17a. The two rectifiers, are fed from opposite ends of the secondary of a centre-tapped transformer, so the voltages they receive will be 180° out of phase. Thus, A rectifies the positive halfcycle whilst B rectifies the negative halfcycle.

Although the output current is still a series of pulses, these are closer together as shown in Figure 6.17b. The circuit is called biphase and the arrangement is called a fullwave single-phase rectifier.

As mentioned above, a transformer is often needed to ensure the correct output voltage. However, where this is not the case, the centre-tapped transformer which must be used for biphase fullwave rectification is expensive and bulky. The alternative is to use a bridge-connected circuit.

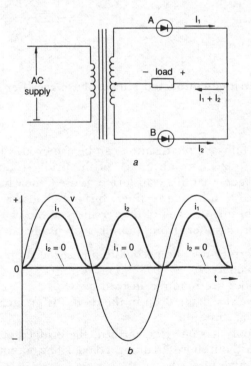

**Figure 6.17   Fullwave rectification**
    *a* Circuit diagram for fullwave (biphase) rectification
    *b* Wave diagrams for fullwave (biphase) rectification

## The bridge rectifier

The bridge connection, shown in Figure 6.18 does not need a transformer but gives a current output similar to that for the fullwave rectifier shown in Figure 6.17*b*. The operation of the bridge rectifier is shown in Figure 6.18, the two smaller circuits indicating the current paths for both positive and negative halfcycles of the supply, the direction of the current in the load being the same in both cases.

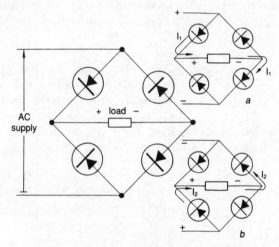

**Figure 6.18   Bridge-connected single-phase-rectifier circuit**

The fact that fullwave rectification can be achieved without the use of a transformer is the main advantage of the bridge-connected rectifier. Another advantage is that, if a transformer is used, the whole of the output voltage is applied to a bridge rectifier, but only half the voltage (due to the centre tap) in turn to each of the rectifiers of the two-diode fullwave circuit. However, it suffers from a number of disadvantages. These are:

(i)   A transformer may be needed to adjust the supply voltage to the correct value.

(ii)   Four diodes are required instead of two.

(iii)   The forward voltage drop in the diodes is greater, since two are in series at any time.

(iv)   If the supply has one side earthed, the bridged output cannot be earthed without short-circuiting a diode. The output therefore must 'float' relative to earth.

These disadvantages do not prevent the very wide use of the bridge-rectifier circuit.

## *Average value of rectified output*

### *Fullwave rectification*

Figure 6.19 shows the output current from a single-phase fullwave rectifier to a resistive load. This current never reverses and is technically a direct current; however, it is far from smooth, falling to zero twice in each cycle of the alternating current from which it was derived. In practice, the pulses of current will not have quite the same shape as halfcycles of sine waves, owing to the nonlinear characteristic of the rectifiers, but for many applications the difference is insignificant.

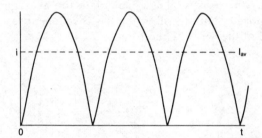

**Figure 6.19    Unsmoothed output from a single-phase fullwave rectifier**

It becomes important to be able to calculate the value of the direct-current output from a rectifier, but this is not straightforward because the current, although not reversing, is changing continuously. Since there are no reversals, the root-mean-square value is unimportant, and the average value is required. It was shown in Section 10.3 of 'Electrical Craft Principles', Volume 1, that the average value of a halfcycle of sinusoidal current or voltage was $2/\pi$ times the maximum value, or

$$I_{av} = \frac{2I_m}{\pi} = 0 \cdot 637 I_m$$

Since this is the average value of each of the pulses making up the direct current, it is also the average value of the whole of the direct current. If the voltage drop in the rectifiers is ignored (and this is a reasonable approximation for semiconductor diodes working at normal voltages), the same formula can be used to calculate the average value of output voltage.

For a single-phase fullwave rectifier, ignoring forward voltage drop, leakage current, and the effects of the rectifier charactersistic,

$$I_{av} = 0 \cdot 637 I_m \quad \text{and} \quad V_{av} = 0 \cdot 637 V_m$$

or
$$I_{av} = \frac{I}{1 \cdot 11} \quad \text{and} \quad V_{av} = \frac{V}{1 \cdot 11}$$

where        $I_{av}$ = average value of rectified current (A)

$I$    = RMS value of alternating current (A)

$I_m$  = maximum value of alternating current (A)

$V_{av}$ = average value of rectified voltage (V)

$V$    = RMS value of alternating voltage (V)

$V_m$  = maximum value of alternating voltage (V).

## Example 6.1

A single-phase fullwave bridge rectifier, fed from a transformer with a 50 V output, feeds a 10Ω resistive load. Calculate the average load current. The 50 V transformer output will be the RMS voltage, because the question does not specify otherwise. Therefore,

$$V_m = \frac{50}{0\cdot707} \text{ volts} = 70\cdot7 \text{ V}$$

$$\left(\text{for a sine wave, } V_m = \frac{V_{RMS}}{0\cdot707}\right)$$

On the output side of the rectifier,

$$V_{av} = 0\cdot637 V_m = 0\cdot637 \times 70\cdot7 \text{ volts} = 45 \text{ V}$$

load current,    $I = \dfrac{V}{R} = \dfrac{45}{10} \text{ amperes} = 4\cdot5 \text{ A}$

## Example 6.2

The charger for a 12 V battery has a single-phase transformer with a 32 V centre-tapped winding feeding the fullwave rectifier. If the internal resistance of the battery is 0·24Ω, calculate the charging current.
The voltage to each rectifier will be that due to one half of the transformer winding, i.e. 16 V.

$$V_m = \frac{16}{0\cdot707} \text{ volts} = 22\cdot6 \text{ V}$$

$$V_{av} = 0\cdot637 V_m = 0\cdot637 \times 22\cdot6 = 14\cdot4 \text{ V}$$

This direct-voltage output from the rectifier will oppose the 12 V EMF of the battery, so the effective voltage driving current will be

$$14\cdot4 - 12 \text{ volts} = 2\cdot4 \text{ V}$$

$$I = \frac{V}{R} = \frac{2\cdot4}{0\cdot24} \text{ amperes} = 10 \text{ A}$$

*Halfwave rectification*
For single-phase halfwave rectification, the output current is a series of isolated pulses, each driven by a positive halfcycle of the alternating voltage.

(See Figure 6.20). Each pulse will have the same average value as the pulses making up the fullwave rectified system, but since there are only half as many pulses the average output is halved. Thus, for halfwave single-phase rectification, ignoring forward voltage drop, leakage current and the effects of the rectifier characteristic,

and

$$I_{av} = \frac{0 \cdot 637}{2} I_m = 0 \cdot 319 I_m = \frac{I}{2 \cdot 22}$$

$$V_{av} = \frac{0 \cdot 637}{2} V_m = 0 \cdot 319 V_m = \frac{V}{2 \cdot 22}$$

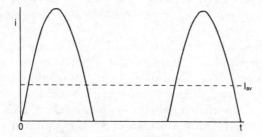

**Figure 6.20**  **Unsmoothed output from a single-phase halfwave rectifier**

## Example 6.3

If the bridge rectifier of Example 6.1 is replaced by a single diode, calculate the current to the $10\Omega$ load.

As in Example 6.1,

$$V_m = 70 \cdot 7 \text{ V}$$

$$V_{av} = 0 \cdot 319 V_m = 0 \cdot 319 \times 70 \cdot 7 \text{ volts} = 22 \cdot 6 \text{ V}$$

load current $\quad I = \dfrac{V}{R} = \dfrac{22 \cdot 6}{10}$ amperes $= 2 \cdot 26$ A

This example shows that the output current of a single-phase halfwave rectifier is half that of the fullwave rectifier.

## Example 6.4

A 50 V emergency-lighting battery has an internal resistance of $2\Omega$ and must be charged at 6 A. If the charger is a single-phase halfwave type, calculate the secondary voltage of the supply transformer, ignoring the forward voltage drop of the rectifier.

effective charging voltage $= IR = 6 \times 2$ volts $= 12$ V

This is the difference between average output of the charger and the battery EMF. Thus,

$$V_{av} = 50 + 12 \text{ volts} = 62 \text{ V}$$

$$V_m = \frac{V_{av}}{0 \cdot 319} = \frac{62}{0 \cdot 319} \text{ volts} = 194 \text{ V}$$

$$V = 0 \cdot 707 V_m = 0 \cdot 707 \times 194 \text{ volts} = 137 \text{ V}$$

## Smoothing

Reference to Figures 6.19 and 6.20 will confirm that the output from the rectifiers so far considered, although not changing direction and thus strictly direct currents, do not have the steady value of the output from a battery, or from a DC generator.

For some applications, such as small battery chargers, the variation of current, called a '**ripple**', is unimportant. For DC motors fed from an AC supply through rectifiers, the pulses of current would give corresponding pulses of output torque, often resulting in harmful vibration. Probably the most usual application of rectifiers is to provide a DC supply to feed electronic equipment from AC mains. Such equipment requires a very smooth supply voltage, free from ripple. The pulsing output of the rectifier must be smoothed to meet this requirement and there are two basic methods of smoothing.

### 1 Capacitor smoothing

A capacitor is connected in parallel with the load across the rectifier as shown in Figure 6.21a. The capacitor charges during the time that the rectifier passes current and discharges when the rectifier voltage is less than the capacitor voltage. With no load at all, the capacitor will charge to the maximum output voltage of the rectifier, and with no discharging current, will maintain this steady ripple-free output voltage. This is shown by the curves marked 1 in Figure 6.21b (for single-phase halfwave rectification) and Figure 6.21c (for single-phase fullwave rectification). If the load current increases, the capacitor discharges more rapidly between charging pulses, the effects of increasingly heavy load currents on output voltage being shown by the curves marked 2 and 3 in Figures 6.21b and 6.21c. An inspection of these wave diagrams will show that:

(a) the ripple on the voltage output has twice the frequency for a fullwave rectifier that it has for a halfwave rectifier;
(b) the average output voltage is greater for a fullwave than for a halfwave rectifier, except on no load;
(c) smoothing becomes progressively worse, and ripple greater, as load current increases.

The bad effect of a heavy load on capacitor smoothing can be reduced by the use of a larger capacitor which stores more energy and discharges less

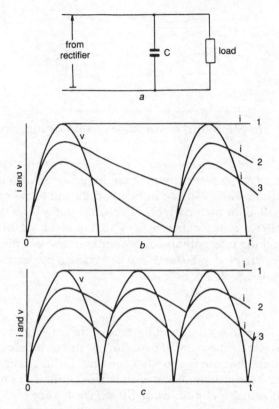

**Figure 6.21  Capacitor smoothing**
   *a* Smoothing capacitor circuit
   *b* Voltage and current waves for no load (1), medium load
       (2) and heavy load (3) for a capacitor-smoothed single-
       phase halfwave rectifier
   *c* Curves as *b* for a capacitor-smoothed single-phase fullwave
       rectifier

between charging pulses. However, the bulk and cost of capacitors, as well
as the effect on the rectifier of the very heavy initial charging current (called
'surge current') puts a limit on the size of capacitor used. Put simply, the
capacitor is an excellent smoothing device for small output currents, but
becomes less effective as current increases.

*2 Choke smoothing*

In Chapter 2, it was shown that a changing current passing through an
inductor induces an EMF which opposes the current change producing
it. This effect, known as Lenz's law, is a very useful one for smoothing.
The induced EMF will try to keep the current flowing as it reduces, and
to prevent its increase as it rises. The greater the rate of change of current,
the greater will be the smoothing effect, so that a large ripple will tend

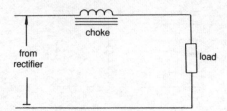

**Figure 6.22   Simple inductor (or choke) smoothing circuit**

to be smoothed more than a small one. It is generally true that ripples tend to be higher in heavy-current circuits than in light-current circuits, so the inductor smoothing circuit shown in Figure 6.22 will be more effective for heavy currents than for light currents. This is the opposite of the smoothing effect of a parallel-connected capacitor. The induced EMF in the coil, and hence the smoothing effect obtained, depends on the self inductance of the coil used. A very high self inductance can be obtained by using an iron-cored coil, usually called a choke.

## Filter circuits

A circuit to provide smoothing and to remove the unwanted ripple is often called a filter circuit. The parallel capacitor or the series choke are simple examples of filters, but the first is effective only for light currents and the second only for heavy currents. A more effective filter is one combining both units and called a **choke input filter** (see Figure 6.23). Better still is the **capacitor input filter**, shown in Figure 6.24, which has capacitors across the rectifier and the load, as well as a series-connected choke. Such a filter will give good smoothing if the capacitor and inductor values are carefully chosen, but has two disadvantages:

 (i) The load voltage falls as the current increases. This is known as 'poor voltage regulation'.
(ii) On light loads, capacitor $C_1$ charges to the maximum value of the supply voltage, and retains most of this voltage during the next halfcycle. The capacitor voltage adds to the maximum supply voltage of the second halfcycle, so the rectifier must have a peak-inverse-voltage rating of twice the maximum value of the supply voltage.

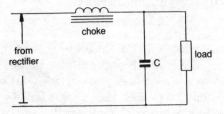

**Figure 6.23   Choke input filter**

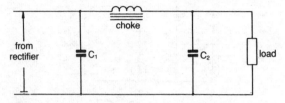

**Figure 6.24 Capacitor input filter**

## *Three-phase rectifier circuits*

If a three-phase supply is available, a very much smoother output voltage can be obtained from the supply mains than with single-phase systems. Figure 6.25 shows circuit and wave diagrams for a three-phase rectifier, each phase supplying the load in turn for as long as its voltage is greater than that of the other two phases. The output voltage of such a rectifier never falls to less than half the maximum value of the supply voltage, and the ripple is comparatively small.

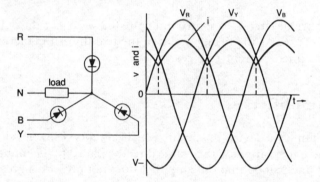

**Figure 6.25 Circuit and wave diagrams for a halfwave three-phase rectifier**

An even lower ripple occurs in the output from six rectifiers connected in a three-phase bridge circuit, the negative voltages being used as well as the positive. Figure 6.26 shows circuit and wave diagrams for such a system, the output voltage of which is never less than 86·6% of the maximum supply voltage.

Three-phase rectifiers are often used to provide high-power DC supplies from AC mains, the degree of smoothing required being very small.

## *Controlled rectification*

We have seen earlier how the thyristor can be used as a halfwave controlled rectifier, the average output current increasing as the firing angle decreases (see Figure 6.8). Two thyristors with two diodes can be connected in a

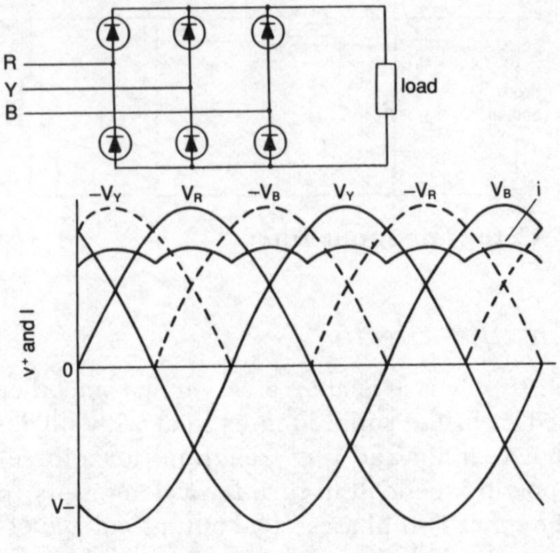

**Figure 6.26   Circuit wave diagrams for a fullwave bridge-connected three-phase rectifier**

controlled bridge circuit, as shown in Figure 6.27*a*, to give the output shown in Figure 6.27*b*. As the firing angles of the two thyristors increase, so the average rectified current reduces.

## 6.6   Amplifiers

An amplifier is a device or circuit which accepts a low-level input and provides a higher output with a value directly related to the input. It should be clearly understood that the amplifier does not provide a greater output than its input in power and energy terms; the apparent increase in energy is provided by the power supply. Amplifiers can be extremely complex devices, and the electrical craftsman is seldom concerned with their circuitry. He can simply treat them as 'black boxes' which are provided with a power supply and an input and which deliver an output which is related to the input but is greater than it. Figure 6.28 indicates the approach used.

The increase in output of an amplifier compared with its input is known as the gain. Clearly, input and output must be in identical units. There is no unit for gain itself:

$$\text{gain} = \frac{\text{output}}{\text{input}}$$

### Example 6.5

An electronic amplifier is fed with a signal of 4·5 mV and provides an output of 6·0 V. Calculate the gain.

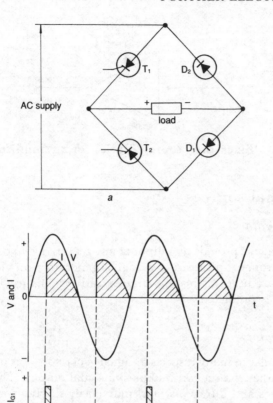

**Figure 6.27  Controlled bridge rectifier**
> *a* Rectifier circuit
> *b* Full-wave controlled rectification

$$\text{gain} = \frac{\text{output}}{\text{input}} = \frac{6 \cdot 0}{4 \cdot 5 \times 10^{-3}} = 1333$$

## Example 6.6

An amplifier is used to increase the output from a photocell and has an output of $2 \cdot 25$ V. If the gain is 560, calculate the input.

$$\text{gain} = \frac{\text{output}}{\text{input}} \quad \text{so} \quad \text{input} = \frac{\text{output}}{\text{gain}} = \frac{2 \cdot 25}{560} = 4 \cdot 02 \text{ mV}$$

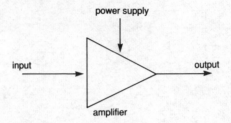

**Figure 6.28** **'Black-box' representation of an amplifier**

## 6.7 Signal sources

### Signal generators

These equipments provide an output at any required voltage and frequency within their range. In most cases, signal generators provide a sinusoidal output and are used for testing circuits and systems to ensure that they will behave as predicted.

### Pulse generators

These are similar in many respects to signal generators, but producing pulses of voltage rather than a continuous sinusoidal output. The pulses are of various shapes and durations, and may occur singly or as a continuous train. Devices of this type are also used for testing purposes, especially for logic circuits, counters and the like.

### Mark–space ratio

This term is used in connection with pulsed systems to indicate the ratio of pulse duration to the time between pulses. Figure 6.29 shows a system where the pulse duration is exactly half the time between pulses, so the mark–space ratio is $0 \cdot 5$.

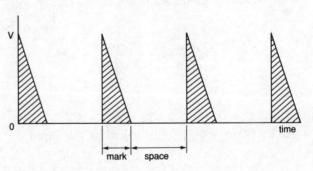

**Figure 6.29** **To illustrate the mark–space ratio**

## Example 6.7

A pulse generator provides rectangular voltage pulses at a frequency of $1 \cdot 6$ kHz. If the mark–space ratio is $0 \cdot 25$, calculate the duration of each pulse.

The duration of each pulse together with its following space is the periodic time $T$ of the system.

$$T = \frac{1}{f} = \frac{1}{1 \cdot 6 \times 10^3} \text{ seconds} = 0 \cdot 000625 \text{ s} = 0 \cdot 625 \text{ ms}$$

If the mark–space ratio is $0 \cdot 25$, four-fifths of the periodic time is the space time, and one-fifth the pulse time.

$$\text{pulse duration} = \frac{0 \cdot 625}{5} \text{ milliseconds} = 0 \cdot 125 \text{ ms}$$

## 6.8 Logic

Logic circuits are in very wide use in electronics, forming the basis of many automatic systems. They are based on the assumption that only one of two possibilities can happen, and can thus be expressed mathematically with a numbering system with only two digits. Such a system is the **binary system** of numbers, where only the digits 0 and 1 are used. In practical terms, the presence of a voltage (often $+ 5$ V) represents the digit 1, and the absence of a voltage (0 V) represents 0. This is called **positive logic**. When the arrangement is reversed (0 V representing 1) the system is called **negative logic**.

The process is carried out using circuits called **logic gates**. The gates may be built up with discrete components, but are much more likely to exist on integrated circuits; we are not concerned here with the circuitry involved, and will consider the gates as 'black boxes', each fulfilling its particular purpose. The type of gate and its interconnection with others will result in an output only when the inputs fulfil certain specified requirements. For example, logic may be used in the starter for an electric motor so that the machine will only start when *all* of the following conditions are met.

(i) The 'stop' system has been reset if used to stop the motor,
(ii) the 'start' button is pressed, and
(iii) starting current does not exceed preset limits.

There are several types of gate, but we will concern ourselves only with the five which are the most usual. Each gate will have only one output, but may have many inputs. To simplify matters, we will limit ourselves to a maximum of two inputs.

### AND and NAND gates

There are several symbols used for logic gates, but here only those approved by the IEC will be used; the symbol for the AND gate is shown in Figure

6.30*a*. This gate is arranged so that an output will only be present when *all* the inputs are fed (output only if the first input AND the second input are fed). Using the binary number 1 to represent an input or an output, and binary 0 to represent their absence, a table can be drawn up to indicate all the possible states of the system. This is called a **truth table** and that for the two-input AND gate is shown in Figure 6.30*b*.

Notice that the only condition for an output is when BOTH inputs are fed (C is 1 *only* when A is 1 AND B is 1).

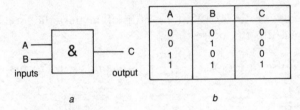

| A | B | C |
|---|---|---|
| 0 | 0 | 0 |
| 0 | 1 | 0 |
| 1 | 0 | 0 |
| 1 | 1 | 1 |

*a*                                         *b*

**Figure 6.30   The AND gate**
    *a* Circuit symbol for an AND gate
    *b* Truth table for an AND gate

The NAND gate behaves in the opposite way to the AND gate (note that NAND means NOT AND). The circuit symbol for the gate and its truth table are shown in Figure 6.31.

Notice that the only condition for an output is when one or both inputs are NOT fed (C is 1 only when A is 0 OR B is 0 or BOTH are 0).

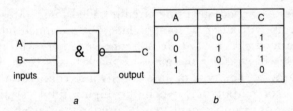

| A | B | C |
|---|---|---|
| 0 | 0 | 1 |
| 0 | 1 | 1 |
| 1 | 0 | 1 |
| 1 | 1 | 0 |

*a*                                         *b*

**Figure 6.31   The NAND gate**
    *a* Circuit symbol for a NAND gate
    *b* Truth table for a NAND gate

## *OR and NOR gates*

The two-input OR gate will give an output when any one or both of its inputs are fed. The circuit symbol for the gate and its truth table are shown in Figure 6.32. Note that an output will occur when either or both of the inputs are fed.

The NOR gate (NOT OR) has the opposite output to the OR gate when its inputs are the same. There will only be an output when neither input is fed. The circuit symbol for the gate and its truth table are shown in Figure 6.33.

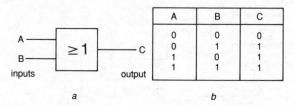

| | A | B | C |
|---|---|---|---|
| | 0 | 0 | 0 |
| | 0 | 1 | 1 |
| | 1 | 0 | 1 |
| | 1 | 1 | 1 |

*a*                    *b*

**Figure 6.32   The OR gate**
      *a* Circuit symbol for an OR gate
      *b* Truth table for an OR gate

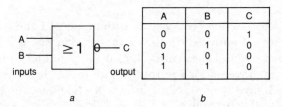

| | A | B | C |
|---|---|---|---|
| | 0 | 0 | 1 |
| | 0 | 1 | 0 |
| | 1 | 0 | 0 |
| | 1 | 1 | 0 |

*a*                    *b*

**Figure 6.33   The NOR gate**
      *a* Circuit symbol for a NOR gate
      *b* Truth table for a NOR gate

## *The NOT gate*

This gate has only one input and one output. Its output is always the opposite of its input; for example, if the input is 1, the output is 0. The circuit symbol for the gate and its truth table are shown in Figure 6.34.

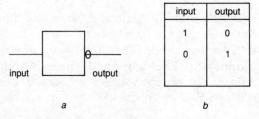

| input | output |
|---|---|
| 1 | 0 |
| 0 | 1 |

*a*                    *b*

**Figure 6.34   The NOT gate**
      *a* Circuit symbol for a NOT gate
      *b* Truth table for a NOT gate

## *Logic circuits*

Logic gates can be connected together to perform logical functions. The following examples will illustrate the process.

## Example 6.8

Logic gates are interconnected as shown in Figure 6.35. If inputs A, B and C are 1, 0 and 1, respectively, deduce the output Z.

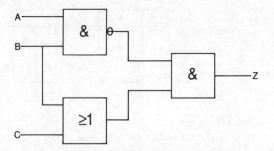

**Figure 6.35   Logic circuit for Example 6.8**

The top left-hand gate is a NAND gate with inputs of 1 and 0. Reference to the truth table of Figure 6.31*b* shows that the output will be 1. This could be reasoned without the truth table; if the gate had been an AND gate, it would only have had a 1 output if both inputs had been 1. Thus, the output would have been 0. Since it is a NAND gate the output will be the opposite to 0, and will be 1.

The bottom left-hand gate is an OR gate, which gives a 1 out if either of its inputs is 1. Although the B input is 0, C is 1 so the output is 1. The inputs to the right-hand (AND) gate are thus both 1, the condition for the output to be 1. Thus, output Z under the given conditions will be 1.

## Example 6.9

Logic gates are interconnected as shown in Figure 6.36. If inputs A, B, C and D are 1, 0, 1 and 1, respectively, deduce the output Z. The logic circuit is reproduced again as Figure 6.37 marked with the status of each part.

The top left-hand gate is an OR gate and, since one if its inputs is fed, the output is 1. At the bottom left-hand is a NOT gate, and since the input is 1 the output is 0. In the middle is an AND gate, and since inputs are not both 1, the output is 0. The right-hand gate is NAND. Had it been AND with one input 0, the output would have been 0; but NAND gives the opposite output, so Z is 1.

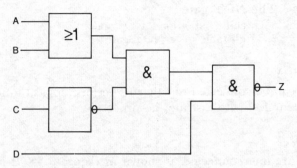

**Figure 6.36   Logic circuit for Example 6.9**

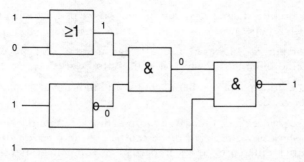

**Figure 6.37** **Logic circuit of Figure 6.36 with the status of each part marked**

## 6.9 Summary of formulas for Chapter 6

See text for definition of symbols

For fullwave single-phase rectifiers:

$$I_{av} = 0 \cdot 637 I_m \qquad I_{av} = \frac{I}{1 \cdot 11}$$

$$V_{av} = 0 \cdot 637 V_m \qquad V_{av} = \frac{V}{1 \cdot 11}$$

For halfwave single-phase rectifiers:

$$I_{av} = 0 \cdot 319 I_m \qquad I_{av} = \frac{I}{2 \cdot 22}$$

$$V_{av} = 0 \cdot 319 V_m \qquad V_{av} = \frac{V}{2 \cdot 22}$$

For an amplifier:

$$\text{gain} = \frac{\text{output}}{\text{input}} \qquad \text{output} = \text{input} \times \text{gain} \qquad \text{input} = \frac{\text{output}}{\text{gain}}$$

## 6.10 Exercises

1 List the advantages of alternating current compared with direct current. List and discuss three applications where direct current is used.

2 List three different methods of obtaining a DC supply, and compare their advantages.

3 What is the purpose of a transducer? Explain, with the aid of a block diagram, how the temperature of an industrial process could be monitored so that an alarm would be given if the temperature became excessive.

4 With the aid of a diagram explain in general terms how a light-dependent resistor (LDR) operates.

5 What is the purpose of a strain gauge? Show how a typical gauge is arranged and one kind of circuit in which it is likely to be used.

6 Explain the principle of the thermocouple and indicate an application where it could be used.

7 How does a Zener diode differ from normal diodes? With the aid of a circuit diagram, show how the Zener diode can be used in a circuit to ensure that the potential difference across a particular component never exceeds a predetermined level.

8 With the aid of circuit and wave diagrams, describe how the thyristor is used for controlled rectification. Explain the terms 'phase control' and 'burst control'. Show which would be used with a filament-lamp-dimming system. Suggest a suitable application for the other method of control.

9 Why is it sometimes difficult to use a thyristor (or thyristors) to control the speed of a direct-current motor? With the aid of circuit and wave diagrams, show how a component can be used to overcome the difficulty.

10 Sketch the waveform of the voltage of a single-phase supply. Using different colours for each curve, show on the diagram the current from such a supply after (*a*) halfwave rectification, and (*b*) fullwave rectification.

11 Draw a circuit diagram to show a single-phase fullwave-rectifier circuit, using two semiconductor diodes.

12 Draw the circuit of a bridge-connected semiconductor-rectifier circuit and explain its operation.

13 List the advantages and disadvantages of a fullwave bridge-connected rectifier when compared with a fullwave rectifier using two diodes and a centre-tapped transformer.

14 A resistive load of $30\Omega$ is connected to a 240 V, 50 Hz supply through a silicon diode, the circuit being the same as that shown in Figure 6.16*a*. Calculate the average PD across the load and the load current, neglecting the forward voltage drop in the diode.

15 The load of Exercise 14 is connected to a transformer with a 240 V centre-tapped secondary winding through two diodes connected as in Figure 6.17*a*. Calculate the average PD across the load and the load current. Neglect the forward voltage drop of the diodes.

16 The $30\Omega$ load of Exercises 14 and 15 is connected to a 240 V, 50 Hz supply through a bridge-connected silicon rectifier (Figure 6.18). Neglecting the forward voltage drop of the diodes, calculate the average PD across the load and the load current.

17 A $50\Omega$ resistive load is required to dissipate a power of 200 W when connected to an AC supply through a transformer and a rectifier. Calculate the output voltage of the transformer if the rectifier is
(*a*) a single-phase halfwave type, and
(*b*) a single-phase bridge.

18  A 200 V stationary battery has an internal resistance of $2\Omega$ and is to be charged from a 240 V AC supply through a fullwave bridge rectifier. Calculate the charging current.

19  The 6 V alkaline battery of a resistance tester is to be charged overnight at a current of 100 mA through a single-phase halfwave rectifier and transformer unit. If the internal resistance of the battery is $0 \cdot 6\Omega$, calculate the output voltage of the transformer.

20  Why is it often necessary to smooth the output of a rectifier before use? Explain, using circuit diagrams, the operation of a typical smoothing filter. What are the disadvantages of this circuit?

21  Show by diagram the meaning of 'rectification' of an alternating current. Describe one form of single-phase rectifier giving a connection diagram, and explain its operation.

22  Show by means of diagrams what you understand by rectification of an alternating current. Describe with sketches one form of rectifier element and show how you would connect four such elements to give 'fullwave' rectification.

23  A charging device is required for 12 lead–acid secondary cells. The mains supply available is single-phase AC at 240 V, 50 Hz. Write down two possible methods of providing the necessary direct current. Describe and explain the operation of one of these methods, using sketches or diagrams to illustrate your answer.

24  Describe with the aid of suitable sketches a device used for rectification of a single-phase alternating current.

25  Draw circuit diagrams for single-phase, fullwave rectification of alternating current

(*a*) using bridge-connected rectifier elements, and

(*b*) using a centre-tapped transformer,

Include a single stage of smoothing.

Explain briefly how the smoothing is achieved.

26  (*a*) State what is meant by rectification of an alternating current.

(*b*) Name one type of device which is commonly used for the purpose.

(*c*) Draw two complete cycles of an AC sine wave. Sketch on the same diagram:

  (i) the waveform of the same current after halfwave rectification

  (ii) the waveform of the same current after fullwave rectification.

27  Draw the circuit symbol for a triac. Explain the operation of the device with the aid of circuit and wave diagrams.

28  How does a GTO (gate-turn-off) thyristor differ from normal thyristors?

29  Draw the circuit symbol for an amplifier. If the output power of such a system is greater than its input power, does this mean that the circuit is providing 'something for nothing'? Explain the reasons for your answer. An amplifier has an input of 25 µV and an output of $5 \cdot 0$ V. Calculate the gain.

30  Use a wave diagram as a basis for explaining the term 'mark–space ratio' as applied to a pulse generator. A pulse generator provides a string of pulses at

a frequency of 15 kHz and a mark–space ratio of 1. Calculate the duration of each pulse.

31  Draw the circuit symbol and the truth table for a two-input AND gate. In your own words, indicate the input conditions resulting in an output.

32  Draw the circuit symbol and the truth table for a two-input NAND gate. In your own words, indicate the input conditions resulting in an output.

33  Draw the circuit symbol and the truth table for a two-input OR gate. In your own words, indicate the input conditions resulting in an output.

34  Draw the circuit symbol and the truth table for a two-input NOR gate. In your own words, indicate the input conditions resulting in an output.

35  What is a NOT gate? Illustrate your answer with the circuit symbol and the truth table.

36  Logic gates are connected as shown in Figure 6.38 with the inputs shown. Deduce the state of the output Z.

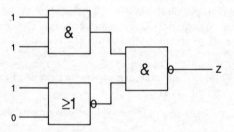

**Figure 6.38   Diagram for Exercise 36**

37  If the output of the logic gate system of Figure 6.39 is 1 when the inputs are as shown, what will be the type of the unmarked gate?

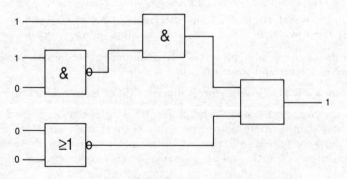

**Figure 6.39   Diagram for Exercise 37**

## 6.11 Multiple-choice exercises

6M1 A transducer can be defined as:
  (*a*) a person who listens to speech in one language and says the same thing in a different language
  (*b*) a device which converts a physical quantity (such as force or vibration) from one form to another (such as a potential difference)
  (*c*) a person who directs a film or a play
  (*d*) an electrical device used to amplify small signals.

6M2 The device whose symbol is shown below is:
  (*a*) a triac               (*b*) a diode
  (*c*) an integrated circuit (*d*) a photo transistor.

**Figure 6.40  Diagram for Exercise 6M2**

6M3 A strain gauge depends for its operation on:
  (*a*) the change in resistance of a conductor as it is stretched or compressed
  (*b*) the use of accurate weighing machines
  (*c*) the physical strength of the user
  (*d*) the incident lighting level.

6M4 The thermocouple depends for its operation on:
  (*a*) the two junctions being at the same temperature
  (*b*) having an accurate kilovoltmeter
  (*c*) the Seebeck effect
  (*d*) the operation of an alarm if temperature becomes too high.

6M5 A resistor designed to change in resistance with change in temperature is called a:
  (*a*) thermistor          (*b*) temperature-coefficient device
  (*c*) transistor          (*d*) diode

6M6 A device designed to break down and pass reverse current at its rated voltage but to recover when the voltage is removed is called:
  (*a*) a triac               (*b*) a Zener diode
  (*c*) a fail-safe device     (*d*) a transistor.

6M7   The device whose symbol is shown below is a:

    (*a*) diode       (*b*) triac       (*c*) thyristor     (*d*) diac

**Figure 6.41   Diagram for Exercise 6M7**

6M8   Controlling a thyristor by switching it fully on for a number of halfcycles and then off for several cycles is known as:
    (*a*) close control       (*b*) burst control
    (*c*) on/off control      (*d*) phase control.

6M9   When the firing angle of a phase-controlled thyristor is increased, the average direct current will:
    (*a*) reduce           (*b*) remain unchanged
    (*c*) increase         (*d*) fall to zero.

6M10  The problems encountered when switching a thyristor controlling an inductive load are reduced by connection of a:
    (*a*) contactor        (*b*) triac
    (*c*) flywheel diode    (*d*) shunt field.

6M11  A GTO device is:
    (*a*) a thyristor or a triac which can be turned off as well as on by a signal to its gate
    (*b*) part of a computer system designed to find a particular piece of information
    (*c*) a high-powered motor vehicle
    (*d*) made with germanium–titanium oxide.

6M12  An electronic circuit where components are made on or in a single chip of silicon is called:
    (*a*) a solid circuit     (*b*) a reliable circuit
    (*c*) expensive        (*d*) an integrated circuit.

6M13  A circuit using a centre-tapped transformer feeding two diodes is called a:
    (*a*) bridge rectifier     (*b*) fullwave (biphase) rectifier
    (*c*) three-phase rectifier  (*d*) halfwave rectifier.

6M14 The smoothing system shown in Figure 6.42 is called a:
   (*a*) choke smoothing circuit      (*b*) choke input filter
   (*c*) capacitive smoothing circuit    (*d*) halfwave rectifier.

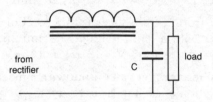

**Figure 6.42   Diagram for Exercise 6M14**

6M15 The circuit shown in Figure 6.43 is for a:

   (*a*) halfwave rectifier          (*b*) fullwave (biphase) rectifier
   (*c*) bridge rectifier            (*d*) three-phase rectifier.

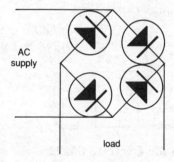

**Figure 6.43   Diagram for Exercise 6M15**

6M16 The number of diodes required for a fullwave three-phase bridge
   rectifier circuit is:
   (*a*) six          (*b*) two          (*c*) three          (*d*) four.

6M17 The circuit shown below is for a:
   (*a*) controlled bridge rectifier    (*b*) halfwave rectifier
   (*c*) diode bridge rectifier        (*d*) fullwave biphase rectifier.

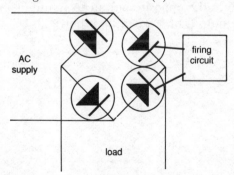

**Figure 6.44   Diagram for Exercise 6M17**

6M18  A device which accepts a low-level input and provides a higher-level output directly related to the input is called:
    (*a*) a transducer               (*b*) a multiplier
    (*c*) a rectifier                 (*d*) an amplifier.

6M19  The output from a transducer is at a level of $1 \cdot 4$ mV. If it feeds an amplifier with a gain of 1200, the amplifier output will be:
    (*a*) $1 \cdot 4$ V      (*b*) $1 \cdot 68$ V      (*c*) $1 \cdot 17$ mV      (*d*) $1 \cdot 17$ μV.

6M20  If a pulse generator gives an output with a mark–space ratio of $0 \cdot 2$ and the duration of each pulse is 15 μs, the frequency of the output will be:
    (*a*) $66 \cdot 7$ kHz      (*b*) 50 Hz      (*c*) 90 μs      (*d*) $11 \cdot 1$ kHz.

6M21  A logic system in which $+5$ V represents the digit 0 is referred to as:
    (*a*) safe to work on          (*b*) negative logic
    (*c*) a nand system         (*d*) positive logic.

6M22  The logic gate shown below is a:
    (*a*) AND gate             (*b*) NOT gate
    (*c*) NOR gate             (*d*) NAND gate.

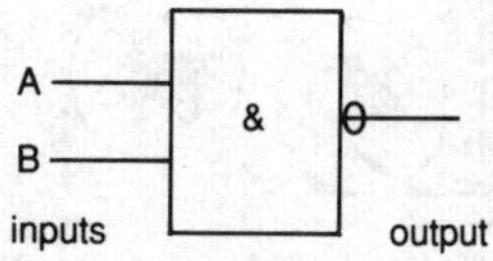

**Figure 6.45   Diagram for Exercise 6M22**

6M23  The logic gate which will give an output when both or one of its inputs is fed is:
    (*a*) an AND gate          (*b*) a NOT gate
    (*c*) an OR gate           (*d*) a NAND gate.

6M24  The NOT gate:
    (*a*) gives an output of 1 when its input is 0
    (*b*) is seldom used in logic circuits
    (*c*) never gives the opposite output to its input
    (*d*) gives an output of 0 when its input is 0.

## Chapter 7

# The transformer

## 7.1    Introduction

Transformers are very important in electrical engineering, because almost all of its many branches make use of them. The efficient transmission and distribution of electricity would be impossible without power transformers. Electronic equipment in industry uses transformers in very large numbers. Communications systems, including television and telephony, rely on transformers for their operation.

Although transformers differ in size and in application, all rely on the principle of mutual inductance for their operation.

## 7.2    Principle of the single-phase transformer

In Section 2.4 it was established that, if two coils are arranged so that the magnetic flux produced by current in one links with the other, mutual inductance exists between the coils and a change of current in one induces an EMF in the other. If the two coils are arranged on a core of magnetic material, this will increase the amount of magnetic flux set up by one coil and will make sure that most of it links with the other coil. In this way mutual inductance is increased.

Consider two coils wound on a simple magnetic circuit as shown in Figure 7.1. This actual construction is not very efficient and is only used for cheap transformers (see Section 7.5), but it serves to illustrate the principle. The arrangement is a simple double-wound transformer, which is represented in the circuit diagram shown in Figure 7.2. The winding fed with current is called the 'primary winding', and the other the 'secondary winding'. All quantities associated with the primary are identified with the subscript 1 (e.g. $N_1$ for primary number of turns), and those associated with the secondary with 2 (e.g. $N_2$ for secondary number of turns). Each winding must be made with insulated conductors to prevent short circuits within

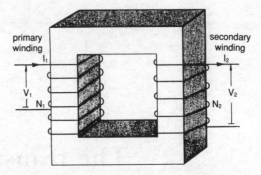

**Figure 7.1   Arrangement of simple transformer**

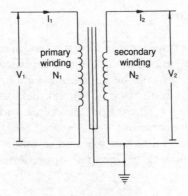

**Figure 7.2   Circuit symbol for transformer**

the winding itself, or to the magnetic circuit or core, which is usually earthed for safety on power transformers.

Alternating current in the primary winding will set up an alternating magnetic flux in the core, the self inductance of the winding inducing in it an EMF opposing the supply voltage. This EMF will be almost the same in value as the applied voltage, and for practical purposes the two may be assumed to be equal. If all of the changing magnetic flux set up by the first winding links with the second, the EMF induced in each turn will be the same regardless of whether it forms part of the primary winding or of the secondary winding. Thus, the voltage per turn of the two windings will be equal.

$$\text{Primary volts per turn} = \frac{\text{primary volts}}{\text{primary turns}} = \frac{V_1}{N_1}$$

$$\text{secondary volts per turn} = \frac{\text{secondary volts}}{\text{secondary turns}} = \frac{V_2}{N_2}$$

and

$$\frac{V_1}{N_1} = \frac{V_2}{N_2}$$

rearrangement gives
$$\frac{V_1}{V_2} = \frac{N_1}{N_2}$$

or
$$\text{voltage ratio} = \text{turns ratio.}$$

This important relationship is not exactly true, since we have started by assuming that primary back EMF and primary voltage are equal, but the error is small in power transformers and may be ignored at this stage.

## Example 7.1

A transformer with 1000 primary turns and 250 secondary turns is fed from a 240 V AC supply. Calculate the secondary voltage and the volts per turn.

$$\frac{V_1}{V_2} = \frac{N_1}{N_2} \quad \text{so} \quad V_2 = V_1 \times \frac{N_2}{N_1}$$

$$V_2 = 240 \times \frac{250}{1000} \text{ volts} = 60 \text{ V}$$

$$\text{volts per turn} = \frac{V_1}{N_1} = \frac{240}{1000} = 0 \cdot 24$$

or

$$\text{volts per turn} = \frac{V_2}{N_2} = \frac{60}{250} = 0 \cdot 24$$

It should be noted that the transformer will only function properly when connected to an AC supply. The steady value of a direct current sets up an unchanging magnetic flux, which induces no EMF into the secondary winding.

A transformer with an output voltage greater than its input is called a step-up transformer, whereas a step-down transformer has a lower output voltage than its input. If a voltage or turns ratio is quoted for a transformer, this is always put in the order input:output, which is primary:secondary.

## Example 7.2

A transformer with a voltage ratio of 11 000:415 has 300 secondary turns. Calculate the number of primary turns.

$$\frac{V_1}{V_2} = \frac{N_1}{N_2} \quad \text{so} \quad N_1 = N_2 \times \frac{V_1}{V_2}$$

$$N_1 = 300 \times \frac{11\ 000}{415} \text{ turns} = 7950 \text{ turns.}$$

## Example 7.3

A neon-sign transformer has an output of 4500 V and is fed at 240 V. If the secondary has 2000 turns, calculate the number of primary turns.

$$\frac{V_1}{V_2} = \frac{N_1}{N_2} \quad \text{so} \quad N_1 = N_2 \times \frac{V_1}{V_2}$$

$$N_1 = 2000 \times \frac{240}{4500} \text{ turns} = 106 \cdot 7 \text{ turns.}$$

In practice, it is not possible for a winding to have part of a turn, so this solution must be rounded off to a whole number. 107 turns would increase the output voltage slightly, whilst 106 turns would give a small decrease.

If we assume that our transformer is 100% efficient, then

power input = power output,

$$V_1 I_1 \cos \phi_1 = V_2 I_2 \cos \phi_2$$

where $I_1$ and $I_2$ are primary and secondary currents, respectively, while $\cos \phi_1$ and $\cos \phi_2$ are the primary and secondary power factors. Assuming that the two power factors are equal, they will cancel from the equation to give

$$V_1 I_1 = V_2 I_2$$

or

$$\frac{V_1}{V_2} = \frac{I_2}{I_1}$$

voltage ratio = inverse of current ratio

Neither of the assumptions made is strictly true but, since the error involved is small for a power transformer, the resulting expression is a useful one.

## Example 7.4

A 50 kVA transformer has a voltage ratio of 3300:400 V. Calculate the primary and secondary currents.

$$\text{kVA} = \frac{V_1 I_1}{1000} \quad \text{so} \quad I_1 = \frac{\text{kVA} \times 1000}{V_1}$$

$$I_1 = \frac{50 \times 1000}{3300} \text{ amperes} = 15 \cdot 2 \text{ A}$$

$$\frac{V_1}{V_2} = \frac{I_2}{I_1} \quad \text{so} \quad I_2 = I_1 \times \frac{V_1}{V_2}$$

$$I_2 = 15 \cdot 2 \times \frac{3300}{400} = 125 \text{ A}$$

An alternative method for the second part of the calculation would be:

$$\text{kVA} = \frac{V_2 I_2}{1000} \quad \text{so} \quad I_2 = \frac{\text{kVA} \times 1000}{V_2}$$

$$I_2 = \frac{50 \times 1000}{400} \text{ amperes} = 125 \text{ A}$$

Notice that, if the transformer reduces the voltage, the current increases, so that input and output power remain equal (but see Section 7.9, which shows that this is not strictly true).

The expressions for ratios of voltage, current and turns can, of course, be combined to give:

$$\frac{V_1}{V_2} = \frac{N_1}{N_2} = \frac{I_2}{I_1}$$

## Example 7.5

The single-phase transformer feeding a soil-warming system is supplied at 240 V, 50 Hz, and must provide a 20 V output. The full-load secondary current is 180 A, and the secondary has 45 turns. Calculate

(a) the output kVA of the unit,
(b) the number of primary turns,
(c) the full load primary current,
(d) the volts per turn.

(a) Output kVA $= \dfrac{V_2 I_2}{1000} = \dfrac{20 \times 180}{1000}$ kilovolt-amperes $= 3 \cdot 6$ kVA

(b) $\dfrac{V_1}{V_2} = \dfrac{N_1}{N_2}$ so $N_1 = N_2 \times \dfrac{V_1}{V_2}$

$N_1 = 45 \times \dfrac{240}{20}$ turns $= 540$ turns.

(c) $\dfrac{V_1}{V_2} = \dfrac{I_2}{I_1}$ so $I_1 = I_2 \times \dfrac{V_2}{V_1}$

$I_1 = 180 \times \dfrac{20}{240}$ amperes $= 15$ A

(d) Volts/turn (primary) $= \dfrac{V_1}{N_1} = \dfrac{240}{540} = 0 \cdot 444$

or volts/turn (secondary) $= \dfrac{V_2}{N_2} = \dfrac{20}{45} = 0 \cdot 444$

## Example 7.6

A 75 kVA transformer has a step-down ratio of 12:1, with 2400 primary turns and a primary voltage of $3 \cdot 3$ kV. Calculate

(a) the number of secondary turns,
(b) the secondary voltage,
(c) the volts per turn,
(d) the full load primary and secondary currents.

(a) $\dfrac{V_1}{V_2} = \dfrac{N_1}{N_2}$   so   $N_2 = N_1 \times \dfrac{V_2}{V_1}$

$N_2 = 2400 \times \dfrac{1}{12}$ turns = 200 turns

(b) Voltage ratio $\dfrac{V_1}{V_2} = \dfrac{12}{1}$   so   $V_2 = \dfrac{V_1}{12}$

$V_2 = \dfrac{3300}{12}$ volts = 275 V

(c) Primary volts/turn = $\dfrac{V_1}{N_1} = \dfrac{3300}{2400} = 1\cdot38$

or   secondary volts/turn = $\dfrac{V_2}{N_2} = \dfrac{275}{200} = 1\cdot38$

(d) $I_1 = \dfrac{\text{kVA} \times 1000}{V_1} = \dfrac{75\,000}{3300} = 22\cdot7$ A

$I_2 = \dfrac{\text{kVA} \times 1000}{V_2} = \dfrac{75\,000}{275} = 273$ A

## 7.3   Tapped windings

Section 7.2 has shown that the voltage ratio of a transformer depends on its turns ratio, and hence on the numbers of primary and secondary turns. If more than one secondary voltage is required of a transformer, one solution is to provide more than one secondary winding. Such arrangements are quite common in television sets, where one secondary winding provides a high voltage which is rectified and used for operation of the electron gun in the cathode ray tube (see Section 11.19), while a much lower voltage from a separate secondary winding feeds the cathode heater for the tube. All windings are placed on the same magnetic core, the circuit diagram for a transformer with two secondary windings being shown in Figure 7.3.

In some cases only a single output is required, but the voltage may need to be adjusted to more than one value. This can be achieved by use of a tapped primary or secondary winding. Tappings are connections made to the windings which are brought out to a connection box so that the effective number of turns on the winding can be altered by changing connections. Circuit diagrams for transformers with tapped primary and tapped secondary windings are shown in Figures 7.4 and 7.5, respectively. It is unusual for both windings on the same transformer to be tapped, primary tappings being favoured where voltages are high and on-load tap changing is necessary. Transformers with small outputs are usually tapped on the secondary winding.

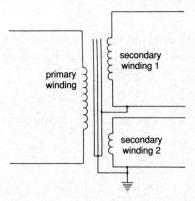

Figure 7.3   Circuit symbol for transformer with two secondary
windings

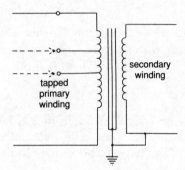

Figure 7.4   Circuit symbol for transformer with a tapped primary
winding

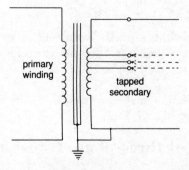

Figure 7.5   Circuit symbol for transformer with a tapped secondary
winding

## Example 7.7

A welding transformer is fed at 415 V and has 20 secondary turns. The primary winding has 300 turns, and is tapped so that the number of turns can be reduced in steps of 20 down to 240 turns. Calculate the secondary voltages obtainable.

$$\frac{V_1}{V_2} = \frac{N_1}{N_2} \quad \text{so} \quad V_2 = V_1 \times \frac{N_2}{N_1}$$

(a) $V_2 = 415 \times \dfrac{20}{300} = 27 \cdot 7$ V

(b) $V_2 = 415 \times \dfrac{20}{280} = 29 \cdot 6$ V

(c) $V_2 = 415 \times \dfrac{20}{260} = 31 \cdot 9$ V

(d) $V_2 = 415 \times \dfrac{20}{240} = 34 \cdot 6$ V

The secondary voltages obtainable in this case are $27 \cdot 7$ V, $29 \cdot 6$ V, $31 \cdot 9$ V and $34 \cdot 6$ V. Notice that, with a tapped primary fed at a constant potential, the output voltage is highest for the lowest number of primary turns.

## Example 7.8

A bell transformer is to be wound for use from a 240 V single-phase supply. Outputs of 24 V, 18 V and 12 V are required. Calculate the total number of secondary turns and the tapping positions required if the primary has 200 turns.

$$\frac{V_1}{V_2} = \frac{N_1}{N_2} \quad \text{so} \quad N_2 = N_1 \times \frac{V_2}{V_1}$$

Total number of secondary turns

$$N_2 = 200 \times \frac{24}{240} = 20 \text{ turns.}$$

Since the output voltages of 18 V and 12 V are three-quarters and one-half, respectively, of the largest secondary voltage, the secondary must be tapped in these proportions, i.e. ¾ of 20 = 15 turns, and ½ of 20 = 10 turns.

The transformer is shown diagrammatically in Figure 7.6.

## 7.4   Principle of three-phase transformers

A three-phase transformer is effectively the same as three single-phase transformers connected in a three-phase arrangement. In some cases, in

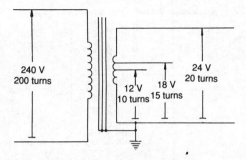

**Figure 7.6   Circuit diagram for Example 7.8**

fact, three separate single-phase transformers may be used instead of a three-phase transformer, although the cost and the total weight of the resulting unit will be greater than that of a purpose-built three-phase unit.

The six windings, three primaries and three secondaries, are interconnected as shown in Figure 7.7 or in Figure 7.8. Figure 7.7 shows both primary and secondary star-connected, while in Figure 7.8 a delta connection is used on both input and output sides. A common set of abbreviations is Y for star and D for delta primaries, with y and d for secondaries. These connection methods are known as star–star (or sometimes wye–wye or Yy) and delta–delta (or mesh–mesh or Dd), respectively. Another possible connection is star–delta (or wye–mesh or Yd), where the primary winding is star-connected as in Figure 7.7, and the secondary is delta-connected as in Figure 7.8. The last of the four possible basic connection methods is delta–star (or mesh–wye or Dy), with a delta-connected primary as in Figure 7.8 and a star-connected secondary as in Figure 7.7.

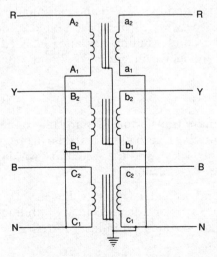

**Figure 7.7   Three-phase transformer, star–star connection**

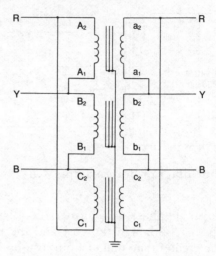

**Figure 7.8   Three-phase transformer, delta–delta connection**

Transformers with delta-connected secondaries are seldom used to supply consumer's loads, since there is no position on a delta winding for connection of the neutral or for the earth. Such transformers are, however, commonly used in the high-voltage substations associated with supply systems.

If the primary and secondary windings are connected similarly (i.e. star–star or delta–delta), calculations are the same as those for single-phase transformers as long as the special relationships between voltages and currents for three-phase systems are followed (see Section 4.8).

The standard method for marking three-phase transformer windings is to label the three primary windings A, B and C in capital (upper case) letters, and secondary windings as a, b and c in small (lower case) letters. Each winding has two ends, which are labelled 1 and 2, so that, for example, the primary of the second winding has ends labelled $B_1$ and $B_2$. The standard markings are shown in Figures 7.7 and 7.8.

## Example 7.9

A 660 kVA three-phase transformer is star–star connected, and each of the three primary windings has 6500 turns. The primary windings are fed with a line voltage of 11 kV, and the secondary windings are to provide a line voltage of 415 V. Calculate:

(*a*) the primary and secondary phase voltages,
(*b*) the number of turns on each secondary winding,
(*c*) the primary and secondary full-load currents.

(*a*)   $V_P = \dfrac{V_L}{\sqrt{3}} = \dfrac{11\ 000}{1 \cdot 73}$ volts $= 6350$ V (primary)

$$V_P = \frac{V_L}{\sqrt{3}} = \frac{415}{1 \cdot 73} \text{ volts} = 240 \text{ V (secondary)}$$

(b) $\dfrac{V_1}{V_2} = \dfrac{N_1}{N_2}$  so  $N_2 = N_1 \times \dfrac{V_2}{V_1}$

Since the transformer is star–star-connected, the phase-voltage ratio strictly should be used, although line voltages have the same ratio.

$$N_2 = 6500 \times \frac{240}{6350} \text{ turns} = 246 \text{ turns.}$$

(c) In Section 5.13, it was shown that the power in a three-phase balanced load is given by

$$P = (\sqrt{3}) V_L I_L \cos \phi$$

It follows that the voltamperes can be calculated by dividing the power by the power factor, so

$$\text{VA} = \frac{P}{\cos \phi} = \frac{(\sqrt{3}) V_L I_L \cos \phi}{\cos \phi} = (\sqrt{3}) V_L I_L$$

$$I_L = \frac{\text{VA}}{(\sqrt{3}) V_L} = \frac{660 \times 10^3}{1 \cdot 73 \times 11 \times 10^3} \text{ amperes} = 34 \cdot 7 \text{ A (primary)}$$

$$I_L = \frac{\text{VA}}{(\sqrt{3}) V_L} = \frac{660 \times 10^3}{1 \cdot 73 \times 415} \text{ amperes} = 918 \text{ A (secondary)}$$

## Example 7.10

A delta–delta connected transformer with a rating of 300 kVA has a turns ratio for each set of windings of 3180:200. If the transformer is required to provide an output of 415 V line, calculate:

(a) the primary line voltage,
(b) the primary and secondary full-load line currents,
(c) the current rating of each primary and secondary winding.

(a) $\dfrac{V_1}{V_2} = \dfrac{N_1}{N_2}$  so  $V_1 = V_2 \times \dfrac{N_1}{N_2}$

$$V_1 = 415 \times \frac{3180}{200} \text{ volts} = 6600 \text{ V}$$

(b) $I_L = \dfrac{\text{VA}}{(\sqrt{3}) V_L}$ (as in Example 7.9) $= \dfrac{300 \times 10^3}{1 \cdot 73 \times 6600}$ amperes

$\qquad = 26 \cdot 2$ A (primary)

$$I_L = \frac{\text{VA}}{(\sqrt{3}) V_L} = \frac{300 \times 10^3}{1 \cdot 73 \times 415} \text{ amperes} = 417 \text{ A (secondary)}$$

(*c*) Since each winding is delta-connected, it must be rated at the phase current of the three-phase system corresponding to the full-load line current.

$$I_P = \frac{I_L}{\sqrt{3}} = \frac{26 \cdot 2}{1 \cdot 73} \text{ amperes} = 15 \cdot 1 \text{ A (primary)}$$

$$I_P = \frac{I_L}{\sqrt{3}} = \frac{417}{1 \cdot 73} \text{ amperes} = 241 \text{ A (secondary)}$$

Where primary and secondary have different types of connection, the overall turns ratio of the transformer is more complicated. For example, consider a three-phase transformer with a 1:1 turns ratio, so that the input and output voltages for the windings are the same. If the transformer is connected in star–delta, and has a primary line voltage of $V$, each of the star-connected primary windings will receive the phase voltage of the system, which is $V/\sqrt{3}$. Each secondary winding will then have this same voltage induced in it, and since these windings are delta-connected, the voltage $V/\sqrt{3}$ will be the secondary line voltage. Thus, with a 1:1 turns ratio, a star–delta transformer provides a $\sqrt{3}:1$ step-down line-voltage ratio. For a star–delta connected transformer.

$$\frac{N_1}{N_2} = \frac{V_1/\sqrt{3}}{V_2}$$

or

$$\frac{N_1}{N_2} = \frac{V_1}{V_2\sqrt{3}}$$

The arrangement is shown in a simplified diagrammatic form in Figure 7.9.

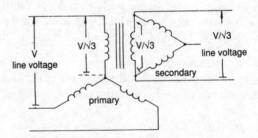

**Figure 7.9   To illustrate the $\sqrt{3}:1$ stepdown ratio of a star–delta-connected three-phase transformer in addition to the effect of turns ratio**
A turns ratio of 1:1 assumed for the diagram

## Example 7.11

A three-phase star–delta transformer is required to have a voltage ratio of 415:1200 V, each primary winding having 300 turns. Calculate the number of turns on each secondary winding. If losses are neglected and the secondary provides a load of 50 kVA, calculate the line and winding currents for both primary and secondary windings.

$$\frac{N_1}{N_2} = \frac{V_1}{V_2\sqrt{3}} \quad \text{so} \quad N_2 = N_1 \times \frac{V_2\sqrt{3}}{V_1}$$

$$N_2 = \frac{300 \times 1\cdot73 \times 1200}{415} \text{ turns} = 1500 \text{ turns}$$

Primary:

As the winding is star-connected, line and winding currents will be the same.

$$I_L = \frac{VA}{V_L\sqrt{3}} = \frac{50 \times 10^3}{1\cdot73 \times 415} \text{ amperes} = 69\cdot6 \text{ A (line and winding)}$$

Secondary:

$$I_L = \frac{VA}{V_L\sqrt{3}} = \frac{50 \times 10^3}{1\cdot73 \times 1200} \text{ amperes} = 24\cdot1 \text{ A (line)}$$

Each delta-connected secondary winding carries the phase current $I_P$

$$I_P = \frac{I_L}{\sqrt{3}} = \frac{24\cdot1}{1\cdot73} \text{ amperes} = 13\cdot9 \text{ A (winding)}$$

For a delta–star transformer, a similar effect occurs, but there is a $1{:}\sqrt{3}$ step-up ratio for line voltage in addition to the effect of the turns ratio. The reason for this is shown for a 1:1-ratio transformer in Figure 7.10.

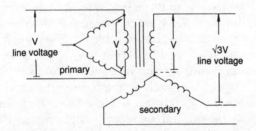

**Figure 7.10**    **To illustrate the $1{:}\sqrt{3}$ step-up ratio of delta–star-connected three-phase transformer in addition to effect of the turns ratio**
A turns ratio of 1:1 is assumed for diagram

Thus, for the delta–star-connected transformer,

$$\frac{N_1}{N_2} = \frac{(\sqrt{3})V_1}{V_2}$$

## Example 7.12

A 3300/415 V delta–star-connected three-phase transformer delivers 100 kVA on full load. Calculate:

  (*a*) the turns ratio, primary to secondary,
  (*b*) the full-load current in each secondary winding,
  (*c*) the full-load primary line current,
  (*d*) the full-load current in each primary winding.

  (*a*) $\dfrac{N_1}{N_2} = \dfrac{(\sqrt{3})V_1}{V_2} = \dfrac{1\cdot73 \times 3300}{415} = \dfrac{5710}{415} = 1142\!:\!83$

  (*b*) For the star-connected secondary winding, line and phase currents are the same, so winding current

$$I_L = \frac{VA}{(\sqrt{3})V_L} = \frac{100 \times 10^3}{1\cdot73 \times 415} \text{ amperes} = 139 \text{ A}$$

  (*c*)  $I_L = \dfrac{VA}{(\sqrt{3})V_L} = \dfrac{100 \times 10^3}{1\cdot73 \times 3300} \text{ amperes} = 17\cdot5 \text{ A}$

  (*d*)  $I_P = \dfrac{I_L}{\sqrt{3}} = \dfrac{17\cdot5}{1\cdot73} \text{ amperes} = 10\cdot1 \text{ A}$

### *Three-phase transformers in parallel*

There are very many cases where it is necessary to operate three-phase transformers in parallel. For example, if a load is fed by two or more transformers in parallel, one unit can be removed in the event of it becoming faulty or for maintenance, without the supply to the load being lost altogether. There may also be considerable tariff savings by disconnecting transformers on very light load, and thus saving the costs associated with iron losses, which are constant and independent of load (see Section 7.9). For example, if a large load, such as a department store, is fed by several paralleled transformers, all but one could be switched out at night when loads are likely to be limited to security lighting and so on, without disconnecting whole circuits. The disconnected transformers will have no iron losses, and the remaining transformer, being more heavily loaded, will operate at higher efficiency.

  Examples 7.11 and 7.12 indicate that if the primary and secondary windings are not both connected in the same way (star–star or delta–delta) there will be a voltage change from primary to secondary in addition to the effects of turns ratio. It thus follows that two transformers which are identical other than that they are differently connected can never be connected in parallel; unless the voltages produced by the secondary windings are identical, heavy circulating currents will result in the event of parallel connection. It would seem, then, that by choosing suitable turns ratios to offset the effects of differing connection methods, paralleling can be carried out.

Unfortunately this is not the case. As well as the voltage difference following from different connection methods, there is also a phase displacement. Figure 7.11 shows how input and output line voltages of a star–delta-connected transformer have a phase difference of 30°, with secondary voltage lagging primary voltage.

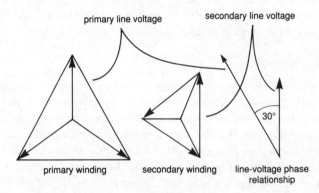

primary line voltage

secondary line voltage

30°

primary winding    secondary winding    line-voltage phase
relationship

**Figure 7.11    Showing how the 30° phase relationship occurs with the star–delta transformer**

It can be shown similarly that the delta–star-connected transformer has a 30° phase difference between line voltages, with secondary voltage leading primary voltage. To simplify the method, it is common to refer to phase differences as **clock numbers**. The only possible phase differences are 30° (or an additional 180° if windings are reverse connected); there are 12 numbers on a clock face, equally spaced through 360°. Thus, each number is 30° from the next. Hence, a lagging angle of 30° is given the clock number 11, and a leading angle of 30° the clock number 1, an in-phase relationship being 12 o'clock.

In fact, there are other transformer connections which are beyond the scope of this book, but it is always true that only transformers with equal secondary voltages and the same clock numbers may be connected in parallel.

## 7.5    Construction of single-phase transformers

The single-phase transformer will consist of primary and secondary windings mounted on a magnetic core.

### Core materials

Chapter 3 dealt with the magnetic materials of which the core is made. Since it must always be subjected to alternating magnetisation, the core material and construction must be chosen to reduce iron losses to a minimum, or the transformer will not be efficient. Most transformer cores are made from laminated silicon steel, the laminations reducing eddy currents and the silicon steel keeping hysteresis loss to a minimum.

Laminations must be arranged so as to reduce the airgaps in the magnetic circuit (see Figures 3.9 and 3.10). The laminations must be tightly held together by clamping or by taping, or they are likely to vibrate and produce excessive noise, as well as increasing the airgaps at joints. Some small high-frequency communications transformers have cores cast of solid ferroxcube, the eddy-current loss thus being kept to a reasonable level.

## Core arrangements

If a transformer were wound as represented by Figure 7.1 with primary and secondary on separate limbs of the core, a proportion of the magnetic flux produced by the primary winding will not pass through the secondary. This 'leakage flux' reduces the transformer efficiency and results in poor voltage regulation (see Section 7.9), and generally it must be reduced to an absolute minimum. The arrangement shown in Figure 7.12 is called a core-type transformer, the windings being split, with part of each wound on each side of the magnetic circuit to reduce leakage flux.

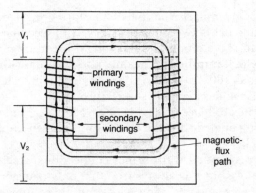

**Figure 7.12   Core-type transformer**

Leakage is reduced still further by using the shell-type circuit shown in Figure 7.13. Both windings are placed on the centre limb, the two outer limbs providing parallel return paths for the magnetic flux.

Since leakage tends to take place at the corners of the magnetic circuit, a ring-type of construction represented in Figure 7.14 is sometimes used for instrument and current transformers (see Section 7.10). Since the core is usually made of continuous-ring laminations, some difficulty is experienced in winding this type of core, but special machines are available

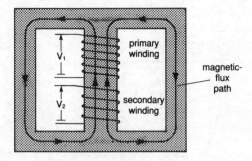

**Figure 7.13   Shell-type transformer**

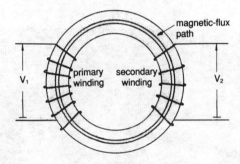

**Figure 7.14   Ring-type transformer**

for the purpose. High cost prevents the use of this construction for general-purpose transformers. Windings can be separately wound on formers for core- and shell-type transformers, the horizontal yoke of the magnetic circuit being made removable to allow the windings to be positioned.

The cross-sections of the cores of small transformers are rectangular, but it is usual to arrange for a near-circular-section core for larger transformers by using differing lamination widths, as shown in Figure 7.15. This reduces the length of copper needed for each winding.

## Windings

Windings are usually made of copper, although some experiments using aluminium as a conductor material have been carried out and there is a likelihood of its wider use.

Cylindrical or concentric windings, where the lower-voltage winding is completely surrounded by the higher-voltage turns, are used mainly for core-type circuits (see Figure 7.16). Sandwich or disc type windings, where the two windings are split into alternately mounted sections, are used generally on shell-type circuits (see Figure 7.17), except for very high-voltage transformers which use the cylindrical type of winding.

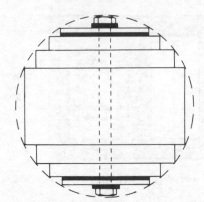

*a*

*b*

**Figure 7.15**   *a* **Typical core cross-section for large transformer**
   *b* **Packs of laminations being laid up to form shaped core**

## Enclosure and cooling

The finished transformer often needs to be protected from dirt and damage
by enclosing it in a box, although this is unnecessary for small units mounted
in electronic equipments. Since the transformer losses (see Section 7.9) result
in heat, cooling becomes necessary. Small transformers are air-cooled, fan
cooling (or 'air-blast cooling') being used on some larger transformers to

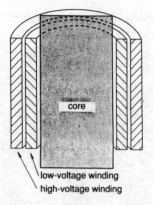

**Figure 7.16 Cylindrical, or concentric, windings**

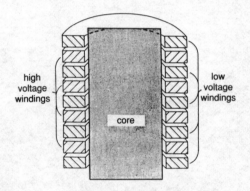

**Figure 7.17 Sandwich, or disc, windings**

reduce cost and weight. Most power transformers are enclosed in an oil-filled tank, the oil improving the insulation of the unit and circulating by natural convection to provide cooling. Cooling pipes are arranged around the tank to allow the oil to cool naturally as it circulates (see Figure 7.18). Unfortunately, the oil is flammable, and a number of fires in industrial and commercial premises have been traced back to its escape from transformers, usually under severe fault conditions. Some years ago, the problem was thought to have been solved by the introduction of polychlorinated benzine (PCB) as a filler to replace the oil, because it is virtually nonflammable and nondegradable. However, PCB has been found to be highly toxic, and as it is nondegrading its disposal is very difficult. Its use to fill transformers ceased some time ago, but large numbers of PCB-filled units are still in operation. It is MOST IMPORTANT that transformers which contain PCB are treated with the greatest care, and that the specialist advice of the Health and Safety Executive is sought concerning disposal of the coolant liquid. UNDER NO CIRCUMSTANCES must such liquid be poured away or buried. Care

**Figure 7.18**   **Completed 11 000/415 V three-phase transformer being lowered into its tank**

must also be taken not to come into contact with it, and to seek medical advice if such an event occurs.

## 7.6   Construction of three-phase transformers

The three-phase transformer is, as stated in Section 7.4, effectively three interconnected single-phase transformers. Considerable savings in cost, size and weight can, however, be made by combining the windings on to a single magnetic circuit. Two arrangements are used:

 (i) the core-type circuit, as shown in Figure 7.19, has three limbs and is the most common method of construction;
(ii) the shell-type five-limb circuit shown in Figure 7.20 is heavier and more expensive to build than the core type, but is sometimes used for very large transformers because it can be made with reduced height. Core materials, windings, enclosure and cooling are much the same as for the larger single-phase units.

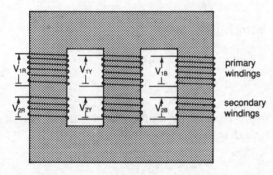

**Figure 7.19    Arrangement of core-type three-phase transformer**

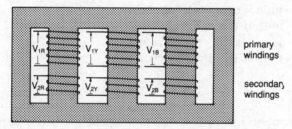

**Figure 7.20    Arrangement of shell-type three-phase transformer**

## 7.7   Autotransformers

An autotransformer has only one tapped winding, which is both the primary and the secondary of the machine. Figure 7.21 shows both step-down and step-up autotransformers. The single winding of the autotransformer is an obvious economy, another saving being due to the reduced current in the part of the winding common to both input and output. For example, the output current of the transformer of Figure 7.21$a$ is $I_2$, but the lower part of the winding carries only the current $I_2 - I_1$, the balance ($I_1$) reaching the output via the upper part of the winding.

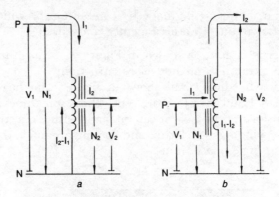

**Figure 7.21   Autotransformers**
    *a* Stepdown autotransformer
    *b* Stepup autotransformer

If losses are neglected, turns, voltage and current ratios are the same as for double-wound transformers, that is

$$\frac{N_1}{N_2} = \frac{V_1}{V_2} = \frac{I_2}{I_1}$$

where   $N_1$ = number of turns connected to input voltage
          $N_2$ = number of turns providing output voltage
          $V_1$ = input (primary) voltage
          $V_2$ = output (secondary) voltage
          $I_1$ = input (primary) current
          $I_2$ = output (secondary) current

## Example 7.13

A step-down autotransformer has a winding of 200 turns, a supply of 240 V being connected across it. An output is taken from the neutral of the supply, and from a tapping 80 turns from the neutral end of the winding. Make a sketch of the arrangement and calculate;

   (*a*) the output voltage,
   (*b*) the current in each part of the winding if the current from the supply
       is 20 A.

A sketch of the arrangement is shown in Figure 7.22.

(*a*) $\dfrac{V_1}{V_2} = \dfrac{N_1}{N_2}$  so  $V_2 = V_1 \times \dfrac{N_2}{N_1}$

$$V_2 = 240 \times \frac{80}{200} \text{ volts} = 96 \text{ V}$$

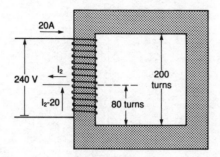

**Figure 7.22    Diagram for Example 7.13**

(b) $\dfrac{I_1}{I_2} = \dfrac{V_2}{V_1}$   so   $I_2 = I_1 \times \dfrac{V_1}{V_2}$

$I_2 = 20 \times \dfrac{240}{96} = 50 \text{ A}$

Current in the common part of the winding,

$$= I_2 - I_1 = 50 - 20 \text{ amperes} = 30 \text{ A}$$

This example shows that the current rating of the upper and lower sections of the transformer can be 20 A and 30 A, respectively, for an output of 50 A. A double-wound transformer to give a similar output would need a 200-turn primary winding rated at 20 A, as well as an 80-turn secondary winding rated at 50 A. This illustrates the major advantage of the autotransformer, which will be smaller, lighter and cheaper than its double-wound counterpart. The disadvantages of the autotransformer are that:

(a) there is a direct metallic connection between the input and the output, whereas the coupling in a double-wound transformer is magnetic only, giving electrical isolation of the two windings;

(b) in the event of an open-circuit fault in the common part of the winding, the input voltage of a step-down autotransformer would appear on the output terminals (see Figure 7.23). Because of this danger, the IEE Wiring Regulations limit the use of autotransformers. However, they are used in high-voltage transmission systems, as starters for induction and synchronous motors, and for voltage control in some types of discharge lamp (see Section 7.11).

## 7.8    Earthing of transformers

Section 12.5 of Electrical Craft Principles Volume 1 explained the use made of earthing of noncurrent-carrying metalwork to prevent it from becoming 'live' relative to the general mass of earth, and to provide a path for fault currents. The IEE Wiring Regulations require that, with certain exceptions,

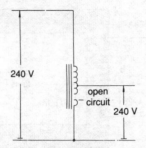

**Figure 7.23   To indicate the danger of input voltages appearing at output terminals of a stepdown autotransformer in the event of an open circuit in the common winding**

the metalwork (other than current-carrying parts) and one point on the secondary winding of a transformer must be connected to earth by means of a suitable protective conductor. As well as providing protection for the transformer itself, this rule ensures that the output of the transformer will have one side earthed so that the circuits and equipments fed from the transformer can also be protected by suitable earthing. Since the double-wound transformer does not provide a direct metallic connection from primary to secondary, the earthing applied to the primary circuit would be absent on the secondary side if not deliberately provided by earthing it. Figures 7.2–7.6 inclusive indicate the necessary earth connections.

Clearly, there is no need to earth the windings of an autotransformer (although the core must be earthed), since the earthing applied to the incoming supply will be carried from the primary winding to the secondary winding due to the direct connection between them. Other special transformers which do not need earthing on the secondary side include those where

  (i)  primary and secondary windings are on different limbs of an earthed core,

 (ii)  an earthed metallic screen is provided between the primary and the secondary windings,

(iii)  the transformer is double-insulated, or

(iv)  the transformer secondary winding is intended to feed only an electric shaver. In this case, the object of the transformer is to make the shaver connection earth-free.

In most electronic applications, the secondary windings of transformers cannot be earthed.

## 7.9   Transformer losses, efficiency and regulation

The assumption made in all the calculations so far carried out in this chapter, that the transformer has no losses, is clearly not true. Losses actually occurring can be considered under two headings. These are:

## (i) Iron (core) losses

Iron losses occur in the magnetic core of the transformer, causing it to heat up. Iron losses can be divided into

(*a*) hysteresis losses (explained in Section 3.5)
(*b*) eddy-current losses (explained in Section 3.6).

Iron losses depend on the frequency of the supply, and the maximum magnetic flux density in the transformer core (see Section 3.7). For power transformers, the supply frequency is almost always constant, and, since the supply voltage is virtually constant, there is very little change in the core flux density. Thus it is reasonable to assume that the iron losses in a transformer remain constant regardless of the load conditions—for example, the iron loss on no load will be the same as that on full load.

## (ii) Copper ($I^2R$) losses

Copper losses are due to the heating effect of the primary and secondary currents passing through their respective windings. Although the resistance of these windings is kept as low as possible, for power transformers it cannot be zero, and copper losses can only be prevented if the current is reduced to zero. In practice, if the secondary winding of a transformer is open-circuited (so that it carries no current), the primary winding will still carry a small current which provides the ampere-turns necessary to set up the magnetic flux in the core. However, this magnetising current is quite small, so that the copper loss due to it may be ignored, and the total copper losses of a power transformer on no load may be assumed to be zero. Power loss in a resistive circuit is given by the expression $P = I^2R$, and since winding resistances are largely constant, copper losses depend on the square of the load current. Thus, a transformer operating on half load will have only one quarter of the copper loss it has when providing full load.

$$\text{copper loss } P_C = \left( \frac{\text{actual load}}{\text{full load}} \right)^2 \times \text{full-load copper loss}$$

## Efficiency

As well as providing for the output power, the input to a transformer must supply the transformer losses. Thus,

$$\text{input power} = \text{output power} + \text{power losses}$$

$$\text{efficiency} = \frac{\text{output power}}{\text{input power}} \times 100\%$$

$$= \frac{\text{input power} - \text{power losses}}{\text{input power}} \times 100\%$$

$$= \frac{\text{output power}}{\text{output power} + \text{power losses}} \times 100\%$$

## Example 7.14

The full-load copper and iron losses for a large power transformer are 16 kW and 12 kW, respectively. If the full-load output of the transformer is 950 kW, calculate the losses and efficiency of the transformer,

(*a*) on full load,
(*b*) on 60% of full load,
(*c*) on half load.

(*a*)   Total loss = copper loss + iron loss

$$= 16 + 12 \text{ kilowatts} = 28 \text{ kW}$$

$$\text{Efficiency} = \frac{\text{output}}{\text{output} + \text{losses}} \times 100\% = \frac{950}{950 + 28} \times 100\%$$

$$= \frac{950}{978} \times 100\% = 97 \cdot 1\%$$

(*b*) At 60% full load, iron loss remains at 12 kW.

$$\text{Copper loss} = \left(\frac{60}{100}\right)^2 \times 16 \text{ kilowatts} = 0 \cdot 36 \times 16 \text{ kilowatts}$$

$$= 5 \cdot 8 \text{ kW}$$

$$\text{Total loss} = 5 \cdot 8 + 12 \text{ kilowatts} = 17 \cdot 8 \text{ kW}$$

$$\text{Efficiency} = \frac{\text{output}}{\text{output} + \text{losses}} \times 100\% = \frac{(0 \cdot 6 \times 950)}{(0 \cdot 6 \times 950) + 17 \cdot 8} \times 100\%$$

$$= \frac{570}{587 \cdot 8} \times 100\% = 97 \cdot 0\%$$

(*c*) At half load, iron loss remains at 12 kW.

$$\text{Copper loss} = (\tfrac{1}{2})^2 \times 16 \text{ kilowatts} = \tfrac{1}{4} \times 16 \text{ kilowatts} = 4 \text{ kW}$$

$$\text{Total loss} = 4 + 12 \text{ kilowatts} = 16 \text{ kW}$$

$$\text{Efficiency} = \frac{\text{output} \times 100\%}{\text{output} + \text{losses}} = \frac{(\tfrac{1}{2} \times 950)}{(\tfrac{1}{2} \times 950) + 16} \times 100\%$$

$$= \frac{475}{491} \times 100\% = 96 \cdot 7\%$$

## *Regulation*

The resistance and inductive reactance of a transformer winding provide an impedance through which output current must pass. A voltage drop thus occurs in the windings of a transformer, its value depending on the effective impedance and on the current ($V = IZ$). As the current taken from a given transformer increases, the voltage drop increases, and since the electromotive force induced in the secondary winding is constant, the output voltage must fall. On no load there will be no secondary current and no voltage drop. On full load the output voltage will fall, and the difference between no-load voltage and full-load voltage, expressed as a percentage of no-load voltage, is called the voltage regulation.

$$\text{regulation} = \frac{\text{no-load voltage} - \text{full-load voltage}}{\text{no-load voltage}} \times 100\%$$

## Example 7.15

A power transformer provides 415 V on no load and 405 V on full load. Calculate the voltage regulation.

$$\text{regulation} = \frac{\text{no-load volts} - \text{full-load volts}}{\text{no-load volts}} \times 100\% = \frac{415 - 405}{415} \times 100\%$$

$$= \frac{10}{415} \times 100\% = 2 \cdot 41\%$$

## Example 7.16

A transformer with a voltage regulation of 4% provides $230 \cdot 4$ V on full load. Calculate its no-load terminal voltage.

Since full-load voltage is 4% below the no-load value, full-load volts = 96% of no-load volts.

$$230 \cdot 4 = \frac{96}{100} \times \text{no-load volts}$$

$$\text{Therefore no-load volts} = 230 \cdot 4 \times \frac{100}{96} \text{ volts} = 240 \text{ V}$$

## 7.10   Instrument transformers

Transformers are used in conjunction with instruments for measurement of high alternating currents or voltages. For measurement of an alternating current of 30 A or more, an ammeter with a range 0–1 A or 0–5 A is connected across the secondary winding of a current transformer whose primary winding is in series with the load circuit (see Figure 7.24). For about 750 V or more, a voltmeter with a range of about 0–110 V is connected across the secondary winding of a voltage transformer whose

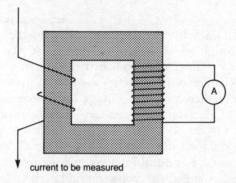

**Figure 7.24   Arrangement of current transformer**

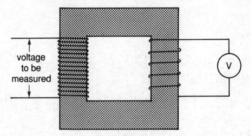

**Figure 7.25   Arrangement of voltage transformer**

primary winding is connected across the voltage to be measured (see Figure 7.25).

## Current transformers

A current transformer is wound on a core of low-loss magnetic material such as laminated Mumetal (see Section 3.3) which has a large cross-sectional area so that the magnetic flux density is low. The primary winding has few turns, in some cases passing straight through the core, so that it has low impedance and does not affect the circuit to be measured. The secondary winding of a current transformer must never be disconnected while the primary is fed, or dangerously high voltages are likely to be induced in it. Figures 7.26 and 7.27 show two types of current transformer, that in Figure 7.27 having a core which will clip over an insulated conductor and so measure the current it carries without the need for disconnection. If, as is sometimes the case, the measuring instrument is situated away from the transformer, the manufacturer should be consulted so that the system is designed to take into account the resistance of the connecting leads.

## Voltage transformers

The voltage transformer is much more like a normal power transformer than is the current transformer, although it has a special core material with

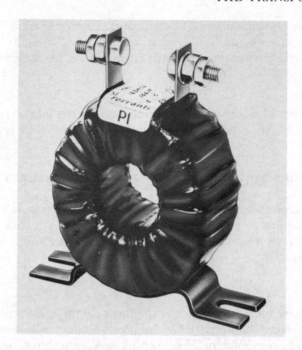

**Figure 7.26  Ring type current transformer**

**Figure 7.27  Digital clamp meter**

a generous cross-sectional area. The windings of a voltage transformer have very low impedance, and care must be taken not to short-circuit the secondary winding while the primary is energised, or it is likely to burn out.

## Wattmeters

Both current and voltage transformers are connected into a type of watt-meter for use on AC supplies. Wattmeters are considered in Section 11.6.

## 7.11   Some applications of autotransformers

The great importance of transformers in electrical systems and apparatus of all types has already been stressed in Section 7.7. This section deals with the application of autotransformers in motor starting and in discharge-lighting circuits.

### Autotransformer motor starter

The three-phase induction motor is the most common electric motor for industrial use (see Chapter 10). One of its disdvantages, however, is that it draws a very heavy current while starting, and various methods are used to reduce it. One of these is to feed the machine with a reduced voltage for starting, and autotransformers are used for this purpose. Figure 7.28 shows the circuit of a simple autotransformer starter for a three-phase induction motor. When the switch is in the 'start' position, the three autotransformers are star-connected to the supply, with the motor stator fed at reduced voltage from tappings. With the switch in the 'run' position, the motor is connected directly to the supply, with the autotransformers disconnected so that no losses occur in them. The transformer tappings are usually adjusted during installation to strike the best balance between low starting current and reasonable starting torque. Autotransformers,

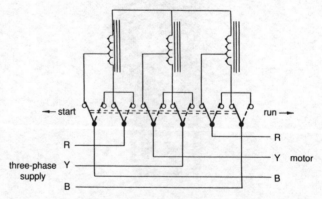

**Figure 7.28   Autotransformer starter for three-phase induction motor**

rather than double-wound transformers, are used because of their lower size, weight and cost.

## Discharge lighting transformers

Discharge lighting is extremely common since lamps of this type are much more efficient than filament lamps. However, a discharge lamp almost always needs special starting and current-limiting circuits, in which auto-transformers are often used. Once again, the low cost of the autotransformer gives advantages: two of the many hundreds of arrangements commonly in use are shown.

Figure 7.29 shows a type of quick-start circuit for a low-pressure mercury-vapour fluorescent lamp, which provides a higher-than-usual voltage to the filament at each end of the lamp for quick heating as well as a high voltage across it for starting purposes. The series choke has a current-limiting effect when the lamp is operating.

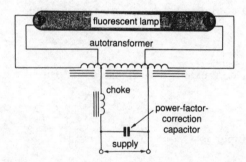

**Figure 7.29  Quick-start fluorescent-lamp circuit with autotransformer**

Figure 7.30 shows a circuit for a low-pressure sodium-vapour lamp of the type in wide use for road lighting. This transformer has a specially designed magnetic circuit which deliberately produces considerable leakage flux so that the high output voltage needed to start the lamp falls to a much

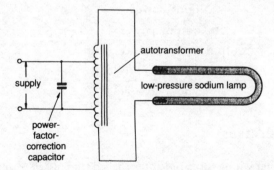

**Figure 7.30  Circuit for low-pressure sodium-vapour lamp**

lower value when the lamp operates. This construction produces a higher regulation than for any other transformer.

## 7.12   Summary of formulas for Chapter 7

See text for definitions of symbols.

For a transformer:

$$\text{primary volts per turn} = \frac{V_1}{N_1}$$

$$\text{secondary volts per turn} = \frac{V_2}{N_2}$$

$$\frac{V_1}{V_2} = \frac{N_1}{N_2} = \frac{I_2}{I_1}$$

For a star–delta-connected three-phase transformer:

$$\frac{N_1}{N_2} = \frac{V_1}{V_2\sqrt{3}}$$

For a delta–star-connected three-phase transformer:

$$\frac{N_1}{N_2} = \frac{V_1\sqrt{3}}{V_2}$$

Transformer copper loss,

$$P_C = \left(\frac{\text{actual load}}{\text{full load}}\right)^2 \times \text{full-load copper loss}$$

$$\text{input power} = \text{output power} + \text{losses}$$

$$\text{efficiency} = \frac{\text{output power}}{\text{input power}}$$

$$= \frac{\text{input power} - \text{power losses}}{\text{input power}}$$

$$= \frac{\text{output power}}{\text{output power} + \text{power losses}}$$

$$\text{regulation} = \frac{\text{no-load voltage} - \text{full-load voltage}}{\text{no-load voltage}}$$

## 7.13   Exercises

1  A transformer with a turns ratio of 5:1 is fed from a 240 V supply. What will be the output voltage?

2 What is the turns ratio of a transformer which provides a 1500 V output from a 240 V supply?

3 A transformer with a turns ratio of 1:10 has 160 primary turns and 1·5 volts/turn. What is the secondary voltage?

4 Calculate the full-load primary and secondary currents of a 60 kVA transformer with a turns ratio of 165:6 fed from a 6·6 kV supply.

5 The input current to the transformer of Exercise 1 is 2 A. What is the output current?

6 A 24 VA bell transformer has an output current of 2 A and a turns ratio of 20:1. Calculate the input current and input voltage.

7 A transformer with 10 000 primary turns and 900 secondary turns is fed at 11 kV. Calculate the secondary voltage available if the transformer primary is tapped at 500-turn intervals down to 8500 turns.

8 A 3300/415 V transformer has 3000 primary turns. Calculate the total number of secondary turns if the unit is required to give the rated voltage output if the input voltage varies by $\pm 2\frac{1}{2}\%$.

9 A transformer has a 900-turn primary winding with a secondary of 300 turns, tapped at 20-turn intervals down to 240 turns. Calculate the available secondary voltages if the primary is fed at 240 V.

10 A three-phase star–star connected transformer has a line-voltage ratio of 240:110 V and 200 turns on each primary winding. Calculate the number of turns on each secondary, and the primary and secondary currents when the secondary provides a load of 20 kVA. Neglect losses.

11 Calculate the full-load line and winding currents for the primary of a 90 kVA delta–delta three-phase transformer with a primary line voltage of 415 V. If the turns ratio is 2:11, calculate the secondary line voltage, as well as secondary line and winding currents.

12 A three-phase 2 MVA transformer is star–delta connected and has a turns ratio of 173:50. The primary winding is fed at 66 kV line. Calculate, neglecting transformer losses,
(a) the primary phase voltage,
(b) the secondary line voltage,
(c) the full-load primary current,
(d) the full-load secondary line current, and
(e) the current in each secondary winding on full load.

13 A three-phase transformer is fed at 6600 V, the primary windings being mesh-connected. The secondary delivers 40 kVA at 400 V, the secondary windings being star-connected. Calculate, neglecting losses,
(a) the current in each secondary phase winding,
(b) the current in each primary phase winding,
(c) the current in each primary line, and
(d) the turns ratio, primary to secondary.

14 A three-phase delta–star transformer has a turns ratio of 275:6. If the primary is fed at 11 kV and the full-load rating is 150 kVA, calculate, neglecting losses,
(a) the secondary line voltage,

(*b*) the primary line and winding currents on full load, and

(*c*) the secondary line and winding currents on full load.

15  Make a simple sketch showing the construction of a single-phase 240/110 V double-wound transformer. Label the parts and state the materials of which they are made.

16  Make a neat diagram or sketch of a simple single-phase double-wound transformer, and with its aid explain the action of the transformer. Calculate the respective numbers of turns in each winding of such a transformer which has a step-down ratio of 3040 to 240 V if the 'volts per turn' are 1·6.

17  (*a*) Describe briefly the essential parts of a single-phase transformer.

(*b*) Why are the iron cores laminated?

18  Give a brief description with sketches, of the main parts of a single-phase transformer, and state the functions of each part.

A step-down single-phase transformer has a ratio of 15:1 and a primary voltage of 3300. Calculate the secondary voltage and the primary and secondary currents when a load of 30 kVA at unity power factor is supplied. Ignore transformer losses.

19  Make a neat sketch or diagram of a single-phase double-wound transformer, and describe its action, clearly explaining the function of each part.

The 'volts per turn' of a certain single-phase transformer are 1·7. The transformer has a step-down ratio of 3825 V to 255 V. Calculate

(*a*) the respective number of turns in each winding, and

(*b*) the secondary current if the primary current is 12 A.

20  Sketch two possible core arrangements for three-phase transformers. In each case, show where the primary and secondary windings for each phase would be placed.

21  Sketch the construction of an autotransformer. Compare the advantages and disadvantages of the autotransformer with those of the double-wound type.

22  A step-up autotransformer has its output connected across its full winding of 800 turns, while the 240 V input is connected across the lower 160 turns. If the transformer provides an output of 20 kVA, calculate

(*a*) the output voltage,

(*b*) the output current,

(*c*) the input current, and

(*d*) the current in the common section of the winding.

23  Why is it necessary to earth a transformer? Sketch a circuit of a double-wound transformer showing which parts must be connected to a suitable protective conductor.

24  Explain the various losses which occur in a transformer and state which of them vary with load.

25  The iron loss for a transformer is 8 kW, and its copper loss on full load is 10 kW. If the full-load output is 600 kW, calculate the total losses and the efficiency:

(*a*) at full load          (*b*) at ⅔ full load          (*c*) at ⅓ full load.

26 A transformer with an output of 392 kW is 98% efficient at full load and has an iron loss of 3·5 kW. Calculate the full-load copper loss.

27 (*a*) What is meant by the statement that a certain transformer has 5% regulation at full load kVA, unity power factor?

   (*b*) A 6600/415 V transformer has tappings on its primary winding. If the tapping which comprises 90% of the whole winding is supplied at 6600 V, what will be the secondary voltage?

28 Calculate the regulation of a transformer which has no-load and full-load secondary terminal voltages of 610 V and 570 V, respectively.

29 The open-circuit terminal voltage of the secondary winding of a transformer is 24 V. Calculate the full-load terminal voltage if the regulation is 8%.

30 On full load, the output voltage of a transformer with a regulation of 4% is 3168 V. Calculate the no-load terminal voltage.

31 What are instrument transformers, and why are they used? Draw circuit diagrams to illustrate your answers.

32 List the special features of current and voltage transformers. What precautions must be taken when using these devices?

33 Indicate why autotransformers are used in a starter for a three-phase induction motor. Draw a circuit diagram to show the arrangement of such a starter.

34 Draw a circuit diagram to show an autotransformer used in conjunction with a discharge lamp.

# 7.14   Multiple-choice exercises

7M1   A transformer will not operate properly when fed from a direct current supply because:
   (*a*) the voltage of the DC system will be incorrect
   (*b*) only step-up transformers are used in a supply system
   (*c*) there will be no changing magnetic flux in the core to induce secondary EMF
   (*d*) the magnetic circuit will not allow flux to be produced.

7M2   It is always true that for a transformer that:
   (*a*) primary volts per turn are equal to secondary volts per turn
   (*b*) secondary voltage exceeds primary voltage
   (*c*) core flux is proportional to load current
   (*d*) voltage ratio is the inverse of turns ratio.

7M3   A single-phase transformer has 1200 primary turns and is fed from a 415 V, 50 Hz supply. The secondary volts per turn will be:
   (*a*) 0·922 V      (*b*) 0·346 V      (*c*) 0·111 V      (*d*) 3·46 V.

7M4   The transformer detailed in Exercise 7M3 has 80 secondary turns, so the secondary voltage will be:
   (*a*) 27·7 V      (*b*) 6·225 kV      (*c*) 15 V/turn      (*d*) 80 V.

7M5   A formula from which the number of secondary turns for a transformer can be calculated is:

(a) $N_2 = \dfrac{N_1 \times V_1}{V_2}$ 　　　　　(b) $N_2 = \dfrac{V_1}{N_1 \times V_2}$

(c) $V_2 = \dfrac{N_1 \times V_2}{N_1}$ 　　　　　(d) $N_2 = \dfrac{N_1 \times V_2}{V_1}$

7M6   A transformer with 1500 secondary turns has its primary fed with a supply at 120 V when the secondary voltage is 360 V. The number of primary turns is:
(a) 455 　　　　(b) 500 　　　　(c) 4500 　　　　(d) 29.

7M7   A 30 kVA transformer has a voltage ratio of 600/240 V. The full-load secondary current will be:
(a) 125 A 　　　　(b) 50 A 　　　　(c) 7200 A 　　　　(d) 13 A.

7M8   The formula relating numbers of turns, voltages and currents for a transformer is:

(a) $\dfrac{V_1}{V_2} = \dfrac{N_1}{N_2} = \dfrac{I_1}{I_2}$ 　　　　　(b) $\dfrac{V_2}{V_1} = \dfrac{N_2}{N_1} = \dfrac{I_1}{I_2}$

(c) $\dfrac{V_1}{V_2} = \dfrac{N_2}{N_1} = \dfrac{I_1}{I_2}$ 　　　　　(d) $\dfrac{V_1}{V_2} = \dfrac{N_1}{N_2} = \dfrac{I_2}{I_1}$

7M9   A single-phase transformer is fed at 240 V, 50 Hz and provides a full-load secondary current of 20 A at 30 V. If the secondary winding has 60 turns, the number of turns on the primary winding will be:
(a) 120 　　　　(b) 600 　　　　(c) 480 　　　　(d) 100.

7M10  The transformer of Exercise 7M9 will have a kVA rating of:
(a) 0·6 kVA 　　　(b) 4·8 kVA 　　　(c) 7·2 kVA 　　　(d) 1·0 kVA.

7M11  The full-load primary current of the transformer of Exercise 7M9 will be:
(a) 25 A 　　　　(b) 2·5 A 　　　　(c) 160 A 　　　　(d) 5 A.

7M12  The volts per turn for the transformer of Exercise 7M9 will be:
(a) 0·5 V 　　　　(b) 1800 V 　　　　(c) 0·25 V 　　　　(d) 0·33 V,

7M13  The turns ratio of the transformer of Exercise 7M9 is:
(a) 1:8 step-up 　　　　　　(b) 240:30 step-up
(c) 8:1 step-down 　　　　　(d) 60:480 step-up.

7M14  A three-phase transformer consists of:
(a) three single-phase transformers with their magnetic circuits linked
(b) a three-limb or five-limb magnetic circuit with a primary and a secondary winding for each phase on separate limbs
(c) a single-phase transformer with a third winding
(d) three single-phase transformers with their primary and secondary windings connected in series.

7M15  A delta–star three-phase transformer has its:
(a) primary winding star-connected and its secondary winding delta-connected

(b) primary and secondary windings interconnected

(c) primary and secondary windings both connected in delta–star

(d) secondary winding star-connected and its primary winding delta-connected.

7M16 A three-phase star–star transformer has a primary line voltage of 6·6 kV, a secondary line voltage of 415 V and a rating of 60 kVA. If each primary winding has 5280 turns, each secondary winding will have:

(a) 3770 turns    (b) 145 turns    (c) 332 turns    (d) 416 turns.

7M17 The primary line current of the transformer of Exercise 7M16 on full load will be:

(a) 9·1 A    (b) 5·25 A    (c) 83·5 A    (d) 8 A.

7M18 The secondary phase voltage of the transformer of Exercise 7M16 will be:

(a) 240 V    (b) 34·6 kV    (c) 3·81 kV    (d) 720 V.

7M19 Three-phase transformers are often used in parallel because:

(a) one transformer will not be able to carry the required load

(b) eddy-current losses may be reduced on light load by switching out some of the transformers

(c) they work better that way

(d) the load can be balanced between them, reducing copper losses.

7M20 Three-phase transformers may only be safely connected in parallel when:

(a) they are identical in all respects

(b) they have exactly the same losses

(c) they have the same secondary voltages and clock numbers

(d) the permission of the electricity supply company is obtained.

7M21 Eddy-current losses in transformers are kept to a minimum by:

(a) making the core of silicon steel

(b) ensuring that there are no short-circuits in the windings

(c) making sure that there are no airgaps in the magnetic circuit

(d) making the core of laminations which are insulated from each other.

7M22 Hysteresis losses in transformers are kept to a minimum by:

(a) ensuring that there no short-circuits in the windings

(b) making the core of laminations which are insulated from each other

(c) making the core of silicon steel

(d) making sure that there are no airgaps in the magnetic circuit.

7M23 A shell-type transformer has a core with:

(a) three limbs    (b) no laminations

(c) two limbs    (d) five limbs.

7M24 A cylindrical, or concentric, transformer winding has:

(a) the primary and secondary windings in multiple layers

(b) the high-voltage winding surrounding the low-voltage winding

(c) no separate primary and secondary windings

(d) the primary and secondary windings on separate cores.

7M25 To improve cooling, power transformers are often enclosed in a tank containing;

(a) liquid sulphur    (b) water

(c) carbon tetrachloride    (d) oil.

7M26 An autotransformer is one with:
(a) automatic voltage-regulation equipment
(b) two windings connected in parallel
(c) a single tapped winding
(d) a step-down ratio.

7M27 The full-load copper and iron losses of a transformer are 7 kW and 6 kW, respectively. If the output of the transformer on full load is 480 kW, its efficiency at half load will be:
(a) 97·4%    (b) 96·9%    (c) 50%    (d) 94·9%.

7M28 The transformer for an electronic power supply has a no-load output voltage of 37 V, falling to 34 V on full load. The voltage regulation of the transformer is:
(a) 8·1%    (b) 0%    (c) 8·8%    (d) 91·9%.

7M29 The secondary winding of a current transformer should never be disconnected whilst the primary is fed because:
(a) it is likely to overheat and burn out
(b) the primary impedance will increase
(c) dangerously high voltages will be induced in it
(d) the indicating instrument will read zero.

7M30 Autotransformers are sometimes used:
(a) in substations feeding domestic consumers
(b) in starters for three-phase induction motors
(c) as voltage-regulation devices
(d) instead of three-phase transformers.

# Electrical-generator principles

## 8.1   Introduction

Chapter 8 of 'Electrical Craft Principles' Volume 1, dealt with electric cells and batteries, which store energy in chemical form and convert it to electrical energy as required. Such devices are of importance for portable equipments and for use as standby supplies, but the amount of energy stored for a given size and weight is low. Our present electrically dependent civilisation would not be possible if we had to rely on supplies obtained from batteries, because they are bulky and expensive. For example, the battery of a car is extremely useful to provide for parking lights and for engine starting, but, without the engine-driven generator which takes over (and also charges the battery), it would become discharged quite quickly.

For most of our electrical needs we rely on the supply mains, which provide electrical energy from electrical generators driven by mechanical prime movers. The generator is not really correctly named. It does not generate electrical energy, but converts mechanical energy to electrical form.

## 8.2   Simple loop generator

The simple generator consisting of a rectangular loop of wire rotated between a pair of magnetic poles was considered in Section 9.4 of 'Electrical Craft Principles' Volume 1. The EMF is induced in the loop because it cuts magnetic flux as it rotates, so it can be seen that there are three requirements for dynamic induction.

These are:

  (i) a conductor system
 (ii) a magnetic field, and
(iii) relative motion between the conductors and field.

The faster the loop rotates, the greater the rate at which flux is cut and the greater the induced EMF. Similarly, if the magnetic field is made stronger, the induced EMF increases. Thus

$$E \propto \Phi n$$

where   $E$ = induced EMF
   $\Phi$ = magnetic-field flux and
   $n$ = speed of rotation of loop

## Example 8.1

When a loop is rotated at 10 r/s in a magnetic field with a total flux of 60 mWb, the induced EMF is 8 V. Calculate the induced EMF if the speed increases to 16 r/s and the total flux falls to 50 mWb.

$$E \propto \Phi n \quad \text{so} \quad \frac{E_1}{E_2} = \frac{\Phi_1}{\Phi_2} \times \frac{n_1}{n_2}$$

$$E_2 = E_1 \times \frac{\Phi_2}{\Phi_1} \times \frac{n_2}{n_1} = 8 \times \frac{50}{60} \times \frac{16}{10} \text{ volts} = 10 \cdot 7 \text{ V}$$

Note that, since both speed and flux are expressed as ratios, the units used for each term must be the same.

In theory, there is no reason why the loop should not remain stationary while the magnetic-field system revolves. In practice, this method is usual for alternating-current generators, which are called alternators (see Sections 8.4 and 8.5).

While the concept of a loop rotating between the widely spaced poles of a magnet may be all very well in theory, the very large airgap introduced into the magnetic circuit by the pole spacing will reduce the flux, and hence the induced EMF, to a very low value indeed. There must be an airgap between the loop and the poles to allow for rotation but, if the loop is wound on an iron core, the magnetic circuit will be improved considerably, with consequent improvement in the values of magnetic flux and of induced EMF. Since this iron core will, like the loop, be rotating in the magnetic field, it too will have an alternating EMF induced in it. The resulting eddy currents are, in practice, reduced to reasonable levels by making the core of laminations (see Section 3.6). Hysteresis loss, owing to the effects of alternating magnetisation as parts of the core alternately pass north and south magnetic poles, is reduced by making the laminations of silicon steel (see Section 3.5).

## 8.3   Direct-current generator principles

Since the loop of the simple generator alternately passes north and south poles of the magnet system, the EMF induced is alternating. To obtain a direct output, the polarity of the wires connecting to the loop must be reversed each time the induced EMF falls to zero. This switching is carried out by means of brushes bearing on a commutator, as described in Section 9.5 of 'Electrical Craft Principles' Volume 1. To synchronise the action of the commutator with the movement of the loop, the two components are mounted on a common shaft and rotate together. For this reason, the system of conductors in which the EMF of a DC generator is induced (called the armature winding) must rotate in a stationary magnetic-field system.

The construction of DC machines and the operation of the DC generator will be considered in more detail in Chapter 9. In practice, such a machine will have a large number of loops on its rotating conductor system, as well as a large number of commutator segments. However, the machine still obeys the rule that induced EMF is proportional to both rotational speed and magnetic flux.

## Example 8.2

A certain DC generator has an EMF of 350 V when its speed is 1750 r/min and its flux/pole is 90 mWb. What must be the flux/pole if it is to generate 250 V at a speed of 1500 r/min?

$$E \propto \Phi N \quad \text{so} \quad \frac{E_1}{E_2} = \frac{\Phi_1}{\Phi_2} \times \frac{N_1}{N_2}$$

$$\Phi_2 = \Phi_1 \times \frac{N_1}{N_2} \times \frac{E_2}{E_1} = 90 \times \frac{1750}{1500} \times \frac{250}{350} \text{ milliwebers} = 75 \text{ mWb}$$

Note that the symbol $n$ is used for speed in r/s, and $N$ for speed in r/min.

## 8.4 Single-phase alternator principles

The simple loop generator has an alternating EMF induced in it, and is thus a generator of alternating current, or alternator. The arrangement of the loop is shown in Figure 9.5 of 'Electrical Craft Principles' Volume 1 and described in Section 9.4 of the same book. In practical cases, it is more usual, for both single-phase and three-phase alternators, for the wire-loop system to remain stationary and for the magnetic system to rotate within it, as justified in Section 8.2. Some of the more important reasons for rotating the magnetic field system instead of the loop are:

(i) the conductor system in which the EMF is to be induced is usually much heavier than the magnet system, which is thus more easily rotated,

(ii) the conductor system will not be subjected to centrifugal forces if stationary, and can be more easily braced to withstand the electromagnetic forces which will result from overload or short circuit,

(iii) the conductor system is more easily insulated for high-voltage operation if stationary,

(iv) the field system of an alternator is seldom a permanent magnet, usually being an electromagnet fed with direct current. This 'exciting' current is much smaller than that taken from the main winding, so the sliprings feeding it are lighter than would be necessary to serve the main winding,

(v) a three-phase alternator would require three heavy-current high-voltage sliprings to serve a rotating main winding, but only needs two low-voltage sliprings for a rotating magnetic-field system.

Figure 8.1 shows a permanent magnet rotating within a simple loop system, as well as the resulting induced EMF. In practice, the permanent magnet in most cases will be replaced by a DC-fed electromagnet, which will result in a higher value of magnetic flux, and very many conducting loops will be let into the surface of a stator of magnetic material; hence the induced EMF will be high. The construction of the single-phase alternator, apart from the obvious difference in the main winding, is similar to that of the three-phase machine described in Section 8.5.

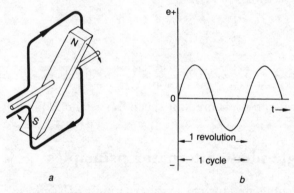

**Figure 8.1    Arrangement and induced EMF of simple loop alternator with two-pole rotor**

There is no reason why a two-pole magnet system must be used, Figure 8.2 showing the arrangement of, and the EMF induced in, a four-pole system. The two-pole system of Figure 8.1 must make one complete revolution to enable north and south poles to pass each conductor and thus to induce one complete cycle of EMF. The frequency in cycle/s (or Hz) of the EMF induced for the two-pole system is thus equal to its rotational speed in r/s. The four-pole system of Figure 8.2 induces one cycle in half a revolution, or two cycles per revolution, whilst a six-pole machine would induce three cycles per revolution.

Generally,   $f = pn$ or $f = \dfrac{pN}{60}$

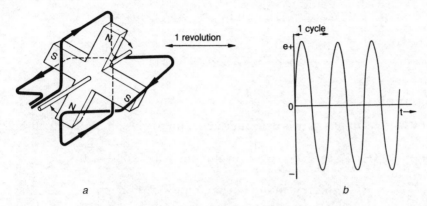

**Figure 8.2   Arrangement and induced EMF of simple loop alternator with four-pole rotor**

where   $f$ = frequency of the EMF induced (Hz)
$p$ = number of pairs of poles on the magnet system
$n$ = rotational speed (r/s)
$N$ = rotational speed (r/min).

Increasing the speed of an alternator will thus increase not only its induced EMF, but also its frequency.

## Example 8.3

Calculate the output frequency of an eight-pole alternator running at 750 r/min.

$$f = \frac{pN}{60} = \frac{8}{2} \times \frac{750}{60} \text{ hertz} = 50 \text{ Hz}$$

## Example 8.4

An alternator running at 20 r/s has an output of 140 V at 60 Hz. How many poles has it? What will be the output voltage and frequency at 32 r/s?

$$f = pn \quad \text{so} \quad p = \frac{f}{n} = \frac{60}{20} = 3 \text{ pairs of poles}$$

The alternator has six poles.

$V \propto n$ if magnetisation is constant, so

$$\frac{V_1}{V_2} = \frac{n_1}{n_2}$$

$$V_2 = V_1 \times \frac{n_2}{n_1} = 140 \times \frac{32}{20} \text{ volts} = 224 \text{ V}$$

$$f = pn = \frac{6}{2} \times 32 \text{ hertz} = 96 \text{ Hz}$$

## Example 8.5

At what speed must a two-pole alternator run to give an output at 50 Hz?

$$f = pn \quad \text{so} \quad n = \frac{f}{p} = \frac{50}{1} \text{ revolutions per second} = 50 \text{ r/s}$$

$$= 50 \times 60 \text{ revolutions per minute} = 3000 \text{ r/min}.$$

## 8.5   Three-phase alternator principles

A three-phase alternator has a stator with three sets of windings arranged so that there is a mutual phase displacement of 120°. These windings are connected in star or in delta to provide a three-phase output. The three-phase alternator is used exclusively in power stations in the UK, with the output from a single machine being as high as 1200 MW.

Small alternators for remote supplies and standby purposes are often driven by diesel engines, but the large alternators used for generating public supplies require more power than can conveniently be obtained by these means. For public supplies, one of three methods is used.

(i) Steam turbines, which are fed with steam from boilers heated by coal, gas or nuclear energy. Such turbines are run at a standard speed of 50 r/s (3000 r/min) in the UK to give a 50 Hz supply from a two-pole alternator, and provide the drives for the great majority of alternators.

(ii) Gas turbines, which are modified aircraft jet engines, are used to provide a high-velocity gas stream, which drives a turbine at 50 r/s. These units are comparatively small, but are useful for meeting peak-load demands because they can be started and connected to the grid system very quickly.

(iii) Water turbines are used in hydroelectric schemes. These turbines are slow-speed devices, and are not widely used in the UK because of the scarcity of wet, mountainous areas. They are common in countries with suitable terrain.

Alternators for use with steam and gas turbines rotate at 50 r/s, and their rotors have a diameter of up to about 2 m. The necessary magnetic flux per pole is obtained by making the two-pole rotor up to about 10 m long. Windings for the two-pole field system are let into the surface of the rotor, this surface being smooth to reduce air friction. A simplified drawing of a typical two-pole cylindrical drum rotor is given in Figure 8.3.

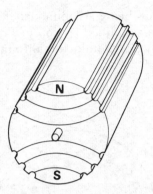

**Figure 8.3** **Simplified diagram of a two-pole drum rotor for an alternator, showing arrangement of field winding**

Since a water-turbine-driven alternator rotates at low speed, it must have a large number of poles to produce the required frequency. The reduced centrifugal forces at low speeds allow large-diameter rotors to be used, and these are usually of the salient-pole type, shown simply in Figure 8.4.

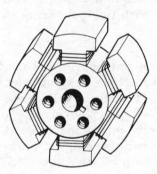

**Figure 8.4** **Simplified diagram of six-pole salient-pole rotor for an alternator**

With both types of rotor, the field winding is excited by direct current, usually supplied by a DC generator called an exciter, which is sometimes mounted in the same shaft as the alternator. Current is fed to the rotor windings through sliprings. The 'brushless alternator' allows the generator commutator, the alternator sliprings and two sets of brushes to be omitted, using rectifiers in the connections from the generator to the alternator within the hollow shaft common to both machines.

The stators of both types of alternator are constructed on the same principles, the hollow cylinder for the water-turbine-driven type having a much larger diameter, but shorter length, than the high-speed steam- and gas-turbine-driven machines. The cylinder is formed of silicon-steel laminations, the windings being accommodated in slots cut into its inner surface as shown in the simplified section of Figure 8.5. Large modern alternators often have hollow conductors through which a coolant, such as hydrogen, is pumped.

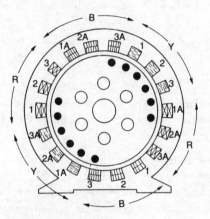

**Figure 8.5   Simplified diagram showing section of three-phase alternator with two-pole drum rotor**

As mentioned earlier in this section, the machine windings can be connected in star or in delta. The star connection has the advantage that the output line voltage is $\sqrt{3}$ times the voltage of each winding, although the individual windings each carry line current. For the delta connection, the line current is $\sqrt{3}$ times the current in each winding, but the line voltage is equal to the voltage of each winding. If a neutral is required, a star-connected machine must be used, although a delta-connected alternator feeding through a delta/star transformer (see Section 7.4) will also give a position for the connection of the fourth wire (see Figure 8.6).

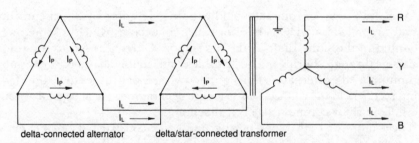

delta-connected alternator      delta/star-connected transformer

**Figure 8.6   Circuit of delta-connected alternator feeding a delta–star output transformer**

## Example 8.6

A water turbine is to drive an alternator, and runs most efficiently at 300 r/min. How many poles must the alternator have if it is to generate at 50 Hz?

$$f = \frac{pN}{60} \quad \text{so} \quad p = \frac{60f}{N} = \frac{60 \times 50}{300} = 10 \text{ pairs of poles.}$$

The alternator must have 20 poles.

## Example 8.7

Each of the three windings of a three-phase alternator is rated at 11 kV and can carry a full-load current of 6000 A. Calculate the line voltage and the maximum line current if the machine is connected (*a*) in star, and (*b*) in delta.

(*a*) $V_L = (\sqrt{3})V_P = \sqrt{3} \times 11$ kilovolts = 19·1 kV

For the star connection, line and phase currents are the same, so the maximum line current is 6000 A.

(*b*) For the delta connection, line and phase voltages are the same, so line voltage is 11 kV.

$I_L = (\sqrt{3})I_P = \sqrt{3} \times 6000$ amperes = 10 400 A

## Example 8.8

The machine of Example 8.7 has four poles and generates at 50 Hz. Calculate its speed. If the machine is to generate at 60 Hz, calculate the running speed, and the output line voltage for both star and delta connections if the excitation is unchanged.

At 50 Hz,

$$f = pn \quad \text{so} \quad n = \frac{f}{p} = \frac{50}{2} \text{ revolutions per second} = 25 \text{ r/s}$$

At 60 Hz,

$$f = pn \quad \text{so} \quad n = \frac{f}{p} = \frac{60}{2} \text{ revolutions per second} = 30 \text{ r/s}$$

$$V \propto N \quad \text{so} \quad \frac{V_1}{V_2} = \frac{N_1}{N_2} \quad \text{and} \quad V_2 = V_1 \times \frac{N_2}{N_1}$$

For the star connection, where the line voltage was 19·1 kV,

$$V_2 = 19 \cdot 1 \times \frac{30}{25} \text{ kilovolts} = 22 \cdot 9 \text{ kV}$$

For the delta connection, where the line voltage was 11 kV,

$$V_2 = 11 \times \frac{30}{25} \text{ kilovolts} = 13 \cdot 2 \text{ kV}$$

Note that the output voltage of the machine at the higher speed could be returned to its original value by reducing the exciting current so that the magnetic flux/pole becomes 25/30 of its original value. Owing to the nonlinearity of the current/magnetic flux relationship (see Section 3.2), the proportional reduction in current to achieve this change may be greater than the proportional change in flux.

## 8.6   Summary of formulas and proportionalities for Chapter 8

See text for definitions of symbols.

$$E \propto \Phi n \qquad \Phi \propto \frac{E}{n} \qquad n \propto \frac{E}{\Phi}$$

$$\frac{E_1}{E_2} = \frac{\Phi_1}{\Phi_2} \times \frac{n_1}{n_2} \qquad E_2 = E_1 \times \frac{\Phi_2}{\Phi_1} \times \frac{n_2}{n_1}$$

$$\Phi_2 = \Phi_1 \times \frac{n_1}{n_2} \times \frac{E_2}{E_1} \qquad n_2 = n_1 \times \frac{\Phi_1}{\Phi_2} \times \frac{E_2}{E_1}$$

$$f = pn \qquad p = \frac{f}{n} \qquad n = \frac{f}{p}$$

$$f = \frac{pN}{60} \qquad p = \frac{60f}{N} \qquad N = \frac{60f}{p}$$

## 8.7   Exercises

1  The EMF induced by a magnetic flux of $0 \cdot 64$ Wb in a coil rotating within it at a speed of 20 r/s is 160 V. What EMF will be induced if the coil rotates at 25 r/s in a magnetic flux of $0 \cdot 48$ Wb?

2  A DC generator has an induced EMF of 240 V when the speed is 1200 r/min and the flux/pole is 80 mWb. What flux/pole will give 300 V at 1800 r/min?

3  An alternator running at 25 r/s has an output of 400 V at 50 Hz when its flux/pole is $0 \cdot 12$ Wb. At what speed will it give 600 V if the flux/pole is reduced to $0 \cdot 09$ Wb? Calculate the new output frequency.

4  At 15 r/s with a flux/pole of 200 mWb a DC generator has an induced EMF of 300 V. Calculate the induced EMF at 20 r/s and 150 mWb/pole.

5  When its flux/pole is 12 mWb, a small DC generator has an induced EMF of 15 V at 3000 r/min. Calculate the speed to give an induced EMF of 16 V with a flux/pole of 10 mWb.

6  A DC tachogenerator has a permanent-magnet field system with a fixed flux/pole of 7 mWb. If the induced EMF is 160 V at 24 r/s, calculate the value at 30 r/s.

7 Calculate the frequency generated by a 12-pole alternator when running at: (*a*) 5 r/s, (*b*) 8 r/s, (*c*) 12·5 r/s.

8 If the machine of Exercise 7 provides an EMF of 120 V at 5 r/s, calculate the EMF at the other two speeds if the excitation is constant.

9 Calculate the number of poles of the alternators which, when running at 1200 r/min, will generate at the following frequencies: (*a*) 40 Hz, (*b*) 80 Hz, (*c*) 20 Hz.

10 A six-pole single-phase alternator generates an EMF of 220 V at 50 Hz. Calculate its speed. Calculate the terminal voltage and the frequency when the machine runs at 20 r/s with unchanged excitation.

11 A 12-pole alternator runs at 4000 r/min. What is the frequency of its output?

12 A special-purpose alternator is required to generate 100 Hz at a speed of 6000 r/min. How many poles must it have?

13 At what speed must an eight-pole alternator run to give an output of 60 Hz?

14 A three-phase alternator is required to have an output line current of 32 000 A at 16 kV. Calculate the voltage and current ratings of each winding if it is connected (*a*) in delta, and (*b*) in star.

15 Each of the three windings of a three-phase alternator is rated at 415 V, 300 A. Calculate the rated line voltage and line current if the machine is (*a*) star-connected, and (*b*) delta-connected.

16 A star-connected four-pole alternator runs at 25 r/s, when its output line voltage is 2·5 kV with an exciting current of 120 A. Calculate the frequency at which it generates.
   The machine is required to operate at 36 Hz for experimental purposes. Calculate the new speed, and the new exciting current for an output at 2·5 kV if a linear current/magnetic flux relationship is assumed.

17 An American three-phase alternator generates 200 V at 60 Hz when running at 1200 r/min. It is required to generate at 50 Hz. Calculate the speed and the generated voltage if the excitation remains the same.

18 A four-pole AC generator produces 400 V at 50 Hz. If the speed is altered to 900 r/min (15 r/s) and the flux is increased by 50%, what voltage and at what frequency will the machine generate?

19 (*a*) What is the purpose of the commutator on a DC generator? Draw sketches of a single wire loop rotating between the poles of a permanent magnet, its ends being connected to a simple two-part commutator. By showing the loop in at least THREE different positions, explain how the commutator achieves its purpose.
   (*b*) When a DC generator is run at 1200 r/min with a constant field current of 2 A, the no-load terminal voltage is 200 V. What will be the voltage if the speed increases to 1600 r/min? Ignore armature voltage drop.

20 A three-phase AC generator has eight poles. Each phase has 100 conductors in series and each conductor generates 2·54 V at 50 Hz.
   (*a*) What is the generator speed in r/min?
   (*b*) What is the phase voltage?

(c) The generator is slowed down to produce $33 \cdot 3$ Hz. What is the new speed?

(d) What is the new phase voltage?

21 A star-connected generator supplies a balanced delta-connected load. The generator phase voltage is 240 V and the current in the line from the generator to the load is $86 \cdot 6$ A. Calculate:

(a) the line voltage,

(b) the current in each phase of the load, and

(c) the impedance of each phase of the load.

## 8.8   Multiple-choice exercises

8M1 One loop of a generator has an induced EMF of $4 \cdot 5$ V when rotated at 380 r/m in a magnetic field with a total flux of 65 mWb. The induced EMF when the speed increases to 500 r/min and the flux falls to 50 mWb will be:

    (a) $2 \cdot 63$ V        (b) $4 \cdot 55$ V        (c) $7 \cdot 70$ V        (d) $5 \cdot 92$ V.

8M2 A DC generator has an open-circuit output voltage of 210 V when its speed is 24 r/s and its flux/pole is 80 mWb. The flux/pole if it provides an output of 230 V at a speed of 1250 r/min will be:

    (a) 1250 mWb        (b) $63 \cdot 4$        (c) $87 \cdot 6$ mWb        (d) 76 mWb.

8M3 The one statement below which is NOT an advantage of rotating the field system of a generator rather than the conductor system is:

(a) the conductor system is usually heavier than the field system

(b) the conductor system is more easily insulated for high voltages if stationary

(c) no EMF will be induced if the field system rotates

(d) the field system is more easily braced to withstand centrifugal forces than the main conductor system.

8M4 The formula which relates the frequency of the induced EMF $f$ to the rotational speed $n$ and the number of pairs of poles on the field system is:

    (a) $f = \dfrac{p}{n}$        (b) $f = pn$        (c) $f = \dfrac{n}{p}$        (d) $n = fp$

8M5 A four-pole alternator running at 30 r/s induces an EMF with a frequency of:

    (a) 60 Hz        (b) 50 Hz        (c) 120 Hz        (d) $7 \cdot 5$ Hz.

8M6 The salient-pole rotor of an alternator is so called because:

(a) it is used only for small power stations in the centre of areas covered by other supply companies

(b) it is particularly suited to areas where battles have been fought

(c) the surface of the rotor is smooth and causes little windage loss

(d) its field poles project from the shaft.

8M7 Most power-station generators have:

(a) very low-voltage outputs

(b) drum rotors

(c) star-connected rotor windings

(d) salient-pole rotors.

8M8 A system has a windmill which has a constant speed of 50 r/m which drives an alternator through a gearbox which increases the rotational speed by a factor of 20. If the output is at 50 Hz, the number of poles on the alternator will be:

(*a*) six          (*b*) two          (*c*) three          (*d*) 12.

*Chapter 9*
# Direct-current machines

## 9.1   Introduction

Many people seem to think that DC machines are of historic interest only, this opinion probably being based on the fact that DC supply systems are now rare. It is true that, for many applications, the three-phase cage induction motor (Chapter 10) is just as satisfactory as a DC motor, is cheaper to buy and requires less maintenance. However, the increasing complexity of industry is demanding greater flexibility from electrical machines in terms of special characteristics and speed control. It is in this field that DC machines, fed from the AC supply through rectifiers, have made their mark. The use of DC machines is tending to increase, and the competent electrical craftsman must have a knowledge of their operating principles.

All motors and generators are energy-conversion machines. The generator is fed with mechanical energy and gives out electrical energy. The motor takes in electrical energy and provides a mechanical output. In both cases, energy is lost in the process, so that the machines are not 100% efficient; their efficiency at full load is, however, generally better than that of mechanical machines. The energy lost in conversion appears as heat in the machine, and as a result the machine runs at a higher temperature than its surroundings.

Since machines are energy convertors, the energy taken in varies with the energy given out. The current to a motor, for instance increases as the mechanical load applied to it becomes greater. The DC machine is simply an energy convertor, and will run either as a generator or as a motor.

## 9.2   Construction

There is no basic difference in the construction of DC motors and generators. Rotating machines of this type must have three basic features:

(i) a magnetic-field system

(ii) a system of conductors, and

(iii) provision for relative movement between the field and the conductors.

In most DC machines, the magnetic field is set up by the stationary part of the machine, which is called the **field system**. The commutator and the conductors which form the armature winding are mounted on the rotating part of the machine, known as the **armature**.

The body of the machine is a hollow cylinder of cast steel (or sometimes cast iron for small machines) called the yoke, which forms the basis for construction of the machine, and is part of the magnetic circuit. Fixed pole pieces, made of solid steel or iron, or sometimes of laminations, are fixed inside the yoke. They are bolted to it and, since they project inwards, they are called 'salient' poles. Each pole has one or more field windings placed over it to produce a magnetic field. There is always an even number of poles, the polarity being alternately north and south. The field windings are held in position by laminated pole shoes, which are shaped to follow the curvature of the armature.

The magnetic circuit is completed, as shown in Figure 9.1 by the armature, which is constructed in the form of a drum of laminated silicon steel keyed to the steel shaft. The armature, as it moves past alternate north and south magnetic poles, has an alternating EMF induced in it, so it is made of silicon steel and laminated to reduce iron losses (see Section 3.7).

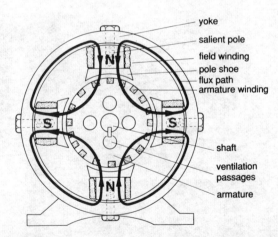

**Figure 9.1   Cross-section of DC machine showing magnetic-flux paths**

Before assembly, each lamination has a series of teeth cut round its edge, so that when a number of laminations are compressed together to form a drum, slots are present in its surface, into which the insulated coils of copper wire or bar can be placed. The windings are wedged into the slots so that they will not move when subjected to severe mechanical stress caused by electromagnetic effects and centrifugal forces.

The armature winding of a DC machine is often very complex. There are two basic types:

(i) wave windings; these tend to be high-voltage low-current windings, and always have two conducting paths in parallel.

(ii) lap windings; these tend to be low-voltage high-current windings, and have as many conducting paths in parallel as there are poles. For instance, a two-pole machine will have two parallel paths, a six-pole machine will have six parallel paths and so on.

The ends of the winding are brought out of the slots at one end, where they are connected to lugs on the commutator by welding, brazing or soldering. A practical commutator will be made up of many hard-drawn copper segments, insulated from each other with mica, the whole being keyed to the shaft. The mica is 'undercut' slightly below the level of the copper segments. This is because the copper wears more quickly than the mica, so that if both were initially level the latter would soon be higher than the copper and would prevent proper brush contact.

The completed armature is mounted on bearings, often supported by end plates mounted on the yoke. The carbon brushes are mounted so that the pressure exerted by them on the commutator can be adjusted, and so that the pressure will remain reasonably constant as the brushes wear. General construction is shown clearly in Figure 9.2, which depicts a DC motor with a split yoke, opened for maintenance.

**Figure 9.2   Steel-mill DC motor with split yoke opened for maintenance**

When a DC machine is running, the following losses in it will give rise to heat:

(a) copper loss: this is the heating effect in the copper conductors due to current flowing through them. These losses occur in both the armature and the field windings. A small loss also occurs owing to the voltage drop across each brush.

(b) iron losses: these losses occur mainly in the armature, and are due to hysteresis and eddy currents.

(c) mechanical losses; these are due to the friction at the bearings and at the commutator, and to windage, which is friction between the rotating parts and the surrounding air.

If the heat due to losses is not removed, the machine temperature will rise to a level where the electrical insulation is likely to be impaired. A fan is often mounted on the shaft to blow cooling air through the machine or to circulate air within the machine enclosure. The temperature at which a machine can be allowed to operate depends on the type of insulation used for the conductors, and can be very much higher for modern silicon and glass-fibre insulations than for cotton and varnishes.

## 9.3   Induced EMF for generators and motors

In Section 8.3 we saw that the EMF induced in a generator depends on the magnetic field flux, and on the rotational speed. Another way of reaching the same conclusion is to go back to the basic equation for the EMF induced in a conductor when moving through a magnetic field (see Section 9.2 of 'Electrical Craft Principles' Volume 1). This is

$$e = Blv$$

where   $e$ = instantaneous value of induced EMF (V)
$B$ = magnetic flux density (T)
$l$ = length of conductor in field (m)
$v$ = velocity of conductor through field (m/s)

The total magnetic flux/pole $\Phi$ for a DC machine depends on the flux density $B$ and on the pole-face area $A$, which in turn depends on the size of the machine ($\Phi = BA$). The rotational speed of the machine determines the velocity of the conductor, in conjunction with machine dimensions. For example, a conductor on the armature of a machine with a large diameter will move faster for a given rotational speed than one on a small machine. Thus,

$$E \propto \Phi n$$

(see Section 8.3) is derived from the basic equation

$$e = Blv$$

It is important to understand that $E$ depends on $\Phi$ and on $n$ for a given machine. The induced EMF will also depend on the number of poles, the number of armature conductors, and the number of parallel paths in the armature.

The equation is

$$E = \frac{2p\Phi nZ}{a}$$

where   $E$ = induced EMF (V)
$p$ = number of pairs of magnetic poles
$\Phi$ = flux/pole (Wb)
$n$ = speed of rotation (r/s)
$Z$ = total number of armature conductors
$a$ = number of parallel paths in the armature

Note that, for a wave-wound machine, $a = 2$
and for a lap-wound machine,        $a = 2p$

For a given machine, the number of field poles ($2p$) and the armature winding ($Z$ and $a$) will be fixed, so once again we come to the result that $E \propto \Phi n$.

## Example 9.1

A four-pole lap-wound DC machine has a flux/pole of 50 mWb and a total of 600 armature conductors. Calculate the EMF induced at 20 r/s.

For the lap-wound machine,

$$a = 2p = 2 \times \frac{4}{2} = 4 \text{ parallel paths}$$

$$E = \frac{2p\Phi nZ}{a} = 2 \times \frac{4}{2} \times \frac{50 \times 10^{-3} \times 20 \times 600}{4} \text{ volts} = 600 \text{ V}$$

## Example 9.2

An eight-pole wave-wound DC machine has a total of 240 armature conductors. At what speed must it run to induce an EMF of 200 V if the flux/pole is 15 mWb?

$$E = \frac{2p\Phi nZ}{a} \quad \text{so} \quad n = \frac{Ea}{2p\Phi Z}$$

For the wave-wound machine, $a = 2$.

$$n = \frac{200 \times 2}{8 \times 15 \times 10^{-3} \times 240} \text{ revolutions per second} = 13 \cdot 9 \text{ r/s}$$

It is important to appreciate that this EMF will be induced in any operating DC machine, no matter whether it is running as a generator or as a motor.

For the generator the induced EMF will provide the terminal voltage and will drive the output current. For the motor, the same EMF will be induced, but will obey Lenz's law (Section 2.3) and will oppose the supply voltage which drives current into the machine. The EMF is known in this case as the 'back EMF'.

## Example 9.3

A DC machine has an induced EMF of 360 V under certain conditions of speed and field. Calculate the induced EMF if the speed increases by 20%, whilst the field flux decreases by 10%.

$$E \propto \Phi n \quad \text{so} \quad \frac{E_1}{E_2} = \frac{\Phi_1}{\Phi_2} \times \frac{n_1}{n_2}$$

and

$$E_2 = E_1 \times \frac{\Phi_2}{\Phi_1} \times \frac{n_2}{n_1}$$

If the speed increases by 20%, $n_2/n_1 = 120/100$
If the field flux decreases by 10%, $\Phi_2/\Phi_1 = 90/100$

$$E_2 = 360 \times \frac{90}{100} \times \frac{120}{100} \text{ volts} = 389 \text{ V}$$

## Example 9.4

The EMF of a DC machine is 300 V at 1200 r/min with a field current of 1·5 A, but falls to 240 V at 900 r/min. Assuming that magnetic flux is proportional to field current, calculate the new value of the latter.

If flux is proportional to field current, $\Phi \propto I_F$

$$E \propto I_F N \quad \text{so} \quad \frac{E_1}{E_2} = \frac{I_{F_1}}{I_{F_2}} \times \frac{N_1}{N_2} \quad \text{and} \quad I_{F_2} = I_{F_1} \times \frac{N_1}{N_2} \times \frac{E_2}{E_1}$$

$$I_{F_2} = 1 \cdot 5 \times \frac{1200}{900} \times \frac{240}{300} \text{ amperes} = 1 \cdot 6 \text{ A}$$

## 9.4  Torque for motors and generators

Torque was explained in Section 3.1 of 'Electrical Craft Principles' Volume 1, as the turning moment of a force. Torque is the force resulting in rotary movement multiplied by the radius at which it acts. In SI units, force is measured in newtons (N) and radius of action in metres (m), so the unit of torque is the newton metre (Nm).

For an electric motor of a given size, the radius of action is the distance from the shaft centre to the armature conductors; this distance is fixed, so torque is proportional to the force on the conductors. It was shown in Section 11.2 of 'Electrical Craft Principles' Volume 1, that the force on a conductor carrying a current and lying in a magnetic field is given by the expression

$$F = BIl$$

$B$ is the magnetic flux density to which the conductor is subjected, and in a DC machine will depend on the flux/pole $\Phi$. $I$ is the current carried by the conductors, which can be assumed to be the armature current $I_A$ for a DC machine. $l$ is the conductor length, which is fixed for a given machine by the length of its armature.

Thus $F \propto \Phi I_A$ and since $T \propto F$

$$T \propto \Phi I_A$$

where   $T$ = torque exerted on the armature of a DC machine
      $\Phi$ = magnetic flux/pole of the machine
      $I_A$ = armature current in the machine

## Example 9.5

A certain DC motor provides a torque of 600 Nm when the flux/pole is 80 mWb and the armature current is 60 A. Calculate the torque if the flux/pole falls to 50 mWb whilst the armature current increases to 100 A.

$$T \propto \Phi I_A \quad \text{so} \quad \frac{T_1}{T_2} = \frac{\Phi_1}{\Phi_2} \times \frac{I_{A_1}}{I_{A_2}}$$

where subscripts 1 and 2 refer to the first and second values, respectively, of torque, flux and armature current.

$$T_2 = T_1 \times \frac{\Phi_2}{\Phi_1} \times \frac{I_{A_2}}{I_{A_1}} = 600 \times \frac{50}{80} \times \frac{100}{60} \text{ newton metres} = 625 \text{ Nm}$$

## Example 9.6

The torque provided by a DC machine is 30 Nm under certain operating conditions. Calculate the torque if the armature current reduces by 10% and the flux/pole increases by 20%.

$$T \propto \Phi I_A \quad \text{so} \quad \frac{T_1}{T_2} = \frac{\Phi_1}{\Phi_2} \times \frac{I_{A_1}}{I_{A_2}}$$

$$T_2 = T_1 \times \frac{\Phi_2}{\Phi_1} \times \frac{I_{A_2}}{I_{A_1}} = 30 \times \frac{120}{100} \times \frac{90}{100} \text{ newton metres} = 32 \cdot 4 \text{ Nm}$$

It must be appreciated that torque can only result in movement, and hence provide an ouput power, if it is large enough to overcome the torque opposing it. For example, a motor with a jammed armature will not provide output power because the shaft is not free to rotate. In a case of this sort there will be no back EMF to oppose the supply voltage, so that a large current will be taken by the armature, burning it out if the protective devices do not function (see Section 9.9). In most cases, of course, torque will result in movement and power will be provided from the shaft.

In Section 9.3 it was shown that EMF is induced in a motor as well as in a generator. Similarly, torque is needed to drive a generator as well as being provided by a motor. The same proportionality ($T \propto \Phi I_A$) applies, the driving torque increasing if field flux and hence induced EMF increases, as well as when the current increases. This is logical, because the generator is simply a convertor of energy, more input power provided by extra torque being necessary to provide a greater output power.

## Example 9.7

A DC generator requires a torque of 150 Nm to drive it at a constant speed with a flux/pole of 10 mWb when delivering 20 A from its armature. What torque will be required to maintain the speed if the flux/pole falls to 8 mWb, but the armature current increases to 30 A?

$$T \propto \Phi I_A \quad \text{so} \quad \frac{T_1}{T_2} = \frac{\Phi_1}{\Phi_2} \times \frac{I_{A_1}}{I_{A_2}}$$

$$T_2 = T_1 \times \frac{\Phi_2}{\Phi_1} \times \frac{I_{A_2}}{I_{A_1}} = 150 \times \frac{8}{10} \times \frac{30}{20} \text{ newton metres} = 180 \text{ Nm}$$

## Example 9.8

A DC generator has an armature current of 90 A when its flux/pole is 110 mWb and it is driven at a constant speed with a torque of 15 Nm. What will be its armature current if the flux/pole falls to 80 mWb and the driving torque to 12 Nm at the same speed?

$$T \propto \Phi I_A \quad \text{so} \quad \frac{T_1}{T_2} = \frac{\Phi_1}{\Phi_2} \times \frac{I_{A_1}}{I_{A_2}}$$

$$I_{A_2} = I_{A_1} \times \frac{\Phi_1}{\Phi_2} \times \frac{T_2}{T_1} = 90 \times \frac{110}{80} \times \frac{12}{15} \text{ amperes} = 99 \text{ A}$$

The answer to this exercise suggests that a reduced input is giving an increased output. In fact, the reduction in flux/pole at constant speed will result in a large fall in induced EMF ($E \propto \Phi n$), so that output power will be reduced even though the armature current increases (power = current × voltage).

## 9.5 Terminal voltage for generators

The armature of a DC machine consists of conductors connected in series–parallel, there always being at least two parallel paths. This conductor system will have resistance, known as the armature resistance; values will vary from a few ohms for a small machine, down to a fraction of an ohm for a large machine. Armature resistance will play an important part in the operation of the machine because the current passing through it gives a voltage drop in the armature. This voltage drop is the difference between the induced EMF and the terminal voltage of a machine, but motors and generators do not follow the same rule.

Figure 9.3a represents the armature of a DC machine operating as a generator. The armature carries a current of $I_A$ amperes and has a resistance of $R_A$ ohms. An EMF of $E$ is induced in it, and this EMF drives the current against the circuit resistance which consists of the external load and the armature resistance. Some of the EMF is dropped inside the armature, so that the terminal voltage is less than the EMF by the amount of the armature voltage drop. From Ohm's law, $V = IR$, so the armature voltage drop is $I_A R_A$ volts.

Thus, for a DC generator,

$$V = E - I_A R_A$$

where 
$V$ = terminal voltage (V)
$E$ = induced EMF (V)
$I_A$ = armature current (A)
$R_A$ = armature resistance ($\Omega$)

Note that, for a generator, the induced EMF is greater than the terminal voltage by the amount of the armature voltage drop. If the armature current increases, making the voltage drop larger, terminal voltage will fall if induced EMF remains constant.

## Example 9.9

A DC shunt generator has an induced EMF of 220 V and an armature resistance of $0 \cdot 2$ $\Omega$. Calculate its terminal voltage when delivering an armature current of (a) 10 A, (b) 40 A, and (c) 100 A.

$$V = E - I_A R_A$$

(a) $V = 220 - (10 \times 0 \cdot 2)$volts = $220 - 2$ volts = 218 V
(b) $V = 220 - (40 \times 0 \cdot 2)$volts = $220 - 8$ volts = 212 V
(c) $V = 220 - (100 \times 0 \cdot 2)$volts = $220 - 20$ volts = 200 V

## Example 9.10

The output of a DC generator is 100 kW at 250 V. If the armature resistance is $0 \cdot 04$ $\Omega$, calculate the induced EMF.

$$P = VI \quad \text{so} \quad I = \frac{P}{V} = \frac{100 \times 10^3}{250} \text{ amperes} = 400 \text{ A}$$

In the absence of information concerning field connections (see Section 9.8) this must be assumed to be the armature current.

$$V = E - I_A R_A \quad \text{so} \quad E = V + I_A R_A$$

$$E = 250 + (400 \times 0 \cdot 04)\text{volts} = 250 + 16 \text{ volts} = 266 \text{ V}$$

The carbon brushes of a DC machine cause a voltage drop which is almost constant (between 1 and 2 V) regardless of load. If this voltage drop is to be taken into account, the terminal-voltage formula becomes

$$V = E - I_A R_A - V_B$$

where $V_B$ = total brush voltage drop (V).

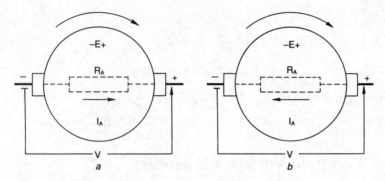

**Figure 9.3    EMF/terminal-voltage relationships**
        *a* Induced EMF/terminal voltage relationship for DC generator
        *b* Induced EMF/terminal voltage relationship for DC motor

## Example 9.11

A DC shunt generator has an induced EMF of 300 V when the armature current is 80 A and the terminal voltage is 274 V. Assuming a brush voltage drop of 2 V, calculate

    (*a*) the resistance of the armature,
    (*b*) the terminal voltage for an armature current of 60 A, and
    (*c*) the generated EMF if the field current is increased to give a terminal voltage of 300 V when delivering 100 A.

(*a*) $\quad V = E - I_A R_A - V_B \quad \text{so} \quad R_A = \dfrac{E - V - V_B}{I_A}$

$$R_A = \frac{300 - 274 - 2}{80} \text{ ohms} = \frac{24}{80} \text{ ohms} = 0 \cdot 3 \ \Omega$$

(*b*) $\quad V = E - I_A R_A - V_B = 300 - (60 \times 0 \cdot 3) - 2 \text{ volts}$

$$= 300 - 18 - 2 \text{ volts} = 280 \text{ V}$$

(*c*) $\quad E = V + I_A R_A + V_B = 300 + (100 \times 0 \cdot 3) + 2 \text{ volts}$

$$= 300 + 30 + 2 \text{ volts} = 332 \text{ V}$$

## Example 9.12

A DC generator has a terminal voltage of 200 V when driven at 30 r/s with an armature current of 40 A. The armature resistance is $0 \cdot 25$ Ω. If the speed is reduced to 25 r/s and the armature current falls to 20 A, with the field flux remaining constant, calculate the new terminal voltage. Let the first value of induced EMF be $E_1$. Brush voltage drop must be ignored since its value is not given.

$$E_1 = V_1 + I_{A_1} R_A = 200 + (40 \times 0 \cdot 25) \text{volts} = 210 \text{ V}$$

If Φ is constant $\dfrac{E_1}{E_2} = \dfrac{n_1}{n_2}$

Therefore

$$E_2 = E_1 \times \frac{n_2}{n_1} = 210 \times \frac{25}{30} \text{ volts} = 175 \text{ V}$$

where $E_2$ is the second value of induced EMF

The new terminal voltage, $V_2 = E_2 - I_{A_2} R_A$

$$V_2 = 175 - (20 \times 0 \cdot 25) \text{volts} = 175 - 5 \text{ volts} = 170 \text{ V}$$

## 9.6   Terminal voltage for motors

The EMF induced in the armature conductors of a motor will obey Lenz's law (see Section 2.3). In opposing the effect responsible for its induction, it will oppose the voltage which is driving current into the armature and is known as 'back EMF'. The effective voltage will be the difference between the terminal voltage $V$ and the induced EMF $E$. Following Ohm's law,

$$I_A = \frac{V - E}{R_A}$$

Rearranging the expression,

$$I_A R_A = V - E$$

and, for a DC motor,

$$V = E + I_A R_A$$

where   $V$ = terminal voltage (V)
$E$ = induced EMF (V)
$I_A$ = armature current (A)
$R_A$ = armature resistance (Ω)

This expression should be compared with that for the DC generator, which is $V = E - I_A R_A$. The difference in sign is most significant, and can be more easily understood by a comparison of Figure 9.3*a* with Figure 9.3*b*. If the generator and the motor both revolve in the same direction, the polarity at the brushes will be the same. However, while the generator drives current out of the positive brush, the supply drives current in at the positive brush of the motor. Some of the EMF induced in a generator is dropped before it becomes the terminal voltage; the terminal voltage of a motor must be larger than the induced EMF because current must be driven against this EMF. For both machines the difference between the terminal voltage and the induced EMF is the armature voltage drop.

If the speed of a motor falls, the induced EMF will fall (assuming constant magnetic flux) so that the terminal voltage will drive extra current into the armature. This extra current will provide extra torque. If the machine speed increases, the induced EMF will rise, the effective voltage applied to the armature $(V - E)$ will fall, and the current will reduce, providing less torque. In practice, the DC machine will reach a steady speed where the difference between voltage and EMF is just enough to drive the current needed to give the correct torque for that speed.

## Example 9.13

A DC motor has an armature resistance of $0 \cdot 15$ $\Omega$ and is fed from a 200 V supply. Calculate the back EMF when the armature current is (*a*) 20 A, (*b*) 50 A, and (*c*) 100 A.

$$V = E + I_A R_A \quad \text{so} \quad E = V - I_A R_A$$

(*a*) $E = 200 - (20 \times 0 \cdot 15)\text{volts} = 200 - 3 \text{ volts} = 197 \text{ V}$
(*b*) $E = 200 - (50 \times 0 \cdot 15)\text{volts} = 200 - 7 \cdot 5 \text{ volts} = 192 \cdot 5 \text{ V}$
(*c*) $E = 200 - (100 \times 0 \cdot 15)\text{volts} = 200 - 15 \text{ volts} = 185 \text{ V}$

## Example 9.14

The armature of a DC motor takes 30 kW from a 250 V supply. If the armature resistance is $0 \cdot 1$ $\Omega$, calculate the back EMF of the machine.

$$P = VI \quad \text{so} \quad I = \frac{P}{V} = \frac{30 \times 10^3}{250} \text{ amperes} = 120 \text{ A}$$

$$V = E + I_A R_A \quad \text{so} \quad E = V - I_A R_A$$

$$E = 250 - (120 \times 0 \cdot 1) \text{ volts} = 250 - 12 \text{ volts} = 238 \text{ V}$$

If the brush voltage drop mentioned in Section 9.5 must be taken into account, the formula becomes

$$V = E + I_A R_A + V_B$$

where $V_B$ = brush voltage drop (V)

## Example 9.15

A DC motor with an armature resistance of $0 \cdot 25$ $\Omega$ and a constant brush voltage drop of $1 \cdot 5$ V has a back EMF of 216 V when the armature current is 10 A. Calculate the terminal voltage.

$$V = E + I_A R_A + V_B$$

$$V = 216 + (10 \times 0 \cdot 25) + 1 \cdot 5 \text{ volts} = 216 + 2 \cdot 5 + 1 \cdot 5 \text{ volts}$$

$$= 220 \text{ V}$$

## Example 9.16

A DC motor with a terminal voltage of 180 V and a back EMF of 173 V has an armature resistance of $0 \cdot 35$ Ω. Calculate the armature current.

$$V = E + I_A R_A \quad \text{so} \quad I_A = \frac{V - E}{R_A}$$

$$I_A = \frac{180 - 173}{0 \cdot 35} \text{ amperes} = \frac{7}{0 \cdot 35} \text{ amperes} = 20 \text{ A}$$

## Example 9.17

A 260 V DC motor has a back EMF of 245 V when the armature current is 300 A. Calculate the armature resistance.

$$V = E + I_A R_A \quad \text{so} \quad R_A = \frac{V - E}{I_A}$$

$$R_A = \frac{260 - 245}{300} \text{ ohms} = \frac{15}{300} \text{ ohms} = 0 \cdot 05 \text{ Ω}$$

## 9.7   Armature reaction and commutation

The operation of DC machines is not as simple as the earlier parts of this chapter would suggest. A number of problems arise in practice, and two of the most severe are considered simply in this section.

### Armature reaction

Our consideration of the DC machine so far has assumed that the only magnetic flux in the machine is due to current in the field windings. This completely overlooks the fact that the armature is carrying a very much heavier current than that in the field winding and will set up its own magnetic flux. In fact, two magnetic fluxes cannot exist in the same place at the same time. If there are two sets of magnetomotive force (ampere-turns) trying to set up two fluxes, the result will be a single magnetic flux due to the combination of the two sets of MMF.

Figure 9.4 represents a DC motor, the number of armature conductors being much less than usual to simplify the diagram. Figure 9.4*a* shows the magnetic field due to the field windings alone, whilst Figure 9.4*b* shows the magnetic field which would be set up by the armature current alone. The brushes are shown at points A and B on the commutator, where in

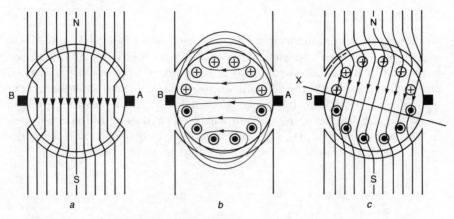

**Figure 9.4   Armature reaction**

    *a* The magnetic field due to the poles alone
    *b* The magnetic field due to the armature alone
    *c* The resulting magnetic field

simple terms they would be in contact with armature conductors passing along the magnetic flux, and thus having no EMF induced in them.

Figure 9.4*c* shows the magnetic flux resulting from the field and the armature. The diagram is drawn assuming the machine to be a motor with an anticlockwise direction of rotation; it can be seen that, if the brushes are to be set at the position where they are in contact with conductors having no EMF induced in them, they must be moved clockwise, that is against the direction of rotation, to axis XY. Had the machine been a generator, induced EMF and current would have both been in the same direction, (opposite to the direction for a motor with the same direction of rotation), reversing the ampere-turns due to the armature, and the brushes would need to be moved anticlockwise, with the direction of rotation.

To sum up, if brushes are to be moved to be in contact with conductors in which no EMF is induced:

    for a motor, move the brushes against the direction of rotation
    for a generator, move the brushes with the direction of rotation

The distortion of magnetic flux does not in itself change the total amount of flux, but the fact that flux will be concentrated in either the leading or the trailing pole tips may lead to magnetic saturation in the pole shoes with a slight overall reduction in flux and hence in a reduction in induced EMF for a generator or increased speed for a motor.

Since the armature current depends on load, the amount of distortion due to armature reaction is variable. Thus, the optimum position for the brushes varies with load. It is not usually practical to keep altering brush position, so a **compensating winding** is sometimes provided at the ends of the pole pieces. This winding is connected in series with the armature winding and offsets the flux distortion due to armature reaction.

## Commutation

As the armature conductors of a DC machine rotate, they pass successively under north and south magnetic poles and have an alternating EMF induced in them. Brushes are situated ideally at the point where no EMF is induced and make contact with the coil ends connected to the commutator segment on which they are bearing. Since the brush carries the current entering or leaving the armature, this current will divide at the commutator segment in contact with the brush, half following coils to the left, and half to the right. This situation is shown in simplified form in Figure 9.5.

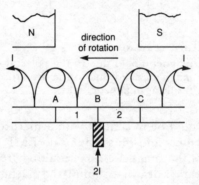

**Figure 9.5   Current relationships in adjacent coils**

It therefore follows that while the brush is in contact with two adjacent commutator segments (1 and 2), the current in the coil connected between the segments (B) must reverse whilst that coil is short-circuited as indicated in Figure 9.6. If, for example, we assume a commutator to have 250 segments and that it rotates at 1200 r/m, the adjacent segments will each be short-circuited for 0.2 ms. In this very short time, the current should completely reverse.

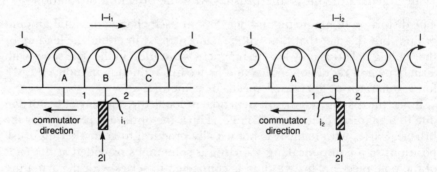

**Figure 9.6   Current reversal in short-circuited coil**

Reference to Chapter 2 will remind the reader that current change in an inductive circuit takes time. It is therefore very unlikely that the current change will be completed while the coil is short-circuited, and the difference between the actual current and the ideal current will flow in the form of an arc between the brush and the segment it has just left. This sparking at the commutator is the outward sign that commutation is taking place.

There are three practical methods of reducing the effects of commutation.

(i) By increasing brush contact resistance. This will reduce the time constant of the short-circuited coil ($L/R$) and hence will reduce the arcing current. Carbon brushes are used in preference to copper gauze because of their relatively high resistance.

(ii) By shifting the brushes forward for a generator or backward for a motor. This will mean that the short-circuited coil will be subject to magnetic flux which will induce an EMF to assist the current change.

(iii) By providing the machine with **commutating poles** (or interpoles) which are small poles mounted between the main poles, their windings being connected in series with the armature. The magnetic field provided by these poles will have a similar effect to that of brush shift. The polarity of commutating poles must be the same as the next main pole ahead in the direction of rotation for a motor, or the previous main pole for a generator.

## 9.8  Methods of excitation

The type of field system of a DC machine and the way in which it is fed have an influence on the behaviour of motors and of generators, as we shall see in Sections 9.11 and 9.12. The methods of connection of the field windings of DC machines fall into five main categories. These are

### (a) Separately excited field

As the name implies, the field is fed from a separate DC source, neither fed from the generator itself (Figure 9.7*a*) nor from the supply system (Figure 9.7*b*). The number of turns used for the field will depend on the current and voltage capabilities of the supply feeding it. If the supply has a fairly high voltage, the field will have many turns of fine wire carrying a small current. For a low-voltage supply, fewer turns of thicker conductor will be used carrying a larger current, because the strength of the magnetic field depends on the ampere-turns of the winding.

### (b) Shunt field

The shunt field is connected across the armature (it 'shunts' the armature). Since it receives the full output voltage of a generator or the supply voltage of a motor, it is generally made of a large number of·turns of fine wire carrying a small field current (see Figure 9.8). Note that, although the field

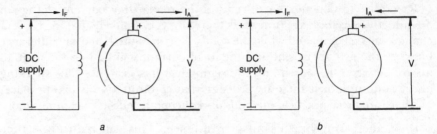

**Figure 9.7   Separately excited DC generator and motor**
  *a* Generator
  *b* Motor

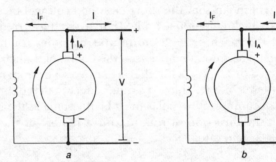

**Figure 9.8   Self-excited shunt-wound DC generator and motor**
  *a* Generator
  *b* Motor

current has the same direction for both machines, if the polarity is the same, the armature and supply currents differ for the motor and the generator. For the generator, the armature current drives the supply current and also feeds the field, so

$$I_A = I + I_F \text{ (generator)}$$

For the motor, the supply provides both armature and field currents,

so,
$$I = I_A + I_F \text{ (motor)}$$

where   $I$ = the current to (or from) the supply (A)
      $I_A$ = the armature current (A)
      $I_F$ = the field current (A)

## Example 9.18

A shunt-connected DC machine is connected to 270 V busbars and has a field resistance of 90 $\Omega$. Calculate the armature current if it is working

  (*a*) as a motor taking 80 A from the busbars, and
  (*b*) as a generator feeding 100 A to the busbars.

The field current obeys Ohm's law, so

$$I_F = \frac{V}{R} = \frac{270}{90} \text{ amperes} = 3 \text{ A in both cases.}$$

(*a*) $I = I_A + I_F$

so $\quad I_A = I - I_F = 80 - 3 \text{ amperes} = 77 \text{ A}$

(*b*) $I_A = I + I_F = 100 + 3 \text{ amperes} = 103 \text{ A}$

## (c) Series field

A series field is connected as shown in Figure 9.9. The armature current (which is also the load current) passes through the field, and since this current is often large, the conductors of the winding have a large cross-sectional area. The magnetic field depends on ampere-turns, so the heavy current results in few turns for a series field. The voltage drop in the field adds to the armature voltage drop and increases the difference between the induced EMF and terminal voltage, so that

$$V = E - I_A R_A - I_A R_F \text{ for a series generator}$$

or $\quad V = E + I_A R_A + I_A R_F \text{ for a series motor}$

where $R_F$ is the resistance of the series field ($\Omega$).

Since armature and field are in series, they both carry full load current, and

$$I = I_A = I_F \text{ (generator and motor)}$$

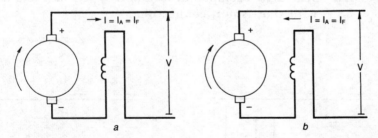

**Figure 9.9    Self-excited series-wound DC generator and motor**
  *a* Generator
  *b* Motor

## Example 9.19

A series-connected DC generator has an induced EMF of 340 V and a terminal voltage of 320 V when delivering 80 A. If the armature resistance is $0 \cdot 15 \ \Omega$, calculate the resistance of the series field.

$$V = E - I_A R_A - I_A R_F = E - I_A(R_A + R_F)$$

$$R_A + R_F = \frac{V - E}{I_A} = \frac{340 - 320}{80} \text{ ohms} = \frac{20}{80} \text{ ohms} = 0 \cdot 25 \ \Omega$$

since $\quad R_A = 0 \cdot 15 \ \Omega, \ R_F = 0 \cdot 25 - 0 \cdot 15 \text{ ohms} = 0 \cdot 1 \ \Omega$

## (d) Compound field

A machine having both shunt and series fields is known as a compound machine, and can be connected as shown in Figure 9.10, which is known as the short shunt connection, or as in Figure 9.11, the long shunt connection. The usual arrangement is for each pole of the machine to carry two windings; the shunt winding has many turns of fine wire, while the series winding has very few turns of heavy conductor.

The magnetic field set up by these windings will depend on the resultant ampere-turns provided by both of them. If the windings are so arranged

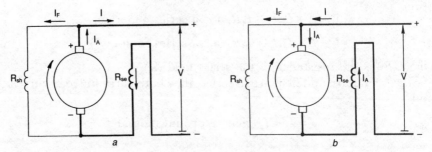

**Figure 9.10   Self-excited compound-wound DC generator and motor with short shunt connection**
*a* Generator
*b* Motor

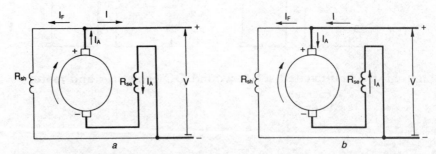

**Figure 9.11   Self-excited compound-wound DC generator and motor with long shunt connection**
*a* Generator
*b* Motor

that the currents in them assist each other in setting up the field, the machine is described as **'cumulatively compounded'**.

If the series winding opposes the shunt winding so that the effective ampere-turns are the difference between the individual values, the machine is said to be **'differentially compounded'**. Either type may be long-shunt or short-shunt connected.

## (e) Field rheostat

It is common to connect a variable resistor (rheostat) in series with the shunt-field winding of a DC machine as shown in Figure 9.12. By variation of resistance the field current, and hence the magnetic flux produced, is altered. For a generator, the induced EMF and hence the terminal voltage, can be controlled in this way ($E \propto \Phi n$). For the motor the rheostat is a speed controller (see Section 9.13).

It is not practicable to connect a resistor in series with a series field winding, but one of two methods of field control is sometimes used in this case. The first is to use a shunt-connected resistor or **'divertor'** as shown in Figure 9.13. This resistor carries part of what would otherwise be the full field current, so that variation of its resistance controls field current. The second method is to use a tapped field as shown in Figure 9.14. By varying the number of turns which carry the current one can alter the ampere-turns. Both methods are expensive, so are not used widely.

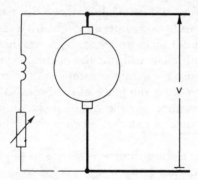

**Figure 9.12   Rheostat in series with shunt field of DC machine**

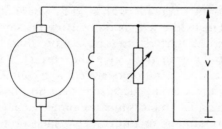

**Figure 9.13   Divertor in parallel with series field of DC machine**

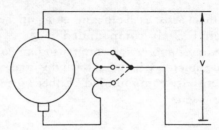

**Figure 9.14   Tapped series field on DC machine**

## 9.9   Starting and protection of DC motors

Assume that a DC motor with an armature resistance of $0 \cdot 1$ $\Omega$ is switched directly, without a starter, on to a 250 V DC supply.

$$I_A = \frac{V - E}{R_A}$$

At the instant of switching on, the motor will be at standstill, so the back EMF $E$ will be zero. Therefore

$$I_A = \frac{V - 0}{R_A} = \frac{250}{0 \cdot 1} \text{ amperes} = 2500 \text{ A}$$

This immense current would certainly damage the armature unless, as is likely, the fuse or circuit breaker operates. The high starting current must be limited while the motor picks up speed to give an increase in back EMF, which will itself be sufficient to limit the current when it becomes high enough. The method used is to add resistance in the armature circuit, the voltage drop across it taking the place of the back EMF and limiting the voltage across the armature. As the motor picks up speed and the back EMF rises, the resistance is reduced, eventually being removed entirely from the circuit.

A typical starter for a shunt motor is shown in Figure 9.15. When the handle is moved to contact 1, the supply is fed to the armature through the resistors $R_1$, $R_2$, $R_3$ and $R_4$ in series. As the motor picks up speed, the arm is moved to contact 2, which cuts out resistor $R_1$. The process is repeated, allowing time for the motor to increase its speed before each movement, until contact is made with stud 5, where all the resistors are cut out of circuit. The field is fed via the starting resistors and the 'no-volt' release coil. In pactice, starters often have more than four resistors, the values being carefully graded to give the correct starting conditions. Automatic starters are push-button operated, and contactors controlled by a timer or by armature current successively cut out the starting resistors.

Like other machines, DC motors must be protected against circumstances which could lead to damage or injury. Two such protective devices are shown on the starter in Figure 9.15.

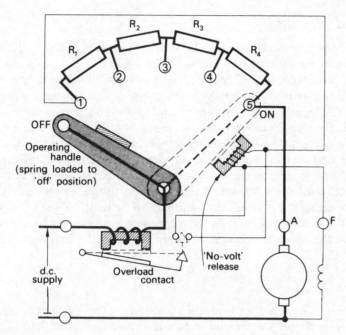

**Figure 9.15    Starter for DC shunt motor with protective devices**

### 'No-volt protection'

The purpose of this device is to return the handle of the starter to the 'off' position in the event of a failure of the mains or of the field circuit. There are three reasons for this protection.

  (i) If the supply fails and the starter remains 'on' the motor will start (or attempt to start) when the supply is restored. This could result in injury to an operator cleaning or adjusting a machine.

 (ii) In the circumstances given in reason 1, the motor will attempt to start without the protection of the starter when the supply is restored, and may damage itself as already indicated.

(iii) If an open circuit occurs in the field circuit of a shunt motor, field current will be cut off and flux will fall to the low value due to remanent magnetism. If provision is not made to stop the motor, a dangerous increase in speed and in armature current will often follow (see Section 9.13).

To protect against these dangers, the handle of a starter is spring-loaded so that it will return to the 'off' position when released. When the motor is running normally, the handle is held in the 'on' position by the 'no-volt' release electromagnet shown in Figure 9.15. If the mains or field circuit fails, the coil, which normally carries the field current, is de-energised, and the handle is released and returns to the off position.

## Overload protection

If a motor is overloaded, or if a fault occurs in its windings, the current taken will exceed the normal value, although it may not be sufficient to blow the protective fuse. Such a current may be high enough to cause a temperature rise in the windings, with resulting damage to the insulation and possibly causing the winding to burn out. To guard against such an overload, the supply current to the motor is fed through a few turns of conductor energising a small electromagnet. If the current exceeds a critical value, an arm is attracted to the electromagnet. When this happens, a pair of contacts is closed and the no-volt coil is short-circuited out. This results in the release of the operating handle and its return to the 'off' position. When such a starter is used, the handle must be held 'on' for a few seconds to allow the running current to fall to normal, when the overload contacts will open, the 'no-volt' coil will be energised, and the handle will be held in position.

## Example 9.20

A DC shunt motor has an armature resistance of $0 \cdot 1 \ \Omega$ and is started by using a shunt motor starter similar to that shown in Figure 9.15. The total value of the resistors $R_1$ to $R_4$ is $9 \cdot 9 \ \Omega$.

(a) Calculate the maximum starting current (when the motor is at standstill) if it is fed from a 240 V supply.

(b) When the motor is running up to speed it has an induced EMF of 110 V, the starter being on stud 3. If resistors $R_3$ and $R_4$ have a combined value of $6 \cdot 4 \ \Omega$, calculate the armature current.

(a) $I_A = \dfrac{V}{R_A + R} = \dfrac{240}{0 \cdot 1 + 9 \cdot 9}$ amperes $= \dfrac{240}{10}$ amperes $= 24 \ \text{A}$

(b) $I_A = \dfrac{V - E}{R_A + R} = \dfrac{240 - 110}{0 \cdot 1 + 6 \cdot 4}$ amperes $= \dfrac{130}{6 \cdot 5}$ amperes $= 20 \ \text{A}$

## 9.10   Self excitation of DC generators

The induced EMF of a generator is proportional to the magnetic flux and the speed of the machine ($E \propto \Phi n$). However fast the machine turns, there can be no EMF if there is no magnetism in the poles. The magnets are energised by the field coils, which are in turn fed by the machine armature, except in the case of the separately excited machine (see Figure 9.7*a*). If the magnetism is produced by the field current which is driven by the induced EMF, how can there be any EMF to begin with?

The answer is that, if there were no field flux at all, there would be no induced EMF and no field current to produce flux; when a DC generator is shut down, however, it retains some remanent magnetism in its poles (remanent flux–see Section 3.5). If the field of the machine is correctly connected relative to the direction of rotation, the small EMF provided by this small amount of flux will drive a small field current which will strengthen the flux. The EMF will then increase, driving more field current which provides a stronger magnetic field to give greater EMF. This process continues until the voltage across the armature (provided by the induced EMF) drives a field current which is large enough to induce the value of EMF providing that voltage; the machine then settles down to provide a steady voltage for a given speed if the field resistance remains unaltered.

A new machine, which has never run before, will often self excite because a very small amount of magnetism, perhaps assisted by the earth's magnetic field, will have appeared in the pole pieces during manufacture. If a machine fails to excite, it can be induced to do so by 'flashing' the field. The method consists of feeding the field with a low-voltage DC supply of the correct polarity for just long enough to start the self-excitation process.

## 9.11  Generator-load characteristics

The load characteristic of a DC generator is a graph of terminal voltage against output current. The behaviour of the machine, and hence its characteristic, depends on the field connection, so the types of machine will be considered in turn.

An effect applying to all generators is an increase in armature voltage drop as load current increases. In most cases this leads to a fall in terminal voltage as load, and hence armature, current increases.

A large armature current will, in fact, affect the terminal voltage in another way. The current-carrying conductors of the armature will try to set up their own magnetic field, so distorting and slightly reducing the magnetic field due to the poles. Because of this effect, known as '**armature reaction**', a reduction in induced EMF occurs when armature current is heavy.

### (a) Separately excited DC generator (see Figure 9.7a)

The load characteristic for one speed and one value of field current (Figure 9.16) shows that terminal voltage falls off slightly as load increases, owing to the combined effects of armature voltage drop and armature reaction. These generators are used where a comparatively steady output voltage is required, but have the disadvantage of requiring a separate DC supply for the field. Field-regulator control will allow the terminal voltage to be kept constant.

### (b) Shunt-excited DC generator (see Figure 9.8a)

The load characteristic for one speed (Figure 9.17) is similar to that of the separately excited machine, but falls more at higher load currents. This

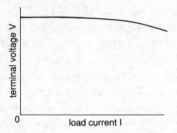

**Figure 9.16   Load characteristic of separately excited DC generator**

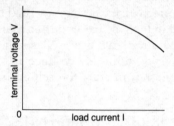

**Figure 9.17   Load characteristic of a self-excited shunt wound DC generator**

is because the field is connected across the armature, so that reducing terminal voltage gives a reduced field current, reducing the magnetic flux and reducing induced EMF. Shunt generators are widely used, particularly where the load resistance is constant, so that terminal voltage also remains constant.

## (c) Series-excited DC generator (see Figure 9.9a)

Since EMF depends on magnetic flux and since this flux is set up by the load current in the field windings, the load characteristic has the shape shown in Figure 9.18. On no load there will be a small terminal voltage due to remanent magnetism in the field poles. At heavy loads, the curve levels off as magnetic flux tends to become constant due to saturation. Series generators are not often used, but an understanding of their principles is helpful when considering compound generators.

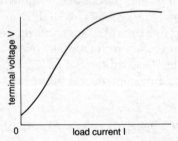

**Figure 9.18   Load characteristic of self-excited series-wound DC generator**

## (d) Cumulatively compounded DC generator (see Figures 9.10a and 9.11a)

Since the series field assists the shunt field, the output voltage will tend to rise as load current increases. The actual characteristic depends on the relative strengths of the shunt and series fields, so a variety of curves is possible (see Figure 9.19). Since the machine can be designed for almost any output characteristic, it is widely used for many purposes.

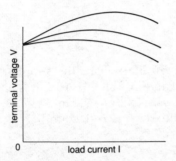

**Figure 9.19** **Load characteristics of a self-excited cumulatively compounded DC generator**

## (e) Differentially compounded DC generator (see Figures 9.10a and 9.11a)

In this machine, the series field opposes and weakens the shunt field, so that terminal voltage falls away quickly as the load current increases in the series winding. The sharply falling voltage characteristic (Figure 9.20) makes this type machine suitable for arc welding, where a high voltage is required to strike the arc, but only a small potential is needed to maintain it.

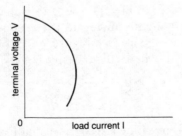

**Figure 9.20** **Load characteristic of a self-excited differentially compounded DC generator**

## 9.12   Motor-load characteristics

The important values in the mechanical output of a machine are its torque and its speed, whilst the input current is also significant. There is more than one way of presenting output characteristics, but the two most usual are curves of speed against torque, and of current, also plotted to a base of torque. To decide the relationships for the two characteristics, we can look again at two of our basic proportionalities.

(i) $E \propto \Phi n$

From this expression

$$n \propto \frac{E}{\Phi}$$

Since the back EMF $E$ differs from the supply voltage $V$ only by the comparatively small armature voltage drop, it is almost true that $n \propto V/\Phi$. As the supply voltage is usually constant, $n \propto 1/\Phi$, or speed is inversely proportional to field flux (see also Section 9.13).

(ii) $T \propto \Phi I_A$

This expression relates all three variables with which we are concerned, i.e. torque, current and magnetic flux.

As with the generator, the characteristics of a motor depend on the way in which its field is fed. The various types are considered in turn.

### (i) and (ii) Separately excited and shunt DC motors (see Figures 9.7b and 9.8b)

In both cases the field is fed from a steady supply so that field current is constant. The two motors therefore behave in the same way. The field flux is constant if armature reaction is ignored, so speed is constant as torque increases, while current is proportional to torque. Armature reaction weakens the field slightly when armature current is heavy, giving an upward curve to the otherwise linear current/torque curve. This curve is for current against load torque, and starts above the zero current level because the machine takes current when running on no load to provide for losses.

If $\Phi$ is constant, $n \propto E$, so speed reduces as armature current and armature voltage drop both increase. Thus the speed/torque curve droops, the fall in speed being reduced slightly owing to the reduction in magnetic flux as armature reaction takes place (see Figure 9.21). Separately excited and shunt motors are used where the comparatively steady speed is an advantage.

### (iii) Series DC motor (see Figure 9.9b)

The field current of this machine is also the armature current. On no load, current, and hence magnetic flux, will be small, so machine speed will be high. As load increases, load current (which is also field current) increases

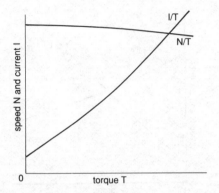

**Figure 9.21**  **Speed/torque and current/torque charactristics for self-excited shunt-wound DC motor**

the magnetic flux and reduces the speed. The characteristics (shown in Figure 9.22) are ideal for use with electric trains and similar applications. At low speeds, torque is high and gives good acceleration. Series motors are also used to drive other fixed loads, such as fans, flywheels etc. Care must be taken, however, not to remove the load. If this happens, the motor speed increases to a high level, sometimes resulting in self destruction due to large centrifugal forces.

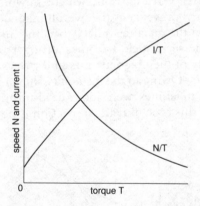

**Figure 9.22**  **Speed/torque and current/torque characteristics for self-excited series-wound DC motor**

## (iv) Cumulatively compounded DC motors (see Figures 9.10b and 9.11b)

These characteristics, shown in Figure 9.23 are a combination of the shunt and series characteristics, both fields assisting as for the generator. The motor is useful for some machine tools, where some reduction in speed due to heavy loads is beneficial.

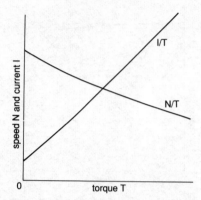

**Figure 9.23   Speed/torque and current/torque characteristics for self-excited cumulatively compounded DC motor**

## (v) *Differentially compounded DC motors (see Figures 9.10b and 9.11b)*

The series field opposes and weakens the shunt field, thus increasing the motor speed as load current increases. This extra speed demands extra torque, which requires extra current, which, flowing in the series field, strengthens it and further weakens the overall magnetic field. This results in extra speed, so that the machine quickly becomes unstable. A point may be reached where the series field becomes stronger than the shunt field, reversing the polarity of the magnetic flux and reversing the motor, often with disastrous results. Owing to the instability shown in the characteristics (Figure 9.24), these machines are seldom used in practice; care must be taken to ensure that this type of field connection is not made by accident.

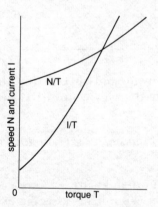

**Figure 9.24   Speed/torque and current/torque characteristics for self-excited differentially compounded DC motor**

## 9.13 Speed control and reversal of DC motors

We have seen in Section 9.12 that the speed of a DC motor varies directly with its back EMF and inversely with its magnetic flux ($n \propto E/\Phi$). Under stable operating conditions, the voltage applied to the armature of a motor is only slightly different from the back EMF, so we do not introduce a very large error if we approximate and say that speed is proportional to the voltage applied to the armature. It follows that the two methods of speed variation are by altering the voltage applied to the armature and by altering the field flux.

### *Speed control by armature-voltage variation*

Motor speed will go up as the voltage applied to the armature rises, and down as it falls, so a variable-voltage DC supply will ideally be required for such a system, the field current being kept constant. Since modern supply systems are almost always AC systems, a common method is to use a controlled rectifier, which will perform the dual role of rectifier and voltage controller. Such a system is shown in simple form in Figure 9.25 and allows full speed variation from zero to maximum. Rectifiers have been considered more fully in Chapter 6. Note that the motor-field system must be fed at a constant voltage, since any change of speed owing to, say, a reduction of armature voltage would be offset by a weakening of field flux.

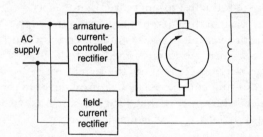

**Figure 9.25    Simple circuit for speed control of DC motor using controlled rectifier**

Another method of speed control is to vary the voltage applied to the armature by putting a variable series resistor in the armature circuit (Figure 9.26). The voltage drop in this resistor reduces the voltage applied to the armature and reduces speed, which can be varied by adjusting the resistor value. This method is only used for small machines, because the resistor carries full armature current and the power losses in it are high. Further, the voltage drop in the resistor varies with load so that speed falls off considerably as load increases if the resistor has a constant value.

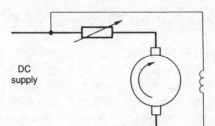

**Figure 9.26   Circuit for speed control of DC motor using variable-armature-circuit resistance**

## *Speed control by variation of field current*

The simplest and most widely used method of varying speed above and below normal over a limited range is by variation of field magnetic flux. A series-connected shunt-field regulator or a parallel-connected series-field divertor can be used to vary field current, and are connected as illustrated in Figures 9.12 to 9.14. As field current increases, magnetic flux increases and speed falls; as field current decreases, magnetic flux decreases and speed rises. The range of speed over which this method is effective is limited. Slowest speed occurs when the field saturates so that flux cannot increase, and upper speed is limited because machines become unstable due to low torque if the magnetic flux becomes too weak.

It is important to appreciate that, unless armature voltage or magnetic flux are altered gradually to increase motor speed, there will be a considerable increase in the current taken from the supply while the speed is changing. If, for example, speed is increased by a sudden reduction in flux, there will be a corresponding immediate reduction in back EMF. The effective voltage driving current to the armature $(V - E)$ will suddenly increase. The resulting current surge will provide extra torque to increase the machine speed, but may damage the machine as well as overloading the supply before it falls to a reasonable value.

## Example 9.21

A 400 V DC shunt motor has a field resistance of 100 $\Omega$ and an armature resistance of $0 \cdot 1 \; \Omega$.

- (*a*) Calculate the back EMF when the motor runs at 15 r/s and takes 84 A from the mains.
- (*b*) At what speed will the motor run if the supply current increases to 104 A due to increased load, without variation of field resistance?

(*a*)  $I_F = \dfrac{V}{R_F} = \dfrac{400}{100}$ amperes = 4 A

$I_{A_1} = I_1 - I_F = 84 - 4$ amperes = 80 A

$$V = E + I_A R_A \quad \text{so} \quad E = V - I_A R_A$$

$$E_1 = 400 - (80 \times 0 \cdot 1) \text{ volts} = 400 - 8 \text{ volts} = 392 \text{ V}$$

(b) $I_{A_2} = I_2 - I_F = 104 - 4$ amperes $= 100$ A

$$n \propto \frac{E}{\Phi} \text{ but } \Phi \text{ is constant,} \quad \text{so} \quad n \propto E \quad \text{and} \quad \frac{n_1}{n_2} = \frac{E_1}{E_2}$$

$$E_2 = 400 - (100 \times 0 \cdot 1) \text{ volts} = 400 - 10 \text{ volts} = 390 \text{ V}$$

$$n_2 = n_1 \times \frac{E_2}{E_1} = 15 \times \frac{390}{392} \text{ revolutions per second} = 14 \cdot 9 \text{ r/s}$$

## Example 9.22

A 300 V DC shunt motor with an armature resistance of $0 \cdot 1$ $\Omega$ is started by using a face-plate starter similar to that shown in Figure 9.15. When the handle is moved to stud 1, a resistance of $9 \cdot 9$ $\Omega$ is connected in series with the armature, which runs to a steady speed of 500 r/min, when it takes a steady current of 15 A. When the handle is moved to stud 2, the armature has $4 \cdot 9$ $\Omega$ in series with it, and again takes a current of 15 A when running at a steady speed. Calculate this speed.

$$V = E + I_A(R_A + R) \text{ where } R \text{ is the starter resistance.}$$

$$E_1 = V - I_A(R_A + R) = 300 - 15(9 \cdot 9 + 0 \cdot 1) \text{ volts}$$

$$= 300 - (15 \times 10) \text{ volts} = 300 - 150 \text{ volts} = 150 \text{ V}$$

$$E_2 = V - I_A(R_A + R) = 300 - 15(4 \cdot 9 + 0 \cdot 1) \text{volts}$$

$$= 300 - (15 \times 5) \text{volts} = 300 - 75 \text{ volts} = 225 \text{ V}$$

$$N \propto \frac{E}{\Phi} \text{ but } \Phi \text{ is constant,} \quad \text{so} \quad N \propto E \quad \text{and} \quad \frac{N_1}{N_2} = \frac{E_1}{E_2}$$

$$N_2 = N_1 \times \frac{E_2}{E_1} = 500 \times \frac{225}{150} \text{ revolutions per minute} = 750 \text{ r/min}$$

## The Ward–Leonard system

If the voltage applied to the armature of a DC motor can be varied from zero to maximum, its speed will also be altered over the full range. A constant-speed motor, which can be an AC motor or a DC motor ($M_1$), drives a DC generator (G) which thus has the same constant speed. The output of this generator is controlled by its field, whose strength can be varied from maximum to very nearly zero by the adjustment of a resistor (R) connected in series with it. This output is used to feed a DC motor ($M_2$) which has a constant field strength, and the speed of which will follow the output voltage of the generator. The arrangement is shown in Figure 9.27.

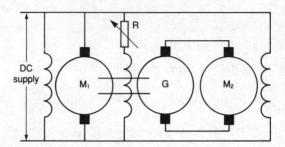

**Figure 9.27   Arrangement of the Ward–Leonard system**

The system shown is fed from a DC supply, and $M_1$ is a constant-speed DC motor. It could just as easily be fed from an AC supply if $M_1$ were an AC motor and the fields of G and $M_2$ were fed through rectifier circuits. While the Ward–Leonard system gives very good speed control over the full range, it does require three machines of the same rating as the final drive motor, and is thus very expensive. A more common method of providing a variable-speed output from a DC motor is to feed it with a variable-voltage DC supply obtained from an AC supply using thyristors (see Figure 9.25).

## *Reversal of rotation*

It is often necessary to reverse the direction of rotation of a DC motor. As we have seen in Chapter 11 of 'Electrical Craft Principles' Volume 1, the direction of the force on a conductor can be reversed by reversing the direction of either the magnetic field or the current in the conductor. Thus the direction of rotation of a DC motor can be reversed by changing over EITHER the armature connections OR the field connections, BUT NOT BOTH. Since a change in polarity of the supply to the motor will reverse both the armature current and the magnetic-field direction, supply polarity is unimportant. A DC motor will therefore run from an AC supply, although its normal construction with the solid poles and yoke would result in heavy eddy-current losses.

Many appliances, such as vacuum cleaners and hairdryers, have small series machines with laminated field systems as their driving motors and can be used from DC or AC supplies. Such machines are known as universal motors.

## 9.14   Summary of formulas and proportionalities for Chapter 9

See text for definitions of symbols.

Induced EMF:

$$E \propto \Phi n \qquad \Phi \propto \frac{E}{n} \qquad n \propto \frac{E}{\Phi}$$

$$E = \frac{2p\Phi nZ}{a} \qquad p = \frac{Ea}{2\Phi nZ} \qquad \Phi = \frac{Ea}{2pnZ}$$

$$n = \frac{Ea}{2p\Phi Z} \qquad Z = \frac{Ea}{2p\Phi n} \qquad a = \frac{2p\Phi nZ}{E}$$

For a wave-wound machine: $a = 2$

For a lap-wound machine: $a = 2p$

$$\frac{E_1}{E_2} = \frac{\Phi_1}{\Phi_2} \times \frac{n_1}{n_2} \qquad E_1 = E_2 \times \frac{\Phi_1}{\Phi_2} \times \frac{n_1}{n_2} \qquad E_2 = E_1 \times \frac{\Phi_2}{\Phi_1} \times \frac{n_2}{n_1}$$

$$\Phi_1 = \Phi_2 \times \frac{E_1}{E_2} \times \frac{n_2}{n_1} \qquad \Phi_2 = \Phi_1 \times \frac{n_1}{n_2} \times \frac{E_2}{E_1}$$

$$n_1 = n_2 \times \frac{E_1}{E_2} \times \frac{\Phi_2}{\Phi_1} \qquad n_2 = n_1 \times \frac{\Phi_1}{\Phi_2} \times \frac{E_2}{E_1}$$

Torque:

$$T \propto \Phi I_A \qquad \Phi \propto \frac{T}{I_A} \qquad I_A \propto \frac{T}{\Phi}$$

$$\frac{T_1}{T_2} = \frac{\Phi_1}{\Phi_2} \times \frac{I_{A_1}}{I_{A_2}} \qquad T_1 = T_2 \times \frac{\Phi_1}{\Phi_2} \times \frac{I_{A_1}}{I_{A_2}} \qquad T_2 = T_1 \times \frac{\Phi_2}{\Phi_1} \times \frac{I_{A_2}}{I_{A_1}}$$

$$\Phi_1 = \Phi_2 \times \frac{T_1}{T_2} \times \frac{I_{A_2}}{I_{A_1}} \qquad \Phi_2 = \Phi_1 \times \frac{I_{A_1}}{I_{A_2}} \times \frac{T_2}{T_1}$$

$$I_{A_1} = I_{A_2} \times \frac{T_1}{T_2} \times \frac{\Phi_2}{\Phi_1} \qquad I_{A_2} = I_{A_1} \times \frac{\Phi_1}{\Phi_2} \times \frac{T_2}{T_1}$$

For a DC generator:

$$V = E - I_A R_A - V_B \qquad E = V + I_A R_A + V_B \qquad V_B = E - V - I_A R_A$$

$$I_A = \frac{E - V - V_B}{R_A} \qquad R_A = \frac{E - V - V_B,}{I_A}$$

For a DC motor:

$$V = E + I_A R_A + V_B \qquad E = V - I_A R_A - V_B \qquad V_B = V - E - I_A R_A$$

$$I_A = \frac{V - E - V_B}{R_A} \qquad R_A = \frac{V - E - V_B}{I_A}$$

For a shunt generator:

$$I_A = I + I_F \qquad I = I_A - I_F \qquad I_F = I_A - I$$

For a shunt motor:

$$I = I_A + I_F \qquad I_A = I - I_F \qquad I_F = I - I_A$$

For a series generator:

$$V = E - I_A R_A - I_A R_F$$

For a series motor:

$$V = E + I_A R_A + I_A R_F$$

For a series generator or motor:

$$I = I_A = I_F$$

## 9.15   Exercises

1  A lap-wound DC machine has six poles with a flux/pole of 80 mWb. It is lap wound with 420 armature conductors, and runs at 1200 r/min. Calculate the induced EMF.

2  A four-pole wave-wound DC machine with a flux/pole of 12 mWb and 300 armature conductors has an induced EMF of 180 V. Calculate its speed in r/s.

3  A two-pole lap-wound DC machine has an induced EMF of 270 V at 900 r/min. If the machine has 500 armature conductors, calculate the flux/pole.

4  A DC machine is required to produce an EMF of 320 V with a flux/pole of 16 mWb at 25 r/s. If the machine is lap wound with four poles, how many armature conductors has it?

5  The EMF in a DC machine is 240 V. If the speed is increased by 30% and the field strength reduced by 20%, calculate the EMF induced.

6  A DC machine has an induced EMF of 200 V under certain conditions of field and speed. What will be the induced EMF if the speed and field flux are both increased by 10%?

7  The EMF of a DC machine is 340 V at 1000 r/min with a field current of 2 A, but falls to 300 V at 600 r/min. Assuming that field magnetic flux is proportional to field current, what will be the new value of the latter?

8  When its speed is 30 r/s and the flux/pole is 36 mWb, a DC machine has an induced EMF of 400 V. Calculate the speed at which 300 V will be induced if the flux/pole is reduced to 30 mWb

9  A DC motor gives a torque of 80 Nm when its armature current is 30 A and its magnetic flux/pole 40 mWb. What torque will it give with an armature current of 40 A and a flux/pole of 20 mWb?

10  A DC shunt motor working from constant-voltage mains delivers a torque of 100 Nm for an armature current of 20 A. What will be the armature current if the load torque increases to 120 Nm?

11  The torque of a DC motor at its rated speed and with an armature current of 33 A is 150 Nm. The speed is increased by weakening the flux to two-thirds

of its previous value, and the load is adjusted to give an armature current of 40 A. What torque is provided?

12 The torque required to drive a DC generator at a given speed is 12 Nm when the armature current is 15 A. The voltage output is increased by raising the flux/pole by 20%, and at the same time the load resistance reduces so that the armature current is 30 A. Calculate the torque needed to keep the speed constant.

13 A DC motor runs at 25 r/s with a flux/pole of 30 mWb and an armature current of 50 A, delivering a torque of 50 Nm. The speed is to be increased by flux weakening so that the armature current is 70 A and the torque is 40 Nm after reduction of load. Calculate the new flux/pole.

14 A DC generator has an armature current of 200 A at a flux/pole of 120 mWb, the driving torque being 200 Nm. If the torque increases to 300 Nm and the flux/pole to 160 mWb, calculate the new armature current.

15 A DC generator has an armature resistance of $0 \cdot 5$ $\Omega$ and an induced EMF of 170 V. What will be its terminal voltage with an armature current of 40 A?

16 The induced EMF and terminal voltage of a DC generator with an armature current of 120 A are 260 V and 230 V, respectively. Calculate the armature resistance.

17 Calculate the induced EMF of a DC generator with an armature current of 25 A at a terminal voltage of 320 V. The armature resistance is $0 \cdot 4$ $\Omega$ and the total brush voltage drop is 2 V.

18 A DC generator with an armature resistance of $0 \cdot 05$ $\Omega$ has an induced EMF of 200 V and a terminal voltage of 185 V. Calculate the armature current.

19 A DC generator delivers 500 kW at 400 V. The armature resistance is $0 \cdot 012$ $\Omega$. Find the value of the EMF.

20 A DC shunt generator has a generated EMF of 250 V, when the armature current is 100 A and the terminal voltage is 240 V.
   (*a*) What will be its terminal voltage when the armature supplies 50 A?
   (*b*) If its field current is increased so that its terminal voltage is 250 V when supplying 100 A, what will then be its generated EMF?

21 A DC generator has armature resistance $0 \cdot 2$ $\Omega$. When driven at 1000 r/min and delivering 50 A its terminal voltage is 240 V. If its speed is raised to 1200 r/min and the current delivered is raised to 100 A, the flux remaining constant, what is then its terminal voltage?

22 A DC shunt motor has an armature resistance of $0 \cdot 2$ $\Omega$ and takes 40 A from 200 V mains. What is the value of the back EMF under these conditions?

23 Calculate the induced EMF of a DC motor having an armature current of 80 A and an armature resistance of $0 \cdot 25$ $\Omega$ if the terminal voltage is 320 V and the total brush voltage drop is 2 V.

24 A DC motor is supplied at 400 V and has an armature resistance of $0 \cdot 012$ $\Omega$. At a certain load the EMF is 388 V. Find the value of the armature current.

25 Calculate the armature resistance of a DC motor which carries an armature

current of 30 A if the terminal voltage and the back EMF are 240 V and 228 V, respectively.

26 Explain the meaning of back EMF of a DC motor, and explain how it is produced. A 460 V DC shunt motor, running on load, has an armature resistance of $0 \cdot 12$ Ω. Calculate:

 (i) the value of the back EMF when the current in the armature is 150 A, and
(ii) the value of the armature current when the back EMF is 452 V.

27 A DC machine is shunt-connected with a field resistance of 160 Ω. It is connected to a system having a constant voltage of 240 V. Calculate the current to or from the supply if the machine operates:

(*a*) as a generator with an armature current of 60 A, and
(*b*) as a motor with an armature current of 40 A.

28 A series-wound DC generator has armature and field resistance of $0 \cdot 1$ Ω and $0 \cdot 2$ Ω, respectively, and an induced EMF of 280 V. What is its terminal voltage when supplying a load of 100 A?

29 A DC shunt generator has an armature resistance of $0 \cdot 25$ Ω and a terminal voltage of 220 V when supplying a load of 48 A. What is the induced EMF if the field circuit resistance is 110 Ω?

30 A series DC generator has armature and field resistance of $0 \cdot 2$ Ω and $0 \cdot 3$ Ω, respectively. Calculate its induced EMF if its terminal voltage on a load of 20 A is 190 V.

31 A series DC motor takes 10 A from a 240 V DC supply. If its back EMF is 236 V and its field resistance is $0 \cdot 25$ Ω, what is its armature resistance?

32 (*a*) Draw a circuit diagram showing the connections for a DC shunt generator supplying current to a bank of lamps. A rheostat should be shown in the field circuit. What is the purpose of the rheostat?

   (*b*) A DC generator delivers $20 \cdot 4$ kW at 240 V to a resistance load. The resistance of the armature is $0 \cdot 125$ Ω and the resistance of the field winding is 80 Ω. Assuming a total brush contact drop of 2 V, calculate (i) the current in the armature, and (ii) the generated EMF.

33 Explain briefly, with the aid of diagrams, the differences between series, shunt and compound DC generators. A DC shunt generator delivers a current of 96 A at 240 V. The armature resistance is $0 \cdot 15$ Ω and the field winding has a resistance of 60 Ω. Assuming brush contact drop of 2 V, calculate:

(*a*) the current in the armature, and
(*b*) the generated EMF.

34 (*a*) Explain briefly, with the aid of diagrams, the differences between series, shunt and compound DC generators.

   (*b*) A shunt-wound DC machine has an armature resistance of $0 \cdot 2$ Ω and a shunt resistance of 115 Ω. It is used (i) as a shunt generator to deliver a current of 150 A to an outside load, and (ii) as a shunt motor driving a mechanical load and taking 150 A from a DC supply. In both cases the terminal voltage of the machine is 460 V. Calculate the currents in both armature and field windings in both cases (i) and (ii).

35 (*a*) Draw diagrams to show TWO methods of connecting the field winding of a DC motor. Include the motor armature and the supply terminals in your diagrams. How will the number of turns and the size of wire used for the field winding differ for the two methods of connection?

(*b*) A shunt-connected DC motor takes a current of 15 A when providing a certain torque. What current will the motor carry if the torque it provides is halved? Assume that the magnetic flux remains constant.

36 (*a*) Show, with the aid of diagrams, three different methods of exciting a DC generator.

(*b*) A shunt-wound DC machine has an armature resistance of $0 \cdot 3$ $\Omega$ and a shunt-winding resistance of 200 $\Omega$. Calculate the armature and field currents when it is used:

(i) as a shunt motor driving a mechanical load and taking a current of 140 A from the 460 V DC supply, and

(ii) as a shunt generator delivering a current of 140 A to an external load at 460 V.

37 What is the meaning of the expression back EMF of a direct-current motor? Explain how the back EMF and the current change during the starting of a DC motor.

38 A 300 V DC motor with an armature resistance of $0 \cdot 25$ $\Omega$ takes an armature current of 20 A on full load. What value of resistance must be connected in series with the armature at the instant of starting to limit armature current to twice its normal value?

39 Explain briefly why a resistance starter is needed for the safe starting of a DC motor. Give a diagram of a starter for a DC shunt motor, with no-volt and overload releases. Include the motor circuits in the diagram.

40 (*a*) Give a complete diagram of a starter for a DC shunt motor with no-voltage and overload releases. Include the motor circuits in the diagram.

(*b*) Describe and explain the use of a variable resistance which may be connected in the shunt circuit of a DC shunt motor.

41 (*a*) Sketch the load characteristics, terminal voltage/load current, of the following types of DC generator:

(i) series-connected, (ii) shunt-connected, (iii) compound-connected.

(*b*) State one purpose for which each is used.

42 Sketch the current/torque and speed/torque characeristics of series, shunt and both types of compounded DC motors. Suggest a suitable use for each type of machine.

43 (*a*) What is meant by the back EMF of a DC motor?

(*b*) A 250 V shunt motor has field resistance of 125 $\Omega$ and armature resistance of $0 \cdot 25$ $\Omega$.

(i) Find the back EMF when the motor is running and taking 42 A from the mains.

(ii) At what speed will it run if the load increases the intake to 62 A?

44 Explain concisely why the speed of a DC shunt motor rises when resistance is inserted in the field circuit and why it drops when resistance is inserted in the armature circuit.

45 Explain with the aid of diagrams, how the speed of a DC shunt motor may be varied (*a*) above normal speed, and (*b*) below normal speed.

　　Why is it desirable that speed changes should be made gradually?

46 (*a*) Why is a starter necessary for a DC shunt motor?

　　(*b*) The first notch of a 240 V DC shunt motor starter puts $11 \cdot 8 \ \Omega$ in series with the armature which has $0 \cdot 2 \ \Omega$ resistance. The motor then runs up to 400 r/min ($6 \cdot 67$ r/s) and takes a steady armature current of 10 A. The second notch leaves $5 \cdot 8 \ \Omega$ in series with the armature. If the current again steadies to 10 A, at what speed will the motor run?

47 Make a drawing showing the connections of a shunt-wound DC motor and starter, complete with the usual protective devices, and explain how you would reverse the rotation of the motor.

48 Describe a type of motor suitable for use on both AC and DC supplies. Explain why the motor will work on either supply. Give examples of where these motors are used commercially.

　　Make a diagram of connections and show how the direction of rotation can be reversed.

## 9.16   Multiple-choice exercises

9M1　The rotating part of a DC machine is called the:

　　(*a*) rotor　　　　(*b*) brushes　　　(*c*) armature　　　(*d*) field.

9M2　The armature of a DC machine is laminated because:

　　(*a*) it is subject to alternating magnetisation

　　(*b*) it is stronger than a solid construction

　　(*c*) eddy-current losses will not occur in a DC machine

　　(*d*) the field system is solid.

9M3　The two basic methods of winding the armature of a DC machine are called:

　　(*a*) progressive and retrogressive　　　(*b*) forward and backward

　　(*c*) drum and salient　　　　　　　　(*d*) lap and wave.

9M4　The armature windings on a DC machine are connected to a synchronised changeover switch called:

　　(*a*) a commutator　　　　　　　　(*b*) an isolator

　　(*c*) a slipring　　　　　　　　　　(*d*) a carbon brush.

9M5　The outer case of a DC machine forms part of its magnetic circuit and is called the:

　　(*a*) case　　　　(*b*) yoke　　　(*c*) stator　　　(*d*) outside.

9M6　The EMF *e* induced in a conductor of a DC machine depends on the magnetic flux density *B*, the conductor length *l* and its velocity *v*. The formula relating these factors is:

　　(*a*) $e = \dfrac{Bl}{v}$　　　(*b*) $e = Blv$　　　(*c*) $B = elv$　　　(*d*) $e = \dfrac{lv}{B}$

9M7  A four-pole wave-wound DC generator has 450 armature conductors, a flux/pole of 40 mWb and runs at 1500 r/m. The induced EMF is:
(a) $1 \cdot 8$ V  (b) 450 V  (c) $2 \cdot 7$ kV  (d) 900 V.

9M8  The EMF induced in the armature of a DC motor is 220 V when running under standard operating conditions. The induced EMF if speed reduces by 15% and field flux increases by 8% will be:
(a) 202 V  (b) $2 \cdot 64$ V  (c) 235 V  (d) 273 V

9M9  The torque produced by a DC motor depends on:
(a) its rotational speed
(b) machine dimensions
(c) armature current and field magnetic-flux strength
(d) armature current and induced EMF.

9M10  The SI unit of torque is measured in:
(a) newton-metres  (b) pounds force-feet
(c) metre-newtons  (d) kilograms.

9M11  If $V$ is the terminal voltage, $E$ is the induced EMF, $I_A$ is the armature current and $R_A$ is the armature resistance, the formula relating them for a DC motor is:
(a) $E = V + I_A R_A$  (b) $V = E - I_A R_A$

(c) $V = \dfrac{E}{I_A R_A}$  (d) $V = E + I_A R_A$

9M12  A DC motor with an armature resistance of $0 \cdot 22$ $\Omega$ and a negligible brush voltage drop has an induced back EMF of 224 V when the armature current is 25 A. The terminal voltage of the motor is:
(a) $5 \cdot 5$ V  (b) $229 \cdot 5$ V  (c) 224 V  (d) $220 \cdot 5$ V.

9M13  The distortion of the field of a DC machine by current in the armature windings is known as:
(a) commutation  (b) field effect
(c) brush movement  (d) armature reaction.

9M14  The distortion referred to in Example 9M13 will have less effect if the brushes of a DC motor are moved:
(a) to line up with the poles  (b) against the direction of rotation
(c) off the commutator  (d) with the direction of rotation.

9M15  A compensating winding to offset the effects of field distortion by the armature current must be connected:
(a) in series with the field  (b) in parallel with the armature
(c) in series with the armature  (d) in parallel with the field.

9M16  Severe sparking at the trailing edges of the brushes of a DC machine is likely to be due to:
(a) commutation  (b) brush pressure too high
(c) armature reaction  (d) the wrong brush type.

9M17  Sparking at the trailing edges of the brushes in a DC machine is usually prevented by fitting:
(a) copper gauze brushes  (b) a compensating winding
(c) a compound winding  (d) commutating (or inter-) poles.

9M18 A series-excited field on a DC machine is connected:
   (*a*) so that it carries the full machine current
   (*b*) directly across the armature terminals
   (*c*) in series with the shunt field
   (*d*) in parallel with the armature.

9M19 A DC machine with compound field windings is one which has:
   (*a*) no speed or EMF variation due to field adjustment
   (*b*) both shunt and series fields
   (*c*) field windings made of special material
   (*d*) is never used.

9M20 If the resistance in series with the shunt field of a DC motor is increased
in value, the motor will:
   (*a*) slow down              (*b*) stop
   (*c*) overheat               (*d*) speed up.

9M21 A starter is necessary for a DC motor because it:
   (*a*) will not rotate without one
   (*b*) prevents the operator from receiving electric shocks
   (*c*) will draw excessive current if one were not used
   (*d*) adds to the cost of the machine.

9M22 A new DC generator which will not self-excite can be induced to do so by:
   (*a*) 'flashing' the field         (*b*) increasing the speed
   (*c*) using a permanent magnet    (*d*) hitting it.

9M23 The generator characteristic shown below is for a:
   (*a*) shunt machine
   (*b*) series machine
   (*c*) differentially compounded machine
   (*d*) cumulatively compounded machine.

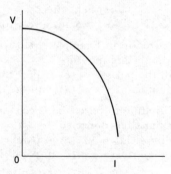

**Figure 9.28   Diagram for Exercise 9M23**

9M24 The speed of a DC motor can be increased by:
   (*a*) reducing the brush pressure
   (*b*) weakening the field strength
   (*c*) strengthening the field
   (*d*) reducing the armature voltage.

9M25 The direction of rotation of a DC shunt motor can be reversed by:
    (*a*) reversing the polarity of the DC supply
    (*b*) reversing the connections of both the armature and the field
    (*c*) changing to a series field
    (*d*) reversing either the armature or the field connections.

*Chapter 10*

# Alternating-current motors

## 10.1  Introduction

The three-phase AC motor has many advantages over its single-phase counterpart. These include a smaller frame size for a given output, and the fact that torque is steady rather than pulsating.

The simplest and most common of the three-phase machines is the induction motor, and most of this chapter is devoted to this very important machine. Like the other three-phase machine considered here, the synchronous motor relies for its operation on the ability of a three-phase winding to set up a rotating magnetic field.

## 10.2  Rotating magnetic field due to a three-phase supply

Consider three coils arranged as shown in Figure 10.1a, each coil setting up a magnetic field in the direction shown if the other two are de-energised. If three-phase currents are fed to all three coils, the directions of the three magnetomotive forces (MMFs) are as shown, and the resulting magnetic field will be due to the resultant of the vector MMFs. Figure 10.1b shows the wave diagram for the currents in the three coils when they are connected to a three-phase supply. Figure 10.1c gives the vector diagrams for six instants of time during a cycle. At 0°, red current is zero, so there is no red-coil MMF. Yellow current is negative, so yellow MMF is opposite in direction to that shown in Figure 10.1a and has a value proportional to the value of current at the instant taken. Blue current is positive, so blue MMF has the same direction as indicated in Figure 10.1a, with a value proportional to the value of instantaneous current. The resultant of these vectors is the horizontal line marked MMF.

The next five diagrams in Figure 10.1c show the situations at successive phase intervals of 60°. From these, it can be seen that the resultant MMF

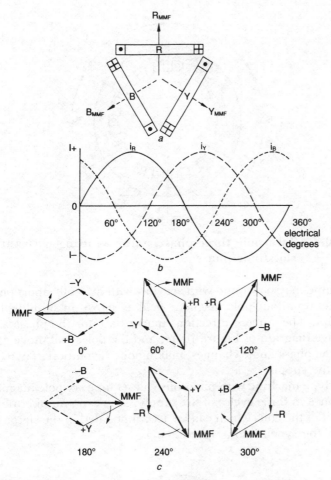

**Figure 10.1** **Rotating magnetic field due to three-phase supply**

has a constant magnitude, and rotates clockwise at constant speed through one revolution for each 360° of the supply.

The speed of rotation of the magnetic field is called the synchronous speed (symbols $n_s$ r/s or $N_S$ r/min) of a machine. For the coil system shown, the synchronous speed in revolutions per second will be the same as the frequency of the supply in cycles per second (hertz), since the field revolves once for each cycle. Such a winding is called a two-pole system, since it sets up one north pole and one south pole.

In practice, the winding does not consist of three coils as shown in Figure 10.1*a*, but of a system of insulated conductors let into slots in the inner surface of a hollow laminated drum which forms the stator. The winding is shown in section in Figure 10.2, the number of slots being reduced to simplify the drawing.

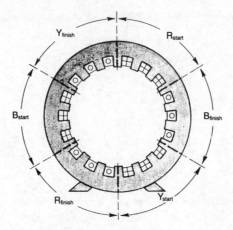

**Figure 10.2    Two-pole three-phase stator winding with three slots/pole/phase**

A machine may be wound with six coils with axes 60° apart physically, and will set up a four-pole magnetic field as shown in simplified form in Figure 10.3. The current directions marked are for the phase angle 30° on the wave diagram of Figure 10.1*b*, and the poles will move round the stator as the phase angle changes, making one revolution in two complete cycles of the supply.

Similarly, a machine with nine coils will set up a six-pole magnetic field rotating once in three cycles of the supply. Any even number of poles up to the limit of the coils which can be accommodated in the stator slots can be set up, the synchronous speed being

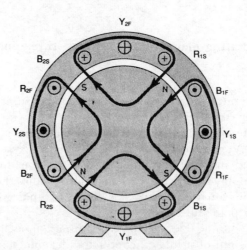

**Figure 10.3    Arrangement of three-phase four-pole winding**
For simplicity, only one conductor/pole/phase has been shown.
Current directions are for 30° instant of Figure 10.1*b*

$$n_s = \frac{f}{p} \quad \text{or} \quad N_s = \frac{60f}{p}$$

where  $n_s$ = synchronous speed (r/s)
$N_s$ = synchronous speed (r/min)
$f$ = supply frequency (Hz)
$p$ = number of pairs of poles

Note that this expression is the same as that given in Section 8.4 for an alternator.

## Example 10.1

Calculate the synchronous speeds, both in r/s and in r/min for two-pole, four-pole and six-pole fields fed from a 50 Hz supply.

Two-pole (one pair of poles):

$$n_s = \frac{f}{p} = \frac{50}{1} \text{ revolutions per second} = 50 \text{ r/s}$$

$$N_s = \frac{60f}{p} = \frac{60 \times 50}{1} \text{ revolutions per minute} = 3000 \text{ r/min}$$

Four-pole (two pairs of poles);

$$n_s = \frac{f}{p} = \frac{50}{2} \text{ revolutions per second} = 25 \text{ r/s}$$

$$N_s = \frac{60f}{p} = \frac{60 \times 50}{2} \text{ revolutions per minute} = 1500 \text{ r/s}$$

Six-pole (three pairs of poles):

$$n_s = \frac{f}{p} = \frac{60}{3} \text{ revolutions per second} = 16 \cdot 7 \text{ r/s}$$

$$N_s = \frac{60f}{p} = \frac{60 \times 50}{3} \text{ revolutions per minute} = 1000 \text{ r/min}$$

## Example 10.2

Calculate the number of poles wound on a stator giving a synchronous speed of 5 r/s (300 r/min) when fed from a 60 Hz supply.

$$n_s = \frac{f}{p}$$

$$p = \frac{f}{n_s} = \frac{60}{5} \text{ pairs of poles} = 12 \text{ pairs of poles, or 24 poles}$$

## *Pole-changing machines*

In due course we shall show that the rotational speed of both synchronous motors and induction motors depends on the speed at which the stator field rotates, i.e. the synchronous speed. It follows that, if a machine operating as, say, a four-pole motor, is changed to a six-pole motor, there will be a 33% reduction in its synchronous speed, and a corresponding reduction in rotor speed (see Example 10.1)

Figure 10.2 shows the arrangement of a two-pole stator, while Figure 10.3 shows a four-pole stator. If, for example, the winding is arranged so that it can be connected as a two-pole or as a four-pole type, a 2:1 speed variation can be obtained. This method is sometimes used to achieve speed control of induction motors, although it must be said that continuous control by frequency variation using power-electronics systems has many advantages (see Section 10.5).

## 10.3   Synchronous motor

Imagine a permanent bar magnet which is free to pivot and is placed inside a two-pole winding setting up a rotating magnetic field (see Figure 10.4). The north pole of the magnet will be attracted to the rotating south pole, while the south pole will try to follow the north pole of the stator field. If free to do so, the magnet will spin and follow the field, keeping in synchronism with it. The magnet will revolve at synchronous speed, or not at all, because if it tends to lag behind the rotating field there will be a repulsion, instead of an attraction of poles, and it will stop.

Practical three-phase synchronous motors seldom have a permanent magnet as a rotor. The rotating part of the machine is an electromagnet, the coil being fed with direct current from an external source by means of sliprings. The synchronous motor is, in fact, the same machine as the

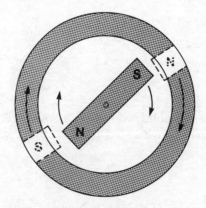

**Figure 10.4   To illustrate principle of synchronous motor**

alternator described in Section 8.5, working as a motor instead of as a generator. Figures 8.3–8.5 inclusive apply to synchronous motors as well as to alternators.

Synchronous motors are uncommon, and are seldom used in small sizes. Usually they have an output in excess of 750 kW; in this size they become competitive with induction motors, and are used for driving compressors, fans and similar loads. These machines will operate at a leading power factor when run on light load with an overexcited field, and may be used for power-factor correction (see Section 5.12).

### Reluctance motors

The simplest of all the synchronous machines is the reluctance motor, having a rotor consisting of a set of iron laminations shaped so that it tends to align itself with the magnetic field set up by the stator. A simple cage rotor may be necessary to enable the machine to run up to a speed approaching synchronous, when the rotor will 'pull in' to synchronism. The rotor speed will be the synchronous speed of the stator field, which will depend directly on the frequency of its supply. The stator is arranged so that parts of it have high reluctance (the magnetic dual of resistance) so that flux is forced to follow the low-reluctance paths. This type of motor is becoming increasingly common, especially when used in conjunction with electronic variable frequency generators, where it offers a means of providing a simple and cost-effective variable speed drive.

## 10.4   Induction-motor principle

The three-phase induction motor is the simplest and most widely used of all electrical motors. The stator is as described in Section 10.2, with a three-phase field winding let into slots cut in the inner face of its laminated structure (see Figures 10.2 and 8.5). The rotor is a laminated cylinder of silicon steel with a system of conductors just beneath its surface. In its simplest and most widely used form, the rotor winding consists of a number of copper or aluminium bars pushed through holes in the rotor drum, the ends of the bars being welded, brazed or keyed to copper or aluminium short-circuiting rings to form a '**cage**' or '**squirrel cage**' as shown in Figure 10.5.

If such a rotor is free to revolve within a wound stator, each conductor will be swept by alternate north and south magnetic poles as the stator field rotates. This will result in an alternating EMF being induced in the rotor conductors, which will cause a current to flow in them. The current acts with the magnetic field to set up a force on the conductors, which results in the rotor moving to follow the magnetic field (see Figure 10.6).

The faster the rotor revolves, the less will be the speed difference between it and the rotating magnetic field. This will result in less induced EMF, less rotor current, and less torque on the rotor. The rotor can never reach

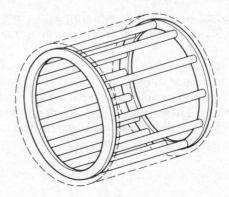

**Figure 10.5   (a) Arrangement of conductor bars in 'squirrel-cage' rotor**

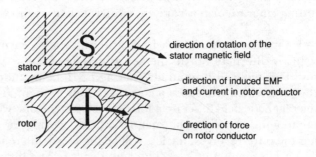

**Figure 10.5   (b) A completed induction rotor**

**Figure 10.6   Induction-motor principle**

synchronous speed, because if it did so, no EMF would be induced and no current would flow. In practice, a 'balance speed' is reached just below synchronous speed, where there is sufficient rotor current to provide the necessary torque. If the torque required of the rotor increases, the rotor speed falls slightly, so that the speed difference between the rotating magnetic field and the rotor increases. This gives greater induced EMF, more current and a higher developed torque. The extra current in the rotor to provide the extra torque results in extra current to the stator by transformer action, so that extra input power is provided to give the additional output.

It is because this type of motor relies on induced EMF in its rotor conductors that it is called an induction motor. Another type of induction motor, the wound-rotor type, has an insulated winding on the rotor, the ends being brought out to sliprings which are short-circuited during normal running.

Cage induction motors are cheaper and smaller than wound-rotor types, but have a lower starting torque. The speed of wound-rotor motors can be reduced by introducing resistance into the rotor circuit by means of the sliprings, but this is a wasteful method owing to the power dissipated as heat in the resistors.

Some induction motors use a 'pole-changing' winding to give step variation of speed range. Such a winding can be reconnected to change its number of poles and hence its synchronous speed.

The direction of rotation of a three-phase induction motor can be reversed by changing over any two of the three stator connections. This will reverse the phase sequence, and reverse the direction of the rotating magnetic field.

The cage induction motor is a virtually constant-speed machine, but is the standard motor for industrial use.

## 10.5   Induction-motor calculations

Section 10.2 shows that the speed of rotation of the magnetic field due to the supply current in the stator winding of a three-phase induction motor is given by the expressions

$$n_s = \frac{f}{p} \quad \text{and} \quad N_s = \frac{60f}{p}$$

In practice, the rotor will revolve at a speed slightly less than synchronous owing to the need for the magnetic field to move past the rotor conductors and so provide the induction effect (see Section 10.4).

### *Slip* s

An induction motor must have slip. If the rotor runs at the same speed as the magnetic field (synchronous speed), it will keep pace with the flux and will not be cut by it. There will be no induced EMF, and no current to drive the rotor.

In practice, an induction motor will settle at a steady speed which is less than synchronous, so that the rotating magnetic field 'slips' past the rotor at a rate just high enough to induce the EMF needed to drive the rotor current to provide the torque for that particular speed.

The difference between rotor speed and synchronous speed divided by the synchronous speed is called slip. It can be expressed as 'per-unit slip' or as 'percentage slip'.

$$\text{Per-unit slip } s = \frac{n_s - n}{n_s} = \frac{N_s - N}{N_s}$$

$$\text{Percentage slip } s = \frac{n_s - n}{n_s} \times 100\% = \frac{N_s - N}{N_s} \times 100\%$$

$$\text{Percentage slip} = \text{per-unit slip} \times 100\%$$

where  $n_s$ = synchronous speed (r/s)
$N_s$ = synchronous speed (r/min)
$n$ = rotor speed (r/sec)
$N$ = rotor speed (r/min)

## Example 10.3

A two-pole 50 Hz three-phase induction motor runs at 2850 r/min. Calculate the per-unit slip and the percentage slip.

$$N_s = \frac{60f}{p} = \frac{60 \times 50}{1} \text{ revolutions per minute} = 3000 \text{ r/min}$$

$$\text{Per-unit slip } s = \frac{N_s - N}{N_s} = \frac{3000 - 2850}{3000} = \frac{150}{3000} = 0\cdot05$$

$$\text{Percentage slip } s = \text{per-unit slip} \times 100\% = 0\cdot05 \times 100\% = 5\%$$

## Example 10.4

A four-pole 50 Hz three-phase induction motor has a per-unit slip of $0\cdot03$ on full load. Calculate the full load speed.

$$n_s = \frac{f}{p} = \frac{50}{2} \text{ revolutions per second} = 25 \text{ r/s}$$

$$s = \frac{n_s - n}{n_s} \qquad sn_s = n_s - n$$

$$n = n_s - sn_s = 25 - (0\cdot03 \times 25) \text{ revolutions per second}$$

$$= 25 - 0\cdot75 \text{ revolutions per second} = 24\cdot25 \text{ r/s}$$

or  $N = 24\cdot25 \times 60 \text{ revolutions per minute} = 1455 \text{ r/min}$

## Example 10.5

A six-pole three-phase induction motor runs at 1152 r/min with 4% slip. Calculate the supply frequency.

$$s = \frac{N_s - N}{N_s} \quad \text{so} \quad \frac{4}{100} = \frac{N_s - 1152}{N_s}$$

$$4N_s = 100(N_s - 1152)$$

$$4N_s = 100N_s - 115\ 200$$

$$115\ 200 = 100N_s - 4N_s = 96N_s$$

$$N_s = \frac{115\ 200}{96} \text{ revolutions per minute} = 1200 \text{ r/min}$$

$$N_s = \frac{60f}{p} \quad \text{so} \quad f = \frac{pN_s}{60}$$

$$f = \frac{1200}{60} \times \frac{6}{2} \text{ hertz} = 60 \text{ Hz}$$

## Example 10.6

A 50 Hz three-phase induction motor runs at $12 \cdot 25$ r/s with $0 \cdot 02$ slip. How many poles has it?

$$s = \frac{n_s - n}{n_s}$$

Transposing the formula,

$$n_s = \frac{n}{1 - s}$$

$$1 - s = 1 - 0 \cdot 02 = 0 \cdot 98$$

$$n_s = \frac{12 \cdot 25}{0 \cdot 98} \text{ revolutions per second} = 12 \cdot 5 \text{ r/s.}$$

$$n_s = \frac{f}{p} \quad \text{sp} \quad p = \frac{f}{n_s} = \frac{50}{12 \cdot 5} \text{ pairs of poles} = 4$$

Number of poles = pairs of poles × 2 = 4 × 2 = 8 poles

## *Rotor frequency*

When the supply is switched on to the stator of an induction motor, the magnetic field will rotate instantly at synchronous speed. The rotor is heavy, and will take time to build up to its normal running speed. When the rotor is at standstill, it will behave like the secondary winding of a transformer (see Chapter 7) with the stator as the primary, and the frequency of the induced EMF and rotor current will be the same as the supply frequency.

As the rotor picks up speed following the stator field, the poles of the rotating magnetic field will sweep past its conductors at a lower rate, inducing an EMF of lower frequency. The frequency of the rotor EMF and current will depend on the difference between synchronous and rotor speeds, and hence on slip.

$$f_r = sf$$

where   $f_r$ = rotor frequency (Hz)
        $s$ = slip (percentage or per-unit)
        $f$ = supply frequency (Hz)

## Example 10.7

Calculate the frequency of the rotor current in the machine of (i) Example 10.3, (ii) Example 10.4, (iii) Example 10.5, and (iv) Example 10.6.

(i) $f_r = sf = 0 \cdot 5 \times 50$ hertz $= 2 \cdot 5$ Hz

(ii) $f_r = sf = 0 \cdot 03 \times 50$ hertz $= 1 \cdot 5$ Hz

(iii) $f_r = sf = 4\% \times 60$ hertz $= \dfrac{4}{100} \times 60$ hertz $= 2 \cdot 4$ Hz

(iv) $f_r = sf = 0 \cdot 02 \times 50$ hertz $= 1$ Hz

## Example 10.8

A four-pole, 50 Hz induction motor runs with a slip of 1% on full load. Calculate the frequency of the rotor current (*a*) at standstill, and (*b*) on full load.

(*a*) At standstill, slip = 1
     $f_r = sf = 1 \times 50$ hertz $= 50$ Hz

(*b*) $f_r = sf = 1\% \times 50$ hertz $= 0 \cdot 5$ Hz

## *Speed control of AC motors*

Historically both induction and synchronous motors have been considered to be single-speed machines because the synchronous (rotating-magnetic-field) speed has always been fixed by the supply frequency, which is constant. Pole-changing motors (see Section 10.2) have been used to give some control of speed, but continuous control is not possible with this method. The speed of a wound-rotor induction motor can be reduced by adding resistance to the rotor circuit via the sliprings, but this is an expensive and inefficient method of speed control. However, starting torque is higher although the wound-rotor motor is inherently less robust than the cage-rotor type. Excluding this method, the speed of an induction motor will be only slightly lower than synchronous speed, so some running speeds are impossible at normal power frequencies. For example, the synchronous speeds of two-pole and four-pole machines at 50 Hz are 3000 r/min and 1500 r/min, respectively, so a rotor speed of 2000 r/min cannot be achieved without the use of belts, chains or gearboxes. Speeds of 2900 r/min and 1450 r/min, respectively, are of the correct order.

The traditional single-speed role of the induction motor has been changed by the use of modern power electronics. Using such methods, it is possible to rectify the three-phase supply to give a direct current, and then to invert this direct supply to provide a three-phase system of any required frequency. Since the synchronous speed of the motor, and hence its running speed, are directly related to supply frequency, electronic frequency control will also give speed control (see Figure 10.7). In some cases battery-driven vehicles use this method, the DC battery supply being inverted to the variable-frequency supply needed to give speed control. Since the induction motor is far more rugged and is likely to need less maintenance than its direct-current counterpart, the advantage is clear.

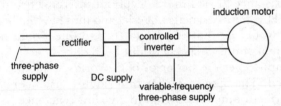

**Figure 10.7   Block diagram for electric speed control of a three-phase induction motor**

## Variation of speed with torque

The torque developed by an induction motor depends on the rotor current and on the strength of the magnetic flux to which the conductors carrying that current are subjected. Owing to rotor reactance, current lags voltage and this means that the maximum values of flux and of current do not coincide. Rotor reactance is a function of rotor frequency; when the rotor is moving slowly (soon after starting) slip, and hence rotor frequency, will be high, resulting in low torque. As the rotor accelerates, rotor frequency reduces and torque increases as maximum flux and current move closer together. Maximum torque will occur when rotor reactance is equal to rotor resistance.

As the machine speeds up further, slip falls and so does the current in the rotor. Because of this, torque also falls, the relationship between torque and speed being shown in Figure 10.8. From the curve, it is clear that starting torque for the cage-rotor machine is low.

## Double-cage induction motors

The starting torque of a normal cage-rotor induction motor is low because its inductive reactance when starting is low by comparison with its resistance.

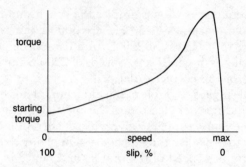

**Figure 10.8   Speed/torque relationship for a cage-rotor induction motor**

If the cage could have lower resistance and higher inductance, starting torque would be better. This could be achieved by using thicker bars to have lower resistance buried more deeply in the rotor drum so that they are surrounded by more magnetic material and thus have higher inductance. The problem would be that, while starting torque would be improved, torque at normal running speed would be reduced.

In an attempt to get the better of both worlds, a double-cage rotor is sometimes used. The outer cage has thin (high-resistance) bars arranged close to the surface of the rotor drum to give low reactance. An inner cage has thicker (lower-resistance) bars buried deeper within the drum to give higher inductance. One possible slot arrangement is shown in Figure 10.9.

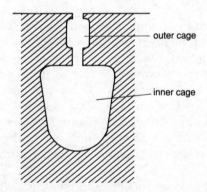

**Figure 10.9   One slot of a double-cage induction motor**

Each cage will make its contribution to total torque. The low-resistance inner cage will provide high starting torque, while the high-resistance outer cage gives good running torque. The torque/speed curves for a double-cage rotor machine are shown in Figure 10.10.

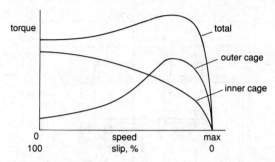

**Figure 10.10 Speed/torque relationship for a double-cage induction motor**

## 10.6 Starting induction motors

Since the magnetic field due to the stator of a three-phase induction motor rotates as soon as the supply is connected, the machine is self starting. The purpose of a starter is not, therefore, to start the machine as the name implies. The starter performs two functions:

 (i)  to provide overload and 'no-volt' protection (see Section 9.8), and
(ii)  to reduce the heavy starting current and the time for which it flows.

There are a number of types of starter, the more important being described as follows.

### Direct-on-line starter

This starter is nothing more than a contactor which switches the supply to the motor. It does nothing to limit starting current, but is used widely for small machines with outputs of up to about 5 kW since it provides protection. Figure 10.11 shows the wiring diagram for a typical starter of this type.

When the 'start' button is pushed, the control circuit is completed so that the operating coil is energised, closing the main contacts and feeding the motor. At the same time, a pair of auxiliary contacts is closed which 'short out' the start-button contacts, thus allowing the start button to be released without opening the contactor and stopping the motor. In the event of the 'stop' button being pressed, or the overload trip operating, the control circuit is broken and the contactor 'drops out'. A starter of this type should normally be mounted as directed in the manufacturer's instructions, to ensure that the contacts are opened by gravity when the control circuit becomes dead. This ensures that the device will stop the motor in the event of an internal failure (it is said to 'fail safe'), and also gives no-volt protection, since once the auxiliary contacts open the starter can only be operated again by the 'start' button. To ensure that an overload is brought to attention, many starters have to be 'reset' by pushing the 'stop' button before they can be restarted.

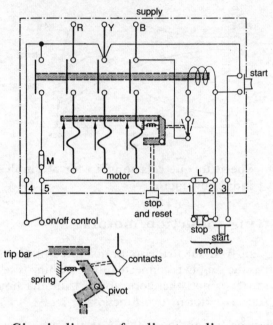

**Figure 10.11   Circuit diagram for direct-on-line starter**

Link L should be removed for remote operation, and link M for on/off control by a float switch or similar device

## Star–delta starter

If three loads are connected in star to a three-phase supply, the line current has one-third the value it has when the same three loads are connected in delta, and only one-third of the power is dissipated (see Example 5.21). If an induction motor has all six ends of its three stator windings available, it can be connected as shown in Figure 10.12 to a star–delta starter which connects the windings in star until the machine is running, when the delta connection is selected. The starter includes overload and no-volt protection, as well as a device to prevent connection in delta before the star connection has been selected, but these items are omitted from the diagram for the sake of clarity.

As the starting current has one-third of the value it would have for the direct-on-line connection, starting torque is low and the time taken for the machine to run up to speed, and hence for the current to fall to the normal running value, is comparatively long. These starters are used for machines with outputs of up to about 25 kW.

## Autotransformer starter

This starter has been described in Section 7.11. Since the tapping can be selected to give the desired starting current and torque values, this starter is to be preferred to the star–delta type. However, it is more expensive and the autotransfomers are bulky.

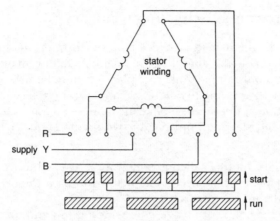

**Figure 10.12  Simplified circuit diagram for star–delta starter**

## Rotor-resistance starter

Some large induction motors have a rotor winding of insulated coils let into slots in the rotor-drum surface, the winding being star-connected with the ends led out to sliprings on the shaft. If the machine is started with resistance added to the rotor circuit via the sliprings, the rotor current, and hence the stator current, is reduced (see Figure 10.13). The improved resistance-to-reactance ratio of the rotor also improves its power factor and improves starting torque. This type of starter is used only for large machines, and may also be used to reduce the speed of the motor, although the method is very inefficient.

**Figure 10.13  Wound rotor for three-phase induction motor, showing sliprings and brushgear**

## 10.7  Starting synchronous motors

The rotating magnetic field provided by a synchronous motor sweeps past the rotor poles so quickly that they are unable to follow an attracting stator pole before a repulsing pole takes its place. It may be helpful to liken the rotor to a passenger trying to board an express train (the stator) as it rushes through a station at full speed. Unlike the induction motor, the synchronous motor is not self starting, and one of two methods may be used to start these machines.

(i) An auxiliary motor, called a 'pony motor' is used to run up the rotor close to synchronous speed. The field winding is then excited and the rotor 'pulls in' to synchronism.
(ii) Solid salient poles, or a cage-type winding in addition to the normal rotor winding, give an induction effect to run the machine up to a speed approaching that of the rotating stator field. Energising the field poles on the rotor with direct current will make the machine 'pull in' to synchronous speed.

Overload and no-volt protection are needed for synchronous motors (see Section 10.6) as well as an interlock to prevent the field current from being switched on before the speed of the rotor approaches synchronous speed. If the latter precaution is not taken, very violent fluctuations in stator current will result.

## 10.8  Single-phase induction motors

We saw in Section 10.2 that a suitable winding fed by a three-phase supply will set up a perfect rotating magnetic field, its rotational speed and the magnitude of its magnetic flux both being constant for a given supply. A single-phase supply will not provide a rotating magnetic flux when applied to a single winding; the flux alternates. However, it can be shown (see Figure 10.14) that such an alternating flux may be considered to consist of two equal fluxes, both rotating at synchronous speed, but in opposite directions.

These two flux components will induce equal and opposite EMFs in a stationary rotor system, and these will cancel out so that there will be no current, and no torque to start the motor. Thus, a single-phase motor of this type cannot be self-starting. However, if the rotor is moving in either direction, it can be shown that EMFs will be induced which will provide current to give torque which makes the rotor continue to follow the component of flux which has the same direction. The machine will then run up to normal speed, which, like the three-phase machine, is less than synchronous speed owing to the necessary slip.

Calculations for single-phase-motor synchronous speed, rotor speed, slip and rotor frequency are exactly the same as for those of the three-phase machine explained in Sections 10.2 and 10.5.

A single-phase induction motor will run perfectly satisfactorily in whichever direction its rotor first begins to move. Thus, we could start these machines by the simple process of twisting them in the correct direction, if this were practicable. Something better is needed, so a starting winding is included. This winding is installed so that any ampere-turns it produces are at an angle to those produced by the main (running) winding. If the current in the starting winding is arranged to be out of phase with that in the running winding by about the same phase angle as the physical displacement of the coils, a rotating magnetic field results.

This field is a very poor imitation of the perfect three-phase rotating field, but is good enough to set the rotor moving in one direction and to provide the extra torque needed for starting. The extra starting winding is normally removed from circuit when the machine runs up to speed, often by means of a centrifugal switch.

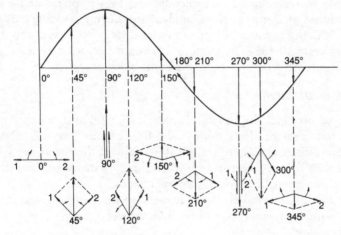

**Figure 10.14   Two rotating fields due to single-phase alternating flux**

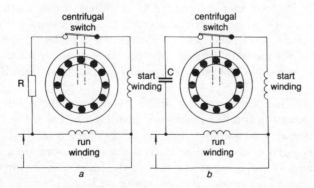

**Figure 10.15   Single-phase motors**
   *a* Resistance start
   *b* Capacitor start

The phase difference between the two windings can be obtained in a number of ways, but the most common are:

### Resistance-start motor

The running winding has a high inductance, so the current in it lags the voltage by a large angle. The starting winding is given a much higher resistance, either by winding it with thinner wire or by adding resistance in series with it. Thus, its current lags voltage by a small angle, and there is a phase difference between the two currents. An arrangement for a resistance-start motor is shown in Figure 10.15a.

### Capacitor-start motor

The current in the start winding is made to lead the supply voltage by connecting a capacitor in series with the start winding. This method is prefereable to resistance start, because the bigger phase angle between lagging running-winding current and leading starting-winding current gives a higher starting torque. An arrangement for a capacitor-start motor is shown in Figure 10.15b. The starting winding is only used for a few seconds, and is usually 'short-time rated' and will overheat if left in circuit.

### Capacitor-run motor

The torque provided by the combination of 'run' and 'start' windings is greater than that of the running winding alone. For this reason, the centrifugal switch is sometimes omitted, so that the machine both starts and runs with the capacitor in circuit. In this case the second winding must of course, be rated for continuous operation.

This machine is strictly called a two-phase machine because it uses two currents which have a phase difference between them.

### Universal series motor

In Section 9.7 we considered types of DC machines, including the DC series motor. If such a machine were connected to an AC supply, the currents in the armature and in the field would reverse together. If both change, the resulting torque would be in the same direction; in other words, the series motor is a universal type, which will operate from both DC and AC supplies.

If we are to use a series motor on an AC supply, there will have to be some changes from the DC machine considered in the previous chapter. The most important change is that the field system must be laminated to reduce eddy-current losses. Most universal series motors use a stator (field system) which is much more like that of the induction-motor stator than the DC machine field system, the field windings being placed in slots around the laminated stator rather than on salient poles. The circuit diagram and the operating characteristics of series motors, both AC and DC, will be found as Figures 9.9b and 9.22, respectively.

Alternating-current series motors, often called AC commutator motors, have been very widely used in the past, particularly for traction purposes, where motors with outputs of up to 2 MW have been used.

## Shaded-pole motor

Shaded-pole motors have poor starting and running torque and are used principally for their cheapness. The rotor of such a machine is of the cage type, and thus depends for its operation on the presence of a rotating magnetic field. A poor rotating field is provided by splitting each of the salient poles into two sections, one of which is surrounded by a heavy, low-resistance shading band of copper or aluminium.

The band has an induced EMF which drives a heavy short-circuit current in it, which forces the magnetic flux in the shaded part of the pole to lag that in the unshaded part. This phase difference results in the poor rotating magnetic field. The arrangement is shown in Figure 10.16. The machine is not widely used except for very small machines used to drive loads such as fans where cost is more important than efficiency.

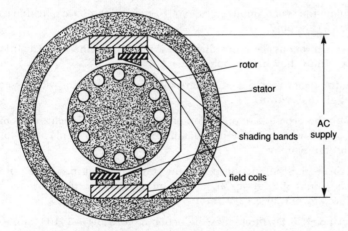

**Figure 10.16   Shaded-pole single-phase motor**
Note that the gap between the stator and the rotor has been enlarged to show pole detail

## 10.9   Summary of formulas for Chapter 10

See text for definitions of symbols.

Synchronous speed:

$$n_s = \frac{f}{p} \qquad f = pn_s \qquad p = \frac{f}{n_s}$$

$$N_s = \frac{60f}{p} \qquad f = \frac{pN_s}{60} \qquad p = \frac{60f}{n_s}$$

Per-unit slip:

$$s = \frac{n_s - n}{n_s} \qquad n = n_s - sn_s \qquad n_s = \frac{n}{1 - s}$$

$$s = \frac{N_s - N}{N_s} \qquad N = N_s - sN_s \qquad N_s = \frac{N}{1 - s}$$

Percentage slip = per-unit slip × 100%

Rotor frequency:

$$f_r = sf \qquad s = \frac{f_r}{f} \qquad f = \frac{f_r}{s}$$

## 10.10   Exercises

1  Show by a series of MMF vector diagrams how a 'rotating' field is produced by a spaced three-phase winding such as in an induction motor.

2  Calculate the synchronous speed of a 10-pole three-phase machine connected to a 50 Hz supply.

3  If the synchronous speed of a three-phase machine connected to a 50 Hz supply is 750 r/min, how many poles has it?

4  What frequency is applied to a four-pole three-phase stator with a synchronous speed of 1800 r/min?

5  Explain the principle of operation of the three-phase synchronous motor. At what speed will this type of machine run? Give an application for a large synchronous motor.

6  In a three-phase induction motor the stator winding produces a flux which travels round the airgap at a constant velocity. Explain briefly how this produces torque in a cage rotor.

7  A four-pole, 50 Hz three-phase induction motor runs at 1410 r/min. Calculate the slip and percentage slip.

8  A six-pole 60 Hz three-phase cage induction motor has a slip of 2%. Calculate its speed.

9  An eight-pole three-phase wound-rotor induction motor runs at 720 r/min with a slip of 4%. What is the supply frequency?

10  (*a*) Explain with the aid of diagrams how a rotating magnetic field produces torque in an induction motor.
    (*b*) State why such a motor always runs below synchronous speed.

11  (*a*) What is meant by the slip of an induction motor?
    (*b*) The rotor of a four-pole three-phase induction motor revolves at 1380 r/min. The slip is 8%. Find the supply frequency.

12  Calculate the frequency of the rotor EMF for each of the machines of Exercises 7, 8 and 9.

13 (*a*) Why cannot an induction motor run at synchronous speed?

(*b*) A six-pole, 50 Hz three-phase induction motor runs with 5% slip. What is its speed? What is the frequency of the rotor current?

14 A four-pole 50 Hz three-phase induction motor runs with 4% slip at full load. What will be the frequency of the current induced in the rotor

(*a*) at starting?

(*b*) at full load?

15 What is the slip of an induction motor? Calculate the speed in r/min of a six-pole three-phase induction motor which has a slip of 6% at full load with a supply frequency of 50 Hz. What will be the speed of a four-pole alternator supplying the motor?

16 Explain the need for no-volt protection and overload protection for an induction motor. Use a circuit diagram to show how these forms of protection are applied to a direct-on-line starter.

17 Explain, with the aid of a circuit diagram, the operation of a star–delta starter for a three-phase cage induction motor. Show how the device reduces starting current.

18 Draw a circuit diagram for an autotransformer starter for a three-phase induction motor and explain its operation.

19 Why will the use of a resistance starter reduce the starting current and improve the starting torque of a wound-rotor induction motor? Sketch a simple circuit diagram for such a starter.

# 10.11   Multiple-choice exercises

10M1   A balanced three-phase supply can be wound in the stator of a machine to set up:

(*a*) an alternating magnetic field

(*b*) a perfect rotating magnetic field

(*c*) a short circuit

(*d*) a poor rotating magnetic field.

10M2   A four-pole winding connected to a 50 Hz three-phase supply will set up a magnetic field which rotates at:

(*a*) 50 r/s      (*b*) 3000 r/min      (*c*) 3600 r/min      (*d*) 25 r/s.

10M3   A three-phase induction motor which sets up a magnetic field rotating at 750 r/min when connected to a 100 Hz supply must have:

(*a*) eight poles   (*b*) two poles      (*c*) four poles      (*d*) one pole.

10M4   The synchronous motor differs from the induction motor because:

(*a*) each pole is provided with a heavy shading band

(*b*) it runs at nearly twice the rotational speed

(*c*) the rotor has a constant magnetic field

(*d*) the rotor is provided with two cage windings.

10M5   The winding on the rotor of the simplest and most usual type of induction motor is called a:

(*a*) lap winding  (*b*) 'squirrel' cage
(*c*) wound rotor  (*d*) wave winding.

10M6   Both the rotor and the stator of an induction motor are laminated because:
(*a*) hysteresis losses would be a problem if not
(*b*) the machine is much more robust when constructed in this way
(*c*) otherwise eddy-current losses would be very heavy
(*d*) it is not a DC machine.

10M7   If $n_s$ is the magnetic field speed in r/s, $f$ is the frequency of the supply and $p$ is the number of pairs of poles on the stator winding of a three-phase machine, the formula relating these factors is:
(*a*) $n_s = \dfrac{f}{p}$   (*b*) $n_s = \dfrac{p}{f}$   (*c*) $p = \dfrac{n_s}{f}$   (*d*) $f = \dfrac{p}{n_s}$

10M8   A two-pole 60 Hz induction motor has a slip of 2% on full load. The full-load speed is:
(*a*) 58·8 r/s  (*b*) 3528 r/min
(*c*) 60 r/s  (*d*) 2880 r/min

10M9   The slip of an induction motor can never be zero in a stable state because:
(*a*) the rotor would be at standstill
(*b*) induced EMF and rotor current would be excessive
(*c*) it would then be a synchronous motor
(*d*) there would be no rotor EMF, rotor current or torque.

10M10   The frequency $f_r$ of rotor currents in an induction motor is related to supply frequency $f$ and slip $s$ by the formula:
(*a*) $f_r = \dfrac{1}{sf}$   (*b*) $f_r = \dfrac{s}{f}$   (*c*) $f_r = sf$   (*d*) $f_s = \dfrac{f}{s}$

10M11   The torque provided by the rotor of an induction motor is at a maximum when:
(*a*) speed is maximum
(*b*) rotor resistance equals rotor reactance
(*c*) the connected mechanical load is at a minimum
(*d*) starting.

10M12   If an electronic speed controller feeds an induction motor, the machine speed depends on:
(*a*) applied voltage  (*b*) rotor current
(*c*) rotor power factor  (*d*) applied frequency.

10M13   A double-cage induction motor is sometimes used because:
(*a*) it gives twice the torque of a single-cage machine
(*b*) the power factor of the machine is better
(*c*) starting torque is much higher than for the single-cage motor
(*d*) it gives twice the speed of a single-cage machine.

10M14   A starter is usually needed for an induction motor:
(*a*) because it is not self starting
(*b*) to give greater safety for the operator
(*c*) because otherwise it will run up to speed too quickly
(*d*) to limit the high starting current.

10M15 A star–delta starter used with an induction motor:
  (*a*) connects the stator in star for starting, then in delta for running
  (*b*) reduces the supply voltage to the machine during starting
  (*c*) connects the stator in delta for starting, then in star for running
  (*d*) connects the rotor in star for starting, then in delta for running.

10M16 A rotor-resistance starter for a three-phase wound-rotor induction motor:
  (*a*) adds external resistance to the wound rotor via sliprings
  (*b*) is connected in series with the machine stator
  (*c*) can only be used in conjunction with a double-cage rotor
  (*d*) applies only to DC motors.

10M17 A large synchronous motor is often started by:
  (*a*) connecting it directly to the three-phase supply
  (*b*) pushing the rotor in the required direction before connecting the supply
  (*c*) running it up to speed using a 'pony motor' before energising the rotor
  (*d*) first switching on the rotor supply then applying stator voltage.

10M18 A single-phase induction motor:
  (*a*) can only be used for loads rated at less than $0 \cdot 5$ kW
  (*b*) will run in whichever direction the rotor is started
  (*c*) must have a capacitor to start it
  (*d*) must never have its frame connected to earth.

10M19 A series motor can be used on AC or DC supplies provided that:
  (*a*) the supply voltage never exceeds 250 V
  (*b*) the load is low
  (*c*) the field system is laminated
  (*d*) only low-frequency AC supplies are used.

10M20 A shaded-pole motor:
  (*a*) can only be used in situations protected from direct sunlight
  (*b*) uses a three-phase supply and feeds very heavy loads
  (*c*) must have a derating factor of $0 \cdot 7$ applied to it
  (*d*) is fed from a single-phase AC supply and is usually of low rating.

## Chapter 11
# Electrical measuring instruments

## 11.1   Introduction

If a carpenter wishes to check his work, he can do so by using a rule, since one of his aids is the measurement of length. Similarly, the greengrocer can use scales to measure the weight of his produce, and the works timekeeper can use a clock. As well as using all these methods, the electrical craftsman is concerned with measurements which have no effect on his senses. The passage of an electrical current cannot be seen, heard, smelt or, we would hope, felt under normal conditions.

The measurements of current is carried out using special instruments which measure one or other of its effects. Such instruments are often much more delicate than the carpenter's rule or the greengrocer's scales, and must be correctly connected to the circuit if both instruments and circuit are to function correctly.

Both basic types of electrical measuring instrument have already been considered in basic principle in 'Electrical Craft Principles' Volume 1. Here we shall consider them in slightly more detail, and show how some of them can be modified to prevent their disadvantages from affecting accurate applications.

## 11.2   Permanent-magnet moving-coil instrument

As explained in Section 11.6 of 'Electrical Craft Principles' Volume 1, this instrument is rather like a very small DC motor with a permanent-magnet field and having a deliberately limited angle of rotation so that no commutator or brushes are required. Beryllium-copper or phosphor-bronze hairsprings provide both the connection for current to the moving coil, and the controlling torque against which the deflecting torque provided by the current moves the coil and its associated pointer. The angle of deflection depends on the deflecting torque if the springs have a linear characteristic, and hence on the current in the coil since the magnetic field due to the permanent magnets is unvarying. The scale over which the pointer moves thus can be calibrated in units of electrical current, and it is an even scale since deflection depends only on current (see Figure 11.7).

The general arrangement of this type of instrument is shown in Figure 11.1, which shows one possible shape for the magnetic circuit. Modern permanent magnets are most easily made in short, relatively straight lengths, so it is usual to complete the often-complicated shape of the magnetic circuit with soft-iron yoke pieces. The bearings must be very fine so as to offer as little friction as possible to the movement. Excellent damping is provided by the induction of eddy currents in the aluminium-coil former, which, obeying Lenz's law, set up a force opposing the movement.

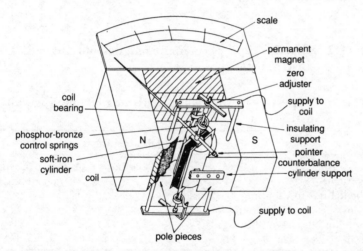

**Figure 11.1   Permanent-magnet moving-coil instrument**

There are several variations on the basic instrument, all intended to improve accuracy or reduce size and weight. For example, the coaxial type, with the permanent magnet inside the coil, and with the magnetic circuit completed by a mild-steel sleeve as shown in Figure 11.2. The taut-band suspension shown in Figure 11.3 has the whole of the moving coil system suspended on a stretched strip of phosphor-bronze which also provides the path for current to the coil and dispenses with bearings completely.

The permanent-magnet moving-coil instrument is undoubtedly the best of the deflecting types, but suffers a very serious disadvantage in that it is unsuitable for measuring alternating currents. Alternating current in the coil sets up alternating forces, so that at very low frequencies the pointer swings first one way and then the other, whilst at normal power frequencies and above, the moving system is unable to follow the variations of torque, and simply stands still, pointing to zero. The properties of this very important instrument are summarised in Table 11.1.

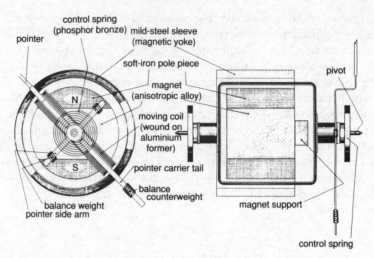

**Figure 11.2   Coaxial type permanent-magnet moving-coil instrument**

**Table 11.1:  Properties of the permanent-magnet moving-coil instrument**

| Advantages | Disadvantages |
|---|---|
| 1 Very accurate | 1 Cannot be used on AC supplies |
| 2 Linear scale | |
| 3 Excellent built-in damping | |
| 4 Very low power consumption | |
| 5 Easily adapted as a multirange instrument (see Section 11.11) | |

## 11.3   Rectifier moving-coil instrument

The advantages of the permanent-magnet moving-coil instrument are so great that it is worth considerable trouble to remove its only real disadvantage, which is its unsuitability in its basic form for AC supplies. Two methods are used for this purpose, the first of which is the rectification of the alternating current before applying it to the instrument. A bridge connection, as shown in Figure 11.4, is usually used, with four rectifiers often of the copper-oxide or selenium types. The instrument reading depends on the average value of the direct current carried, and hence on the average value of a halfcycle of the alternating current. Since the effective (or RMS) value of the AC supply is actually required, this is usually taken into account by calibrating the instrument on the assumption that the current will be sinusoidal. If the waveshape varies from sinusoidal, the ratio of its RMS to average values (the form factor) will vary, and an error will be introduced.

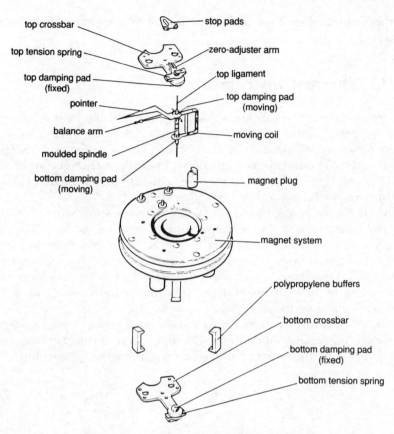

Figure 11.3 **Permanent-magnet moving-coil instrument with taut-band suspension (exploded view)**

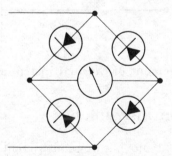

Figure 11.4 **Circuit for rectifier moving-coil instrument**

The rectifier moving-coil instrument is most usually applied as a low-current ammeter, or as an AC voltmeter giving full-scale deflection above about 50 V. Below this level, the forward voltage drop in the rectifiers becomes significant and introduces error. The self capacitance of the rectifiers usually limits use of the instrument to frequencies below about 50 kHz.

## 11.4   Thermal instrument

If wires of different metals are welded together and the joint is heated, a small potential difference is found to occur between them, its value depending on the temperature of the junction. This is known as the **Seebeck effect**, and the dissimilar metal junction is known as a **thermocouple**. This phenomenon is used to convert the standard permanent-magnet moving-coil instrument for use on AC supplies.

The current to be measured is passed through a small wire of resistive material which is heated to a temperature depending on the RMS value of the current. Attached to the wire by a bead of ceramic material, which is a good electrical insulator and a good heat conductor, is a thermocouple with its ends connected to a sensitive instrument as shown in Figure 11.5. The reading of the instrument depends on the heater current, regardless of its waveshape or frequency.

Thermal instruments are not in wide use, usually appearing as ammeters for use on AC supplies of doubtful waveshape and of variable frequency, such as are sometimes found in telephony and radar applications.

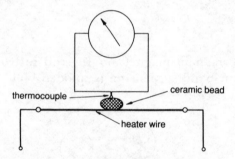

**Figure 11.5   Arrangement of thermal instrument**

## 11.5 Moving-iron instrument

The two basic types of moving-iron instrument, attraction type and repulsion type, are explained in Section 7.8 of 'Electrical Craft Principles' Volume 1. The repulsion type depends on the mutual repulsion of two iron bars or plates when both are magnetised with the same polarity by the current carried by the coil inside which both are placed. The attraction type depends on the attraction of an iron vane into a coil carrying the current to be measured. In both cases the instrument reads for both alternating and direct current; for the repulsion type, this is because both bars or plates change polarity together and continue to repel, whereas for the attraction type the iron vane is attracted by magnetic field of either polarity.

Figure 11.6 shows a moving-iron instrument which combines the repulsion and attraction effects, and thus has a more even scale than usual. Since deflecting torque depends on the square of the current, the instrument reads true RMS values on AC supplies, regardless of waveform, but the scale is very cramped at the lower end, the first 20% of the full-scale reading being almost useless (see Figure 11.7). The power absorbed by a moving-iron instrument, both in terms of copper loss in the coil and iron losses in the bars, plates or vanes is high. The impedance of the coil increases with frequency, giving another source of error.

The moving-iron instrument, unlike the moving-coil type, has no built-in damping, which must be specially arranged by using an air vane in a closed box or a disc moving between the poles of a permanent magnet. Owing to its self inductance, and the change of impedance with frequency, the moving-iron instrument is not suitable for use with shunts (see Section

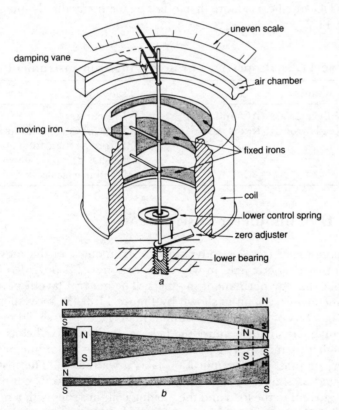

**Figure 11.6  Long-scale combined attraction and repulsion moving-iron instrument**
*a* Shows the general arrangement
*b* Shows the developed shapes of the fixed and moving irons

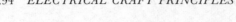

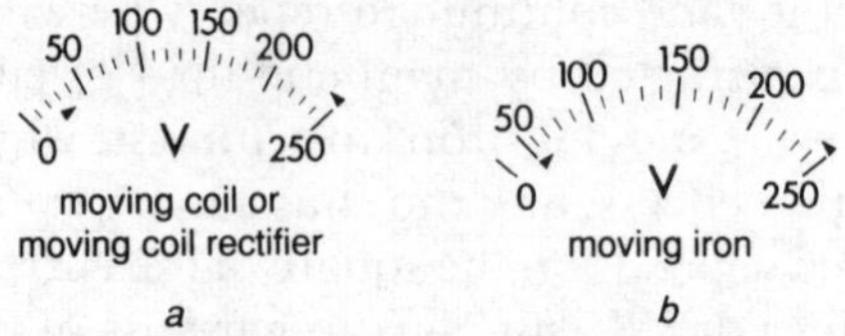

moving coil or
moving coil rectifier

*a*

moving iron

*b*

**Figure 11.7   Comparison of instrument scales**
*The arrow heads indicate the useful parts of the scales*
*a Moving-coil or moving-coil rectifier*
*b Moving-iron*

11.7) and either has a coil wound to take the measured current or is used in conjunction with a current transformer (see Section 7.10). Owing to the need for a minimum magnetic-field strength for operation, the moving-iron instrument cannot be made as a low-current, low-resistance ammeter.

The major advantage of the moving-iron instrument of universal use (AC and DC supplies) is more than offset by the many disadvantages listed in Table 11.2.

**Table 11.2: Properties of the moving-iron instrument**

| Advantages | Disadvantages |
|---|---|
| 1 Universal use (AC and DC) | 1 Inaccurate |
| 2 Reads true RMS on AC | 2 High power consumption |
| | 3 Errors at high frequency |
| | 4 Special damping needed |
| | 5 Nonlinear scale |

## 11.6   Dynamometer instrument

The dynamometer (or electrodynamic) instrument is of the moving-coil type, but the magnetic field in which the coil moves is provided by fixed coils rather than by permanent magnets. The general layout of a typical dynamometer instrument is shown by Figure 11.8, the movement being enclosed in a box of 'Mumetal' or similar material to give shielding against external magnetic fields because the field set up by its fixed coils is weak.

Although this instrument is sometimes used as an ammeter or as a voltmeter, its most usual application is as a wattmeter. The instrument is connected as shown in Figure 11.9, with the fixed coils connected to carry the current to the load and the moving coil, in series with a multiplier (see Section 11.8), connected across the supply. The torque on the moving coil at any instant depends on the values of current in both sets of coils, and, since the moving coil carries a current proportional to the supply voltage, torque depends on the circuit current and voltage values, and hence

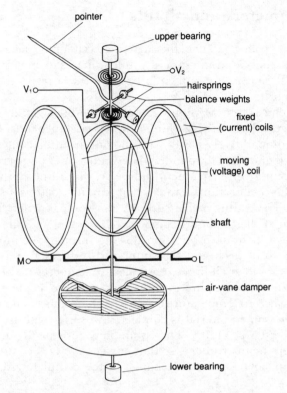

**Figure 11.8 General arrangement of dynamometer instrument**

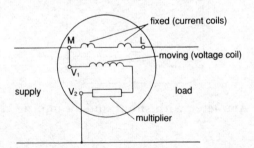

**Figure 11.9 Connection of dynanometer instrument used as wattmeter**

on power. The instrument will thus read average power, taking into account the circuit power factor on AC supplies. The magnetic field due to the fixed coils is too weak to provide eddy-current damping, and an airbox-and-vane damper is often used.

The dynamometer wattmeter is less accurate than the permanent-magnet moving-coil instrument owing largely to the weak magnetic field provided by the fixed coils, but is in very wide use.

## 11.7   Ammeters and shunts

An instrument arranged to read the current in a circuit is called an ammeter, and is usually calibrated in amperes, milliamperes or microamperes. An ammeter must be connected in circuit so that the current to be measured passes through it.

If an ammeter is to carry full-load current without getting very hot, or causing a large voltage drop within itself, its resistance must be low. This is arranged comparatively easily with moving-iron instruments, because the coil is fixed and can be wound with a conductor of large cross-sectional area. The moving-coil instrument, however, has a fine coil with delicate hairsprings for connections, and its maximum load is usually a few milliamperes. To measure higher currents, most of the current flows through a low resistance called a shunt, connected in parallel with the instrument, only a small but definite proportion flowing through the moving coil. If this system is to be accurate, the shunt and instrument resistance must be in the same ratio at all times, regardless of temperature. The shunt $R_2$ is usually made of manganin (an alloy of manganese, copper and nickel), which has a near-zero temperature coefficient of resistance, and a '**swamping resistor**' of the same material is connected in series with the instrument as shown in Figure 11.10. The swamping resistor $R_3$ has a higher resistance than the instrument, so that any change in the latter will result in a much smaller proportional change in the complete circuit.

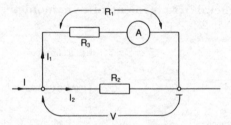

**Figure 11.10   Ammeter with shunt and swamping resistor**

It is often necessary to be able to calculate the resistance of the shunt required for a given instrument coil. Although formulas can be easily deduced for this, it is better to treat the arrangement simply as a parallel circuit, remembering that:

 (i) total current = shunt current plus instrument current,
(ii) the PD across the shunt is the same as that across the instrument (including its swamping resistor if one is used).

## Example 11.1

A moving-coil instrument has a full-scale deflection at 15 mA and a resistance, including its swamping resistor, of 4 $\Omega$. Calculate the resistance of a shunt to make the instrument suitable for a current range of up to 5 A.

The symbols used are those marked on Figure 11.10.

$$I_2 = I - I_1 = 5 - 0 \cdot 015 \text{ amperes} = 4 \cdot 985 \text{ A.}$$

Instrument voltage drop $V = I_1 R_1 = 0 \cdot 015 \times 4 \text{ volts} = 0 \cdot 06 \text{ V}$

This is also the voltage drop across the shunt.

$$R_2 = \frac{V}{I_2} = \frac{0 \cdot 06}{4 \cdot 985} \text{ ohms} = 0 \cdot 01204 \ \Omega \text{ to four significant figures}$$

This example shows that shunts often have very low resistance, and must be carefully made if reasonable accuracy is required; the shunt is often fitted within the case of a single-range instrument.

## 11.8   Voltmeters and multipliers

So far we have considered only measurement of current. Most instruments for measurement of potential difference, called voltmeters, are in fact measuring the current in a circuit of known resistance when connected across the unknown PD. This is not to say that an ammeter can be used as a voltmeter without modification. For instance, if an ammeter with a resistance of $0 \cdot 01 \ \Omega$ were connected across a 240 V supply the resulting current would be 24 000 A and the instrument would be destroyed.

To limit the current to a low value, a high resistance $R_M$ is connected in series with the instrument as shown in Figure 11.11, the scale then being marked off in volts. The series resistance is called a multiplier and, since its resistance must not vary, it is made of manganin. It therefore forms an almost perfect 'swamping resistor' for the instrument movement.

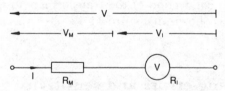

**Figure 11.11   Voltmeter with multiplier**

Multiplier calculations are easily carried out if we treat the instrument as a simple series arrangement, and remember that:

(i)  total PD = multiplier PD + instrument PD
(ii) the same current flows through the instrument and the multiplier.

## Example 11.2

The instrument of Example 11.1 is to be converted for use as a voltmeter with full-scale deflection at 250 V. Calculate the resistance of the multiplier required.

To give full-scale deflection, the instrument, and hence the multiplier, must carry 15 mA. The PD across the multiplier and instrument must be 250 V, so total resistance

$$R = \frac{V}{I} = \frac{250}{0 \cdot 015} \text{ ohms} = 16666 \cdot 67 \ \Omega$$

But instrument resistance is 4 $\Omega$, so multiplier resistance is

$$16\ 666 \cdot 67 - 4 \text{ ohms} = 16\ 662 \cdot 67 \ \Omega$$

Since a voltmeter has high resistance, the current taken by it can be ignored in power circuits. Connection of a voltmeter may, however, affect some circuits as explained in Section 11.10.

## 11.9   Current and voltage transformers

The principles of instrument transformers have been explained in Section 7.10. A current transformer steps down the very high current to be measured to a level acceptable to the measuring instrument, which is usually of the moving-iron type, the connection diagram being shown in Figure 7.24. Similarly, a voltage transformer reduces the potential difference to a level which can be handled by a moving-iron voltmeter as shown in Figure 7.25. A clip-on type ammeter, which uses a heavy current conductor as the primary of a current transformer is shown in Figure 7.27.

Dynamometer instruments are often used for power measurements in high-voltage, high-current circuits in conjunction with voltage and current transformers. The connection of the transformers to the wattmeter and measured circuit is shown in Figure 11.12.

## 11.10   Loading errors and sensitivity

In high-resistance circuits, the connection of a voltmeter to the system may cause changes which prevent correct measurements from being taken. Consider, for example, the simple circuit of Figure 11.13*a*. An apparatus has an internal voltage of 200 V and an internal resistance of 10 k$\Omega$, but, as there is no current, no potential drop occurs in this internal resistance and 200 V appears at the output terminals. To measure this output potential, a voltmeter of resistance 90 k$\Omega$ is connected as shown in Figure

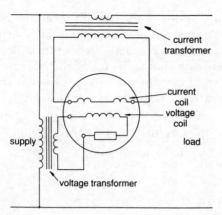

**Figure 11.12   Wattmeter connected to measure power in high-voltage, high-current circuit using instrument transformers**

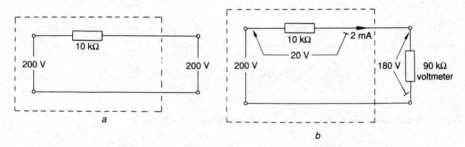

**Figure 11.13   To illustrate effect of voltmeter loading on high-resistance source**

11.13*b*. Total circuit resistance

$$R_T = 10 \text{ k}\Omega + 90 \text{ k}\Omega = 100 \text{ k}\Omega$$

and circuit current

$$I = \frac{V}{R_T} = \frac{200}{100 \times 10^3} \text{ amperes} = 2 \text{ mA}$$

The voltage drop in the internal resistance

$$V_R = IR = 2 \times 10^{-3} \times 10 \times 10^3 \text{ volts } = 20 \text{ V}$$

The PD across the output terminals, and hence across the voltmeter, is then 200 − 20 volts = 180 V.

The voltmeter will thus read 180 V where, before its connection, a PD of 200 V existed. The instrument is not in error, but its connection into the circuit has changed conditions and thus changed the PD across the output terminals.

The circuit change is due to the current taken by the voltmeter. If less current were taken (if the voltmeter resistance had been higher), the error would have been less. For a voltmeter to operate with less current it must have a higher sensitivity; that is, its full-scale deflection must be reached with a lower deflecting current, the current limited by higher voltmeter resistance. The sensitivity of a voltmeter is measured in ohms/volt, which is calculated from

$$\frac{\text{total voltmeter resistance} \quad \text{(ohms)}}{\text{full-scale reading} \qquad \text{(volts)}}$$

## Example 11.3

A voltmeter movement gives full-scale deflection at 1 mA. Calculate its sensitivity in ohms/volt when used with multipliers giving full-scale deflection (*a*) at 40 V, and (*b*) at 250 V.

Total instrument resistance so that full-scale current will result from application of full-scale voltage can be calculated from Ohm's law.

(*a*) $R_T = \dfrac{V}{I} = \dfrac{40}{1 \times 10^{-3}}$ ohms $= 40$ k$\Omega$

$\qquad$ sensitivity $= \dfrac{R_T}{V} = \dfrac{40 \times 10^3}{40}$ ohms/volt $= 1000$ $\Omega$/V

(*b*) $R_T = \dfrac{V}{I} = \dfrac{250}{1 \times 10^{-3}}$ ohms $= 250$ k$\Omega$

$\qquad$ sensitivity $= \dfrac{R_T}{V} = \dfrac{250 \times 10^3}{250}$ ohms/volt $= 1000$ $\Omega$/V

This example shows that sensitivity does not depend on the values of multipliers, and hence on full-scale deflection voltage, but on the current needed by the instrument for full-scale deflection. In fact, the sensitivity of the instrument may be found directly from the full-scale current by taking its reciprocal. For example, the sensitivity of the instrument of Example 11.3 with a full-scale deflection current of 1 mA ($0 \cdot 001$ A) can be found from:

$$\text{sensitivity} = \frac{1}{\text{current at FSD}} = \frac{1}{0 \cdot 001} = 1000 \ \Omega/\text{V}$$

Moving-iron voltmeters generally have low sensitivity, often in the region of 100 $\Omega$/V, which indicates a full-scale deflection current of 10 mA. Moving-coil instruments used on DC supplies are much better, with up to about 20 000 $\Omega$/V (full-scale deflection at 50 $\mu$A); when used with rectifiers on AC supplies, sensitivity falls, typically to 2000 $\Omega$/V, or $0 \cdot 5$ mA for a full-scale deflection.

## Example 11.4

A moving-coil voltmeter with a multiplier has a full-scale deflection of 100 V and a sensitivity of 10 000 $\Omega$/V. Calculate the current needed to give full-scale deflection. The voltmeter is used to measure the output voltage of an equipment with an internal voltage of 100 V and an internal resistance of 50 k$\Omega$. Calculate its reading.

total resistance of voltmeter = 10 000 ohms per volt × 100 volts = 1 M$\Omega$

$$\text{full-scale current } I = \frac{V}{R} = \frac{100}{10^6} \text{ amperes} = 100 \text{ }\mu\text{A}$$

current to instrument

$$= \frac{V}{\text{voltmeter resistance} + \text{internal resistance}}$$

$$= \frac{100}{10^6 + (50 \times 10^3)} \text{ amperes} = \frac{100}{1050 \times 10^3} \text{ amperes} = 0 \cdot 0952 \text{ mA}$$

Voltage drop in internal resistance

$$V = IR = 0 \cdot 0952 \times 10^{-3} \times 50 \times 10^3 \text{ volts} = 4 \cdot 76 \text{ V}$$

Output voltage = voltmeter reading = internal voltage − voltage drop

$$= 100 - 4 \cdot 76 \text{ volts} = 95 \cdot 24 \text{ V}$$

The difficulties over sensitivity of voltmeters arise from the fact that most of them are actually measuring the current driven by the voltage concerned through the resistance of the voltmeter. An electronic voltmeter works on a different principle and actually measures potential difference, using electronic circuits. Electronic instruments are considered in Section 11.12. The cathode-ray oscilloscope (see Section 11.21) can also be used as a highly sensitive voltmeter.

Ammeters connected in low-voltage, low-resistance circuits can cause errors similar to those caused by voltmeters in low-current, high-resistance circuits. Figure 11.14 shows how the series connection of a 0·1 $\Omega$ ammeter in a circuit of resistance 0·4 $\Omega$ fed from a 1 V supply reduces the circuit current from 2·5 A to 2 A.

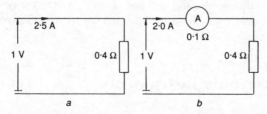

**Figure 11.14    To illustrate effect of ammeter resistance on circuit current**

## 11.11   Multirange test sets

The electrical craftsman would require a very large number of instruments to measure all the widely differing values of resistance, current and voltage with which he is likely to be concerned. In many cases, one instrument with a variety of shunts, multipliers and instrument transformers may be used; in some cases the accessories may be separate units for connection to the instrument as required, while in others they may be incorporated in one case and selected by means of dial switches or plugs. A simplified diagram for a multirange instrument is shown in Figure 11.15, showing the principles of connection of shunts and multipliers. In most cases, a rectifier is included so that AC values may be measured. A dry battery and variable resistor fitted in the case will make the instrument suitable for resistance measurements as described in Section 11.13. Photographs of a very widely used multirange instrument are given in Figure 11.16.

### *Clamp-type instruments*

We saw in Section 11.7 that an ammeter must be connected so that the current it is to measure passes through it. This means that the measured circuit must be disconnected to allow insertion of the ammeter, often a time-consuming process when it is remembered that the ammeter must usually be disconnected after the measurement. If very large currents are to be measured, very heavy cables are involved. Again, disconnection of an earth lead for measurement of the leakage current in it could cause danger in the event of an earth fault while it is disconnected.

Current in a cable sets up a magnetic field proportional to that current, and the clamp-type instrument uses this fact as the basis of its operation. A magnetic circuit is placed round the conductor concerned and an EMF proportional to the magnetic field, and hence to the current, is induced in a coil. Measurement of this EMF is thus a measurement of the current. Most clamp meters operate only on AC supplies using the transformer principle, but a few will also measure direct currents using the Hall effect.

This type of instrument may have an analogue or a digital display, and is particularly used by electricians and by power engineers because of the simplicity and speed of taking readings. Units are often arranged with sockets for connection of leads giving access to voltage, resistance and temperature readings. In the current mode, currents of up to several thousand amperes or as low as 10 μA are possible. Figure 7.27 shows a typical clamp meter.

## 11.12   Electronic instruments

Electronics circuits have not only increased dramatically in their range of applications, but have also become inexpensive and extremely rugged. The

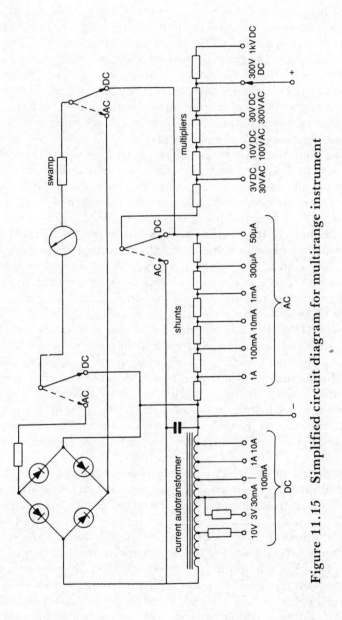

**Figure 11.15  Simplified circuit diagram for multirange instrument**

invention of the integrated electronic circuit (the 'silicon chip') means that modern instruments are capable of measuring and calculating to give facilities hitherto not considered possible. Analogue instruments are those mentioned so far in this chapter, and provide a reading by means of the physical movement of a pointer over a calibrated scale. Most electronic instruments provide a digital readout, giving their readings in figures on a liquid-crystal display (LCD). The light-emitting-diode (LED) type has largely disappeared because of the high current consumption (and hence short battery life) but is sometimes still preferred where there is a need for a reading to be taken where the lighting level is low. However, there are some applications where the analogue readout is preferred, and for such cases the final output of the electronic measuring system is to a permanent-magnet moving-coil display.

Unlike their analogue counterparts, most digital instruments are true voltmeters, measuring a very low value of potential difference across one or more circuit components. Using multipliers, the low value can be made a small but exact proportion of a very much larger voltage, and digital instruments capable of reading direct or alternating voltages of up to 1000 V are commonplace. If a current must be measured, it is passed through a low resistance of known value, when the measured voltage drop across it can be displayed as the value of current, often direct or alternating.

In overall terms it is usually less expensive to combine ammeters and voltmeters in the same instrument, although digital instruments for just one type of measurement are not uncommon (see Figure 7.27). The digital multimeter (DMM) is in very wide use. Types include the switched type, where each range is connected by movement of a rotary switch (see Figure 11.17), or by the operation of combinations of press keys, often on the side of the instrument (see Figure 11.18). A possibility with the electronic instrument, which is not available on its analogue counterpart, is autoranging. The instrument must be set for the type of reading (voltage, current, resistance etc.) and then automatically selects the best range, usually displaying the unit of the measured quantity (e.g. A or mA) (see Figure 11.19).

Most digital multimeters will also measure resistance, usually by the process of measuring the voltage needed across the resistance to allow a certain current to be driven through it. This facility is available on each of the digital multimeters shown in Figures 11.17, 11.18 and 11.19. Some devices also include a buzzer to give an audible tone, if the resistance is lower than a given value, for simple continuity testing. In some cases the testing of semiconductor diodes causes a problem. The silicon diode is the most common type, and requires a forward voltage of about 600 mV to be applied before it will switch on and indicate a low resistance. Some electronic testers provide an output test voltage below this value, so care must be taken to ensure that a good diode does not appear to have high resistance in both forward and reverse directions. Digital instruments often have a diode test setting, which can be relied on to test at a high enough

**Figure 11.16   Commonly used multirange instrument**
*a* External view
*b* View with cover and indicating instrument removed

**Table 11.3: Operating ranges for some instruments**

| Type of instrument | Range as an ammeter (full scale) | Range as a voltmeter (full scale) | Useful frequency range |
|---|---|---|---|
| Permanent-magnet moving-coil | 10 μA–20 mA | 100 μV–200 mV | DC only |
| Permanent-magnet moving-coil with shunts and multipliers | 20 mA–600 A | 200 mV–3 kV | DC only |
| Moving iron | 100 mA–10 A | 50 V–750 V | DC–100 Hz |
| Moving iron with instrument transformers | 10 A–1000 A | 500 V–15 kV | DC–100 Hz |
| Permanent-magnet moving-coil with thermocouple | 10 mA–2 A | – | DC–5 MHz |
| Permanent-magnet moving-coil with rectifier* | 1 mA–50 mA | 10 V–3 kV | 25 Hz–2 kHz |

*Reading is waveform dependent.

**Figure 11.17   Four-and-a-half-digit electronic multimeter with switch selection of ranges**

**Figure 11.18** Three-digit electronic multimeter with key range selection

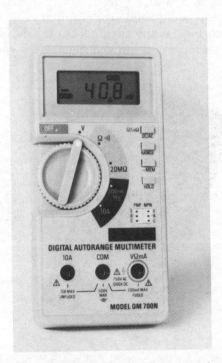

**Figure 11.19** Three-digit autoranging multimeter with memory and hold facilities

voltage to ensure turn-on in the forward direction (see Figures 11.17 and 11.18).

We are all aware of the calculating capabilities of electronics, because we use them in the dedicated miniature computers which we call calculators. These facilities can be built into electronic instruments. For example, the rectifier moving-coil instrument measures the RMS values of alternating currents and voltages by assuming that the applied signal has a sinusoidal waveshape; for any other shape an error will be introduced. A calculating chip in a DMM can determine and display the true RMS value regardless of waveshape. The instrument may also test the correct operation of transistors—the devices shown in Figures 11.17, 11.18 and 11.19 all have this facility. The difficulty of reading a continuously changing digital display can be overcome in some cases by using the 'hold' button to fix the display (see Figure 11.19). A few instruments have the facility to display the average reading over a period. Special instruments can also be made; for example, Figure 11.20 shows a wattmeter which is also capable of reading voltage and current. Power factor can be readily calculated by the division of watts by the product of volts and amperes. Figure 11.21 shows an $L/C$ meter for accurate measurement of inductance and capacitance. Another facility is the ability to 'remember' a reading and to subtract it from a later reading, displaying the difference (see Figure 11.19). This can be useful when measuring resistance with a wander lead, whose resistance must be subtracted from the total reading before the required value can be found.

Electronic instruments have many advantages over the older analogue types, including greater accuracy, improved ruggedness, smaller size and

**Figure 11.20   Digital wattmeter with key selection of functions and ranges**

**Figure 11.21   Digital *L/C* meter with key selection of functions and ranges**

reduced weight. However, they do need a power supply, provided by an internal battery in most cases, although some specialist instruments such as the line–earth loop tester derive their power from the supply being tested. The inclusion of the battery seldom causes problems, because the low current drain of liquid-crystal displays ensures long battery life, and a clear indication (often 'LOBAT' is displayed) gives ample warning when a battery change becomes necessary. Sensitivity is a very important quality of an analogue instrument because it is almost certain to be current-measuring, even when scaled in volts (see Section 11.10). The term has little meaning with electronic instruments because they are true voltmeters and have very high input resistance, usually of the order of ten megohms.

## Comparison of analogue and digital scales

For many years, an argument has raged over the relative advantages of the two scales, and there can be no answer which is correct for all applications and users. If one type of scale feels right for the tester and gives consistently reliable results, this is the type which should be used. The advantages of the two types are listed below.

*Advantages of digital instruments*
   (i)  No skill is required to read a digital instrument. The reading is clearly displayed, so that even an unskilled person will obtain the correct figures; the reader does not require to judge the position of a pointer

on a scale, or to take care that the reading is taken from directly above the scale to prevent parallax error.

(ii) A digital instrument can provide a very precise reading. However, it is important not to confuse precision, which is the ability to read the display accurately, with true accuracy.

(iii) Digital instruments are generally far more accurate than analogue instruments. For a voltmeter, accuracy is quoted at $\pm\ 0 \cdot 1\%$ for a digital instrument and $\pm\ 3\%$ for an analogue instrument in the same price range.

(iv) Many digital instruments can provide a digital output to a data recorder or other computer, which is extremely useful when results must be saved and stored.

(v) Most digital instruments include automatic polarity indication and overload protection, avoiding damage or the need to reconnect.

(vi) Some digital instruments include a microprocessor, making it possible to carry out mathematical operations. For example, if continuity tests are being carried out and it is necessary to subtract the resistance of a previously measured wander lead from each reading, the instrument can be programmed to carry this out automatically.

*Advantages of analogue instruments*

(i) The fluctuations which sometimes prevent readings being taken from a digital instrument appear as needle fluctuations with the analogue instrument, allowing an approximate reading to be taken.

(ii) The analogue instrument is powered from the measured source and does not usually require a battery.

(iii) A digital instrument is incapable of measuring values lying above and below its least significant digit. For example, if a voltage of $24 \cdot 7$ V is to be measured, a three-digit instrument may read 024 or 025, whereas an analogue instrument will give the more accurate reading. Having said this, it must also be stated that the improved accuracy of digital instruments and the ability to choose their operating scale usually make this unimportant.

## 11.13   Resistance measurement — introduction

There are many methods for the measurement of resistance, but most of them require a calculation from instrument readings or from other resistance values. The circuit of a simple direct-reading ohm-meter is shown in Figure 11.22 and consists of a battery, a permanent-magnet moving-coil instrument and a variable resistor all connected in series and usually contained within one case.

The test leads X and Y are first connected together and the resistor R adjusted until the instrument reads full-scale deflection. This point is marked as zero on the resistance scale. When the leads are disconnected, the

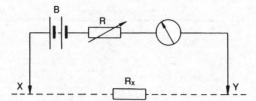

**Figure 11.22   Simple direct-reading ohm-meter circuit**

instrument needle will rest at the left-hand end of the scale, which is marked to indicate infinite resistance. The current through the instrument depends on the unknown resistance $R_X$ (shown broken-lined in Figure 11.22), becoming larger as $R_X$ becomes smaller. The scale can thus be marked off in ohms, but is very cramped at the high (left) end of the scale, and is 'back-to-front', that is with the zero position at the right-hand end of the scale. Any variation in the EMF of the battery is compensated by 'zeroing' the instrument by variation of $R$ with X and Y connected together.

This method of resistance measurement is not very accurate, but gives a reasonable indication of resistance value, and is built into many multirange instruments.

## 11.14   Ammeter-and-voltmeter method

An ammeter and a voltmeter having suitable ranges can be used for the determination of resistance. Figure 11.23 shows the connections of the instruments for this test. The supply is shown as coming from the mains via a double-wound stepdown transformer, but could alternatively be taken from any suitable supply, such as a battery. The resistance of the required unknown ($\Omega$) will then be:

$$\frac{\text{voltmeter reading}}{\text{ammeter reading}} = \frac{\text{(V)}}{\text{(A)}}$$

The variable resistor is used to adjust the current to a suitable value. Many conductors whose resistances must be measured in installation work, such as earth-protective-conductor resistance, are very long, and a lengthy wander lead is needed to reach from one end to the other. The calculation described above will give a result which includes the resistance of this wander lead. A true result can be obtained by measuring the resistance of the wander lead (disconnect from Y and connect to X in Figure 11.23) by the same method, and subtracting this value from that of the complete circuit which included it.

## Example 11.5

The resistance of an earth-continuity conductor is to be measured using a 12 V battery, a variable resistor, a wander lead, an ammeter and a

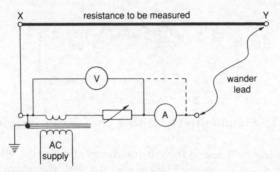

**Figure 11.23   Circuit for ammeter-and-voltmeter method of resistance measurement**

voltmeter. The wander lead is connected to the earth-continuity conductor at the point farthest from the mains, the earthing terminal and the other end of the wander lead being connected to the battery and the ammeter, respectively. When the current is adjusted to 4 A, the voltmeter reads 5·4 V. The wander lead alone is now connected across the test circuit, the voltmeter reading 4·8 V when the current is adjusted to 5 A. What is the resistance of the earth-continuity conductor?

$$\text{total resistance} = \frac{V_1}{I_1} = \frac{5 \cdot 4}{4} \text{ ohms} = 1 \cdot 35 \ \Omega$$

$$\text{wander-lead resistance} = \frac{V_2}{I_2} = \frac{4 \cdot 8}{5} \text{ ohms} = 0 \cdot 96 \ \Omega$$

$$\text{earth-continuity-conductor resistance} = 1 \cdot 35 - 0 \cdot 96 \text{ ohms} = 0 \cdot 39 \ \Omega$$

Figure 11.23 shows alternative connections for the voltmeter. Neither of these gives a strictly accurate result, the connection shown in full lines ignoring the voltage drop in the ammeter, and that in broken lines ignoring the voltmeter current.

If good-quality instruments are used (high-resistance voltmeter and low-resistance ammeter), these errors are likely to be negligible. If a low-sensitivity voltmeter is used, its effect may be taken into account as shown in Example 11.6.

## Example 11.6

The ammeter-and-voltmeter method is used to measure the value of a resistor, the connections being shown in Figure 11.24. The voltmeter reads 60 V and the ammeter 5 A. Calculate the apparent value of the resistor.

If the voltmeter resistance is 500 Ω, calculate the true resistor value.

$$\text{apparent value} \quad = \frac{V}{I} = \frac{60}{5} \text{ ohms} = 12 \ \Omega$$

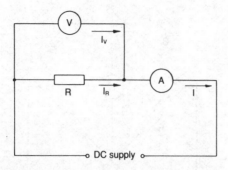

**Figure 11.24   Diagram for Example 11.6**

voltmeter current $\qquad I_V = \dfrac{V}{R_V} = \dfrac{60}{500} = 0 \cdot 12$ A

resistor current $\qquad I_R = I - I_V = 5 - 0 \cdot 12$ amperes $= 4 \cdot 88$ A

true value $\qquad = \dfrac{V}{I_R} = \dfrac{60}{4 \cdot 88}$ ohms $= 12 \cdot 3$ $\Omega$

An alternative method for the second part of this exercise is to consider that the apparent resistor value, 12 $\Omega$, consists of the voltmeter resistance, 500 $\Omega$, in parallel with the unknown resistor $R$.

$$\frac{1}{12} = \frac{1}{500} + \frac{1}{R}$$

$$\frac{1}{R} = \frac{1}{12} - \frac{1}{500} = \frac{500 - 12}{500 \times 12}$$

Therefore $\qquad R = \dfrac{500 \times 12}{500 - 12}$ ohms $= \dfrac{6000}{488}$ ohms $= 12 \cdot 3$ $\Omega$

## 11.15   Wheatstone-bridge principle

This is a measurement method which compares the unknown resistor with others of known values. The circuit is shown in Figure 11.25 and is made up of four resistors, switches, a direct-current supply source such as a battery, and a galvanometer, which is a sensitive centre-zero instrument. Resistors $R_A$ and $R_B$ are usually variable in multiples of ten and are called the 'ratio arms'. $R_C$ is a known variable resistor, and $R_X$ is the unknown resistor. The bridge is 'balanced' by closing switch $S_1$ and quickly tapping switch $S_2$. Any difference of potential across the galvanometer will cause it to deflect. Resistor $R_C$ is varied and the process repeated, until closing

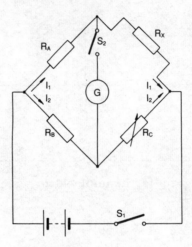

**Figure 11.25   Wheatstone-bridge circuit**

$S_2$ gives no deflection on the galvanometer and indicates no difference of potential across it. In this condition the bridge is said to be 'balanced', and the current $I_1$ in $R_A$ will also flow in $R_X$ since none passes through the galvanometer. Similarly, a second current $I_2$ will flow through both $R_B$ and $R_C$. At balance, the PD across $R_A$ = the PD across $R_B$ since there is no PD across the galvanometer

$$I_1R_A = I_2R_B \qquad (1)$$

Also, the PD across $R_X$ = the PD across $R_C$

$$I_1R_X = I_2R_C \qquad (2)$$

Dividing eqn. 1 by eqn. 2,

$$\frac{I_1R_A}{I_1R_X} = \frac{I_2R_B}{I_2R_C} \quad \text{so} \quad \frac{R_A}{R_X} = \frac{R_B}{R_C}$$

and

$$R_X = \frac{R_A}{R_B}R_C$$

Since the values of $R_A$, $R_B$ and $R_C$ are all known, the value of $R_X$ can be calculated. The ratio arms $R_A$ and $R_B$ are set at values whose ratio will result in the best use of variable resistor $R_C$. Considerable accuracy is possible using a bridge of this type for two reasons. These are:

(i) The accuracy of the deflecting instrument does not affect the accuracy of the bridge, because at balance it carries no current, reads zero and can have no error. For this reason the Wheatstone bridge is called a **'null-deflection'** device.

(ii) Accuracy depends entirely on the accuracy of the three known resistors. Extremely accurate resistors can be bought at low cost.

## Example 11.7

The bridge shown in Figure 11.25 is balanced when $R_A = 5\ \Omega$, $R_B = 50\ \Omega$ and $R_C = 2 \cdot 76\ \Omega$. What is the value of $R_X$?

$$R_X = \frac{R_A}{R_B} \times R_C = \frac{5}{50} \times 2 \cdot 76 \text{ ohms} = 0 \cdot 276\ \Omega$$

Wheatstone bridges are made in many practical forms, one of the simplest and commonest being the slide-wire bridge shown in Figure 11.26. AB is a resistance wire of uniform cross-sectional area stretched beside a rule, the length commonly being 1 m or 50 cm. Since its cross-sectional area does not vary, the resistance of any part of the wire will be proportional to its length. P is a sliding contact on the wire, and the bridge is balanced by adjusting the position of P. At balance, again indicated by zero deflection of the galvanometer when $S_1$ and $S_2$ are closed,

$$R_X = R \times \frac{l_2}{l_1}$$

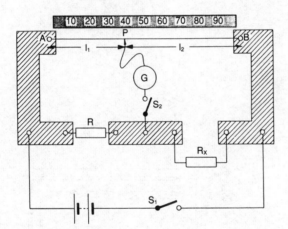

**Figure 11.26    Slide-wire bridge**

## Example 11.8

A one-metre slide-wire bridge as shown in Figure 11.26 is balanced when the sliding contact is 36 cm from the left-hand end. If the resistor $R$ has a value of 5 $\Omega$, what is the value of the unknown resistor?

$$l_1 = 36 \text{ cm}$$

Therefore
$$l_2 = 100 - l_1 = 64 \text{ cm}$$

$$R_X = R\ \frac{l_2}{l_1} = 5 \times \frac{64}{36} \text{ ohms} = 8 \cdot 89\ \Omega$$

The value of the known resistor $R$ should have the same order as the unknown $R_X$. If, for instance, $R_X$ is one hundred times $R$, balance would be obtained less than 1 cm from the end of the scale and accuracy would be poor.

## 11.16   Murray-loop test

A variation of the basic Wheatstone-bridge circuit, known as the Murray loop, is used to calculate the position of an earth fault or of a fault between cores on a multicore cable. The faulty cable is connected as in Figure 11.27 for an earth fault, or as in Figure 11.28 for a fault between cores. $P$ and $Q$ are accurately calibrated resistors, $P$ being a variable resistor.

$P$ is adjusted to give a balance indicated by a null deflection on the galvanometer.

At balance,

$$\text{PD across } P = \text{PD across } Q,$$

so
$$I_1 P = I_2 Q \tag{3}$$

and PD across $R_B$ = PD across $R_A$,

so
$$I_1 R_B = I_2 R_A \tag{4}$$

Dividing eqn. 3 by eqn. 4,

$$\frac{I_1 P}{I_1 R_B} = \frac{I_2 Q}{I_2 R_A} \quad \text{so} \quad \frac{P}{R_B} = \frac{Q}{R_A}$$

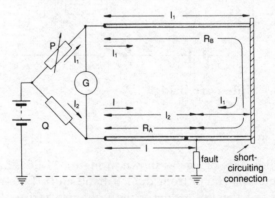

**Figure 11.27   Cable with earth fault connected for Murray-loop test**

If the cable cores all have the same cross-section the core resistance will be proportional to length, and $R_A \propto l$ while $R_B \propto l_1 + (l_1 - l)$.

Substituting these values for $R_A$ and $R_B$ and solving for $l$ gives

$$l = 2l_1 \times \frac{Q}{P + Q}$$

Hence, if the conductor length and the values of $P$ and $Q$ at balance are known, the length $l$ and hence the position of the fault can be calculated. When carefully carried out, this test will pinpoint the position of the fault to within 1 m for a 1000 m cable.

## Example 11.9

A cable 800 m long has an earth fault, and is connected as shown in Figure 11.27 for a Murray-loop test. At balance the resistors $P$ and $Q$ have values of $16 \cdot 45$ $\Omega$ and 10 $\Omega$, respectively. Calculate the distance from the test end of the cable to the fault.

$$l = 2l_1 \times \frac{Q}{P + Q} = \frac{2 \times 800 \times 10}{10 + 16 \cdot 45} \text{ metres} = \frac{16\ 000}{26 \cdot 45} \text{ metres} = 605 \text{ m}$$

Note that the resistance of the earth fault, or the fault between cores, does not form part of any of the four bridge resistors, and does not affect accuracy. If the fault resistance is very high, a high supply voltage may become necessary to drive a sufficiently high current into the bridge to give good sensitivity.

While the methods described above give accurate results and are still widely used, it would be wrong to give the impression that there is no alternative. Increasing use is being made of radio-frequency fault detectors, where a high-frequency signal is fed to the end of the faulty cable. A sensing head like those used for metal detection is then run at ground level above the cable route to find the point where a change in signal strength indicates a fault. Methods such as these have the advantage of directly pinpointing the position of the fault and preventing the need for accurate measurement along the whole route of the cable.

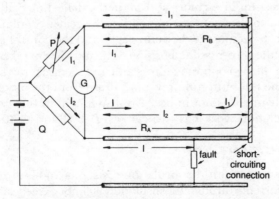

**Figure 11.28   Cable with fault between two cores connected for Murray-loop test**

## 11.17   Direct-reading ohm-meters

The simple ohm-meter described in Section 11.13 suffered from the disadvantages of a cramped, reversed scale and the need to zero the movement before use. A more satisfactory ohm-meter is shown in Figure 11.29; unlike the instrument of Section 11.13 it can be used only as an ohm-meter, and is therefore less common.

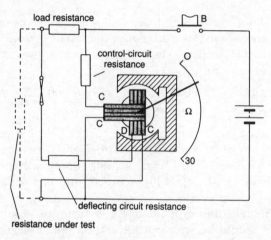

**Figure 11.29   Simplified circuit for a direct-reading ohm-meter**

The direct-reading ohm-meter has two moving coils, which are fixed together and have no controlling torque, current lead-in being by means of fine wire ligaments. Control coil CC is connected directly across the battery and is arranged to make the movement swing to the low end of the scale. Deflecting coil DC is in series with the resistor under test, and tends to swing the movement to the high end of the scale. Thus a high value of test resistor gives more current in the deflecting coil, which swings the movement to the high-reading end of the scale. Similarly, a zero test resistance short-circuits the deflecting coil, and the control coil swings the movement to the zero-reading position. The movement takes up a position depending on the two opposing forces and gives a direct reading of external resistance. The test button B prevents a drain from the battery to the coils when the instrument is not in use. An instrument of this type commonly used in installation work has two ranges, 0–3 $\Omega$ and 0–30 $\Omega$.

### Insulation-resistance tester

The electrician must test the insulation resistance of his circuits to ensure that leakage currents through the insulation will not exceed safe levels. Since the values to be measured are normally very high, of the order of megohms, and since the test voltage can seldom be less than 500 V (and may often be much higher), a special test instrument is used. This is called a **megohm-meter**, and the high test voltage is derived by one of two methods. The first has a hand-wound generator built into the same case as the test unit. A steady output voltage is obtained by using a clutch, which slips at speeds above that for correct output voltage, or by use of a special generator. The second method is to obtain the output voltage from a transistorised amplifier, fed by low-voltage dry or rechargeable batteries. The latter type of tester has the advantage of being easier to operate by one person, who can use one hand for manipulation of the test leads. Digital versions are also available.

A simplified diagram of the transistorised tester is shown in Figure 11.30. The tester must use a direct-current system so that it measures the resistance, and not the impedance, of the insulation (see Chapter 4).

## Earth-fault-loop testers

Wiring systems are subject to *BS 7671* (the 16th Edition of the IEE Wiring Regulations) and must have earth-fault-loop-impedance values measured to check compliance. The value concerned is the impedance met by a short-circuit fault to earth, on which depends the value of the fault current and hence the speed of operation of the protective device. In simple terms, the impedance of the phase-to-earth loop is measured by connecting a resistor (typically 10 $\Omega$) from the phase to the protective conductor as shown in Figure 11.31a. A fault current, usually something over 20 A, circulates in the fault loop, and the impedance of the loop is calculated within the instrument by dividing the supply voltage by the value of this current. The resistance of the added resistor must be subtracted from this calculated value before the result is displayed. An alternative method is to measure the supply voltage both before and while the loop current is flowing. The difference is the voltage drop in the loop due to the current, and loop impedance is calculated from voltage difference divided by the current.

Since the loop current is very high, its duration must be short and is usually limited to two cycles (or four halfcycles) or 40 ms for a 50 Hz supply. The current is usually switched by a thyristor or a triac, the firing time being controlled by an electronic timing circuit. A typical instrument of this kind is shown in Figure 11.31b.

During operation, a heavy current flows from the phase to earth. There is thus an imbalance between phase and neutral which will operate any residual-current device (RCD) upstream of the point of connection. During testing it is therefore necessary to short-circuit all RCDs unless the loop tester in use is of a type designed to limit the rate of rise of current and thus prevent RCD operation.

## Prospective-short-circuit-current (PSC) testers

It is important to ensure that protective devices (fuses and circuit breakers) are capable of breaking the highest fault current to which they may be

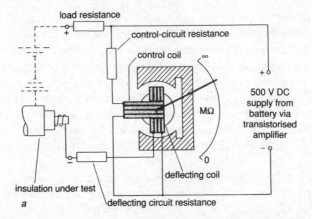

*b*

**Figure 11.30   Measuring insulation resistance**

    *a* Simplified circuit diagram for insulation-resistance tester
      (megohm-meter)
    *b* Insulation and continuity testers

subjected. A small modification to the earth-fault-loop tester makes this possible. The tester (see Figure 11.31*b*) is connected from phase to neutral (instead of phase to earth) and is set to calculate the prospective short-circuit current by dividing the supply voltage by the phase-to-neutral impedance. The result is usually displayed in kA (thousands of amperes).

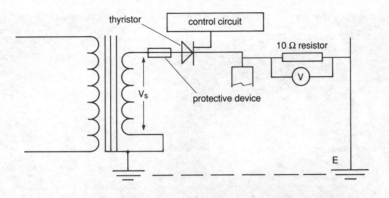

*b*

**Figure 11.31  Earth-fault-loop tester**
 *a* Simplified circuit of an earth-fault-loop tester
 *b* Typical earth-fault-loop and prospective-short-circuit-
   current tester

## Earth-electrode testers

These specialist instruments are used to measure the resistance of an earth
electrode to the general mass of earth. They have four terminals marked
$C_1$, $C_2$, $P_1$ and $P_2$, and are connected as follows:

   $C_1$ and $P_1$ are both connected to the electrode under test,
   $C_2$ is connected to a test electrode arranged outside the resistance area
      of the main electrode, and
   $P_2$ is connected to a test electrode driven halfway between the other
      two.

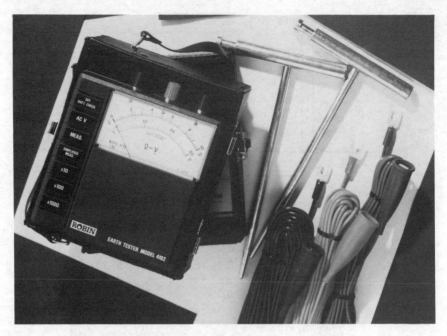

**Figure 11.32   Earth-electrode tester**

The instrument gives a direct reading of main electrode resistance, although the central electrode must be repositioned twice, once nearer to and once further away from the original position to check that resistance areas do not overlap. A typical instrument of this kind is shown in Figure 11.32.

## 11.18   Special instruments

The electrical craftsman is sometimes concerned with the use of special instruments, the principles of some of which are covered in this section.

### *RCD testers*

The RCD, or residual-current device, is very widely used as a protection against electric shock and against fire. Its principle of operation is covered in Section 12.7 of 'Electrical Craft Principles' Volume 1. The IEE Wiring Regulations (*BS 7671*) require that the time taken for the device to open a circuit which is subject to earth leakage above its rated value should be measured. This time is very short, usually less than 20 ms, so a special instrument is required for the purpose.

A number of RCD testers is available from a variety of manufacturers, most of which allow the imposition of a number of earth-leakage currents from 5 mA up to 2·5 A. For each setting, the time taken for the device to trip is displayed digitally. One requirement is that the device should not trip when subjected to half its rated value of earth-leakage current,

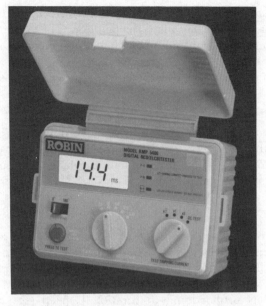

**Figure 11.33   RCD test instrument**

and instruments are provided with a 'half multiplier' setting for this purpose. A typical RCD tester is shown in Figure 11.33.

## Portable-appliance testers

The Electricity at Work Regulations came into force in 1990 in Great Britain and in 1992 in Northern Ireland. They are concerned with the duty of an employer to ensure that the equipment used by employees is safe to use, and one method of demonstrating compliance with the Regulations is to maintain records showing regular and satisfactory test results for appliances.

To comply with the Regulations, many employers (and contractors who carry out tests on their behalf) have taken to using dedicated instruments called 'portable-appliance testers', usually referred to as PATs. Such devices carry out a number of tests which include:

(i) earth-bond test, where the resistance of the earth system is measured by passing a current (which may be as high as 25 A) through it, although much lower currents may be necessary with electronic equipment;

(ii) insulation test, usually at 500 V DC, with pass resistances of 2 M$\Omega$ for single-insulated or 7 M$\Omega$ for double-insulated appliances;

(iii) fuse test, to confirm the continuity of the fuse;

(iv) flash test, at 1·5 kV or at 3 kV, to measure insulation leakage current. Carried out less frequently than the insulation test [(ii) above], the results will give early warning of insulation deterioration;

(v) load test, to measure the resistance of the instrument and hence allow calculation of the current taken in operation. The test cannot be applied to certain types of equipment, such as those incorporating discharge lamps or motors;

(vi) operation test, to check that the current drawn in use is not excessive; and

(vii) earth-leakage test, to measure earth-leakage current during operation.

In many cases the health of the appliance can be judged by comparison of results with those taken earlier. It is thus important that results are stored, and a number of PATs have storage facilities which will also allow transfer of results to a computer so that they may be printed out for display.

## Tachometers

A tachometer is an instrument used to measure speed of rotation. There are many operating principles, perhaps the most common of which is to use a very small generator with constant excitation, such as a permanent-magnet field system, which is driven by the rotation to be measured. Since induced EMF is proportional to rotational speed (see Section 9.3) a voltmeter calibrated in revolutions per minute or per second will give the required reading (see Figure 11.35).

An alternative which imposes no load on the measured rotation is to use a photoelectric method. For example, a bright reflecting strip could be fixed to the rotating shaft, and a light shone on it. A photocell is positioned

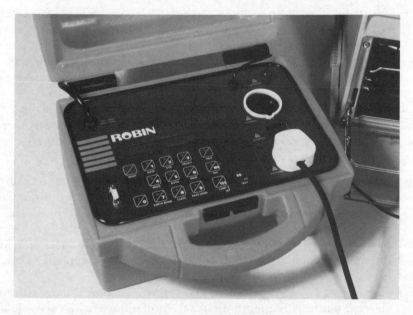

**Figure 11.34   Portable-appliance tester**

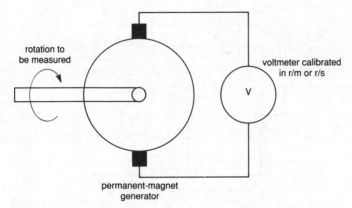

**Figure 11.35  Principle of generator-type tachometer**

to pick up the light reflected from the strip, so it is energised once for each rotation of the shaft, as shown in Figure 11.36. The output from the cell is fed to a modified frequency meter, which can be directly calibrated for rotational speed.

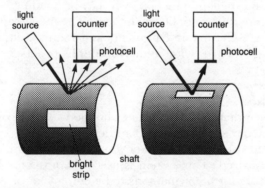

**Figure 11.36  Principle of electronic-counter tachometer**

## *Temperature-measuring instruments*

There are many electrical methods for measurement of temperature, which include:

### *Resistance pyrometers*

The resistance of most conductors varies with temperature, so a change in temperature may be related to a change in resistance. Pure platinum has an almost linear change in resistance with temperature, so is often used as the resistance element. A Wheatstone-bridge circuit is used with the resistance element as one arm as shown in Figure 11.37, the indicating instrument being calibrated directly in terms of element temperature.

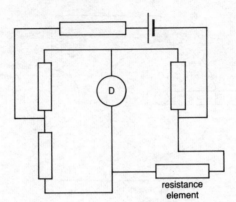

**Figure 11.37   Arrangement of bridge-type resistance pyrometer**

*Thermocouple pyrometers*
If two junctions are made between different metals, a difference in potential will occur if there is a difference in the temperatures of the junctions. This is called the Seebeck effect and the arrangement is often called a thermocouple. A millivoltmeter connected between the two junctions (the 'cold' junction and the 'hot' junction) will have a reading which depends on the temperature difference between them. The meter can be calibrated directly in degrees Celsius or Farenheit, and many multimeters have a setting which allows use with a thermocouple probe for temperature measurement.

*Radiation pyrometers*
Electromagnetic radiation occurs when a hot body 'sees' a colder body; the heat received by the earth from the sun depends on this principle. If the radiated heat is focused onto a thermocouple, the temperature of the hotter body can be measured. This system is used where the hot body is at too high a temperature for other methods to be employed, or is in a position where access for measurement is difficult.

## Frequency measurement

The most usual method of measuring frequency is to use a dedicated electronic meter which is based on an electronic counter. A crystal oscillator similar to that used in a battery watch provides a stable comparison frequency, but this is usually too high for most purposes. Thus the first step is to reduce this frequency using a divider, before putting the resulting output into a logic gate of the AND type, which only gives an output when both its inputs are fed.

The Schmitt-trigger circuit is used to convert the input waveform to a square wave. The number of pulses allowed through the gate each time it opens is equal to the number of times that the frequency to be measured is greater than the divided crystal frequency. Since the latter is accurately known, input frequency is known. The arrangement is shown in Figure 11.38.

An alternative is to use a double-beam oscilloscope for frequency comparison as described in Section 11.21.

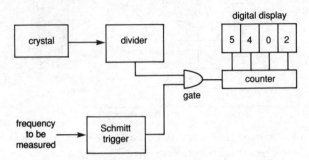

**Figure 11.38 Block diagram for an electronic frequency meter**

## Power-factor measurement

In simple terms, power factor can be defined as the cosine of the phase angle between current and voltage. Thus, while electromechanical power-factor meters have been available for very many years, it is now usually cheaper and more accurate to use electronic methods.

The simplest of these uses a system which measures the phase angle between voltage and current at the point where each passes through zero and calculates the cosine of this angle electronically. This method assumes that waves are perfectly sinusoidal, which is unlikely to be truly the case.

The true definition of power factor is the ratio of power to voltamperes. A better, but more expensive, system is to measure power, voltage and current. These results are fed into a calculating system to divide power by the product of voltage and current. The system can be arranged to include a detector which determines if the current lags or leads the voltage and thus if the power factor is lagging or leading.

## 11.19 Cathode-ray tube

The cathode-ray oscilloscope (CRO) is widely known in its application of displaying waveforms of alternating currents and voltages. In addition to being a display device, the CRO can also be used as a high-sensitivity voltmeter, adapted as an ammeter, and as a frequency-measuring device. The heart of the CRO is its cathode-ray tube.

The cathode-ray tube used for television picture display is well known, and the tube in an oscilloscope uses many of the same basic principles, although the tube shape is longer and more slender, with a smaller tube face or screen. The tube, made of very strong glass, is shaped as shown in Figure 11.39, all of the air being evacuated from the tube so that electrons

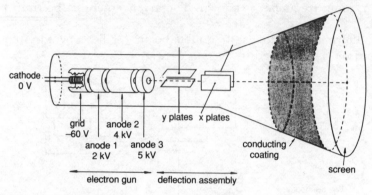

**Figure 11.39    Simplified layout of cathode-ray tube with electrostatic deflection**

passing through it will not collide with atoms of gas. At the narrow end, or neck, of the tube is situated an arrangement of electrodes, called the **electron gun**, which produces a beam of electrons and fires it at the tube face. The gun consists of a coated cathode heated by a current-carrying tungsten heater wire, a grid which is held negative with respect to the cathode so as to begin the beamshaping operation, and two or more anodes, held at varying potentials of up to several thousands of volts more positive than the cathode. The anodes perform the twin functions of accelerating the electrons (and hence of 'firing' them at the screen) and of shaping and focusing the beam so that all the electrons strike the inside of the screen at one very small spot. The inside of the screen is coated with a mixture of fluorescent and phosphorescent powders which glow when excited by the electron beam and produce a spot of light. Electrons from the beam bounce off the inside of the screen, and are collected by a conducting coating round the sides of the tube. This coating is connected to the cathode, so a closed circuit exists round which the electrons circulate. The brilliance of the spot on the screen is dependent on the number of electrons striking the screen, and hence on the negative voltage on the grid; a more negative grid allows fewer electrons to join the beam. The potentials of the anodes with respect to each other vary the beam shape, so controlling focusing.

## Beam deflection

As described so far, the electron gun will fire a beam of electrons to strike the screen centre and to produce a spot of light. For the CRO to be useful, the beam must be capable of deflection so that the position of the spot on the screen can be altered. There are two methods of deflection.

### (i) *Electromagnetic deflection*
The beam of electrons constitutes an electrical current, even though the electrons are moving at high speed through a vacuum instead of slowly in a conductor. Such a beam will therefore be subjected to a force, and hence to deflection, when passing through a magnetic field (see Sections

11.2 and 11.3 of 'Electrical Craft Principles' Volume 1). Fleming's left-hand rule can be applied in this case, although it must be remembered that the conventional current direction assumed by the rule will have the opposite direction to that taken by the electrons.

A magnetic field is set up by currents in coils wound round the outside of the tube as shown in Figure 11.40. This method of deflection is capable of bending the beam very sharply, so it is used for television sets which have short tubes with very large screens. The high currents needed in the deflecting coils rule out electromagnetic deflection for most oscilloscopes, the only exceptions being instruments for display purposes.

### (ii) *Electrostatic deflection*

If the electron beam is made to pass between two parallel conducting plates (see Figure 11.41), it can be deflected by applying a difference of potential between the plates. The negative electron beam is repelled from the negative plate and attracted to the positive plate. The angle through which the beam can be bent using electrostatic deflection is small, so the tubes used in CROs, which employ this method almost exclusively, are comparatively long and narrow, with small screens.

The deflecting plates of oscilloscope tubes are usually shaped as shown in Figure 11.42. A horizontally mounted set of plates deflects the beam vertically. These are called the Y plates, because the beam deflection at the tube face is along the line which is the Y axis of a graph. Vertically mounted X plates give deflection horizontally, along the line corresponding to the X axis of a graph. By using potentials applied to both sets of plates, the spot can be deflected to any part of the screen. For example if the upper Y plate and the left-hand X plate are both made positive, the spot will be deflected towards the upper left of the screen as shown in Figure 11.43.

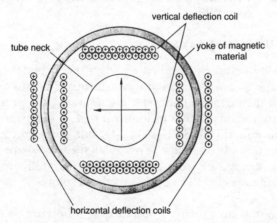

**Figure 11.40   Section through tube neck and yoke to show the arrangement of electromagnetic deflection coils**
The arrows show directions of horizontal and vertical deflection for current directions shown

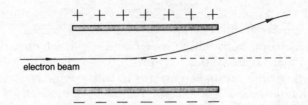

electron beam

**Figure 11.41   To illustrate principle of electrostatic deflection**

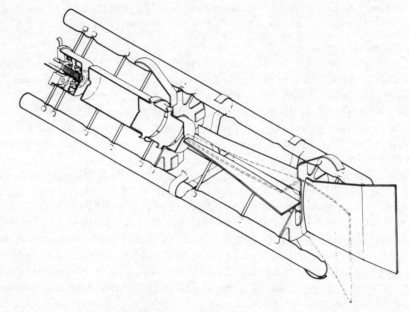

**Figure 11.42   Typical electron gun and electrostatic-deflection assembly**

## 11.20   Tube displays

If an alternating voltage is applied between the Y plates of a CRO, the spot on the tube face will be swept upwards in the positive halfcycle, and downwards in the negative halfcycle, giving a vertical straight line in the middle of the screen as shown in Figure 11.44. This display is useful for voltage measurement (see Section 11.21), but gives no indication of the waveshape of the applied voltage. To display the waveshape, the spot must be moved at a steady rate from left to right horizontally across the screen at the same time as the PD applied to the Y plates is moving the spot vertically.

This horizontal movement is obtained by applying to the X plates a voltage having the waveform shown in Figure 11.45. The voltage is known as the **timebase** of the CRO. Figure 11.46 shows how the timebase applied to the X plates results in a display of the waveform applied to the Y plates. It will be appreciated that the frequency of the timebase must match exactly

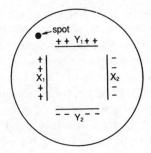

**Figure 11.43   To illustrate spot deflection**

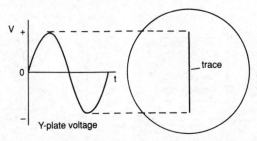

**Figure 11.44   Trace on the screen due to sinusoidal voltage applied to Y plates**

that of the Y plate signal if a single stationary waveform is to appear on the screen. If the timebase frequency is slightly lower than that of the Y-plate signal, the display will move across the screen from right to left; a high timebase frequency gives display movement from left to right.

A timebase with exactly half the frequency of the Y-plate signal will display two cycles of that signal, as shown in Figure 11.47, and any number of cycles can be displayed by making the timebase frequency the correct fraction of the Y-plate signal frequency.

We have seen the necessity of the Y-plate signal and the timebase frequencies being synchronised exactly to provide a steady display. Most timebase circuits have a facility which enables them to lock on to the Y-plate signal. This takes the rising Y-plate signal and uses it to trigger the timebase circuit; the effect may be switched off if not required by a control which is usually labelled 'synch' (synchronising).

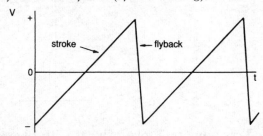

**Figure 11.45   Waveshape of timebase applied to X plates**

It is usual for one of each pair of deflecting plates to be connected together and to the earthed chassis of the CRO, as indicated in Figure 11.48, the signals being applied to the other plates, marked X and Y, respectively, in the figure. Care must be taken when using a CRO to display waveforms of voltages or of currents derived from the mains supply. Since the neutral of the supply is earthed, connection of the phase conductor to the E terminal of the CRO will provide a short circuit. The E terminal must always be connected directly to the neutral conductor unless the supply has been rendered earth-free by use of an isolating transformer (see Section 7.8).

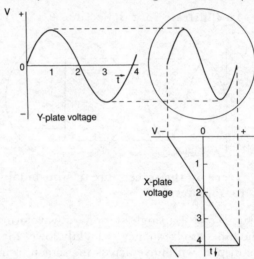

**Figure 11.46    Trace due to X- and Y-plate signals of same freqeuency**

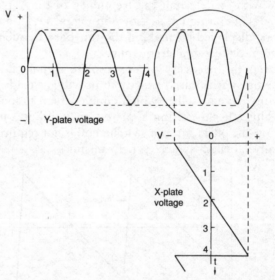

**Figure 11.47    Trace due to Y-plate signal with twice frequency of X-plate signal**

The CRO with electrostatic deflection is capable only of responding to potential differences, and will not directly measure or display currents. This apparent disadvantage may be overcome by passing the current through a resistor. The voltage drop in the resistor will depend on the current it carries, so the PD across it will have the same waveshape as the current, and can be applied to the CRO for display as indicated in Figure 11.49.

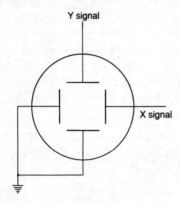

**Figure 11.48 Earthing of deflecting plates**

# 11.21 Measurements with the CRO

As well as being a very useful device for the display of waveforms, the CRO is a highly sensitive measuring device, since its input impedance is the capacitance between its deflecting plates. Although it is capable of very many measurements, only the three most common will be considered here.

## (i) Voltage measurement

In many cases the voltage applied to a CRO will be insufficient to give a deflection large enough to be measured, and amplifiers are built in to multiply the Y input voltage before it is applied to the plates. This will

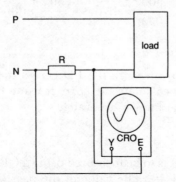

**Figure 11.49 Circuit for displaying current waveform**

tend to reduce input impedance although it remains very high. The number of times the voltage is increased (called the gain of the amplifier) can be selected using a control which is calibrated in terms of the applied voltage needed to move the spot vertically through 1 cm. A typical Y amplifier will have settings such as 100 mV/cm, 5 V/cm etc.

The height of the vertical trace can be measured against a graticule, which is like graph paper with a 1 cm square size printed on a clear sheet of material placed in front of the screen. It must be remembered that the vertical height of a trace due to the application of an alternating voltage depends on the potential difference between the positive and negative maximum (peak-to-peak) values of voltage. If the applied voltage is sinusoidal, the effective or RMS value can be found by dividing the peak-to-peak value by $2\sqrt{2}$ (see Figure 11.50 and Section 10.4 of 'Electrical Craft Principles' Volume 1).

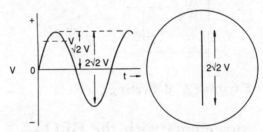

**Figure 11.50   To illustrate measurement of RMS value of voltage using CRO**

## Example 11.10

An unknown sinusoidal alternating voltage is applied to the Y plates of a CRO. When the timebase is switched off and the Y amplifier control is set to 50 mV/cm, a vertical trace 6 cm high appears on the screen. What is the RMS value of the applied voltage?

Peak-to-peak voltage = 6 cm × 50 mV/cm = 300 mV.

$$\text{RMS voltage} = \frac{300}{2\sqrt{2}} \text{ millivolts} = \frac{300}{2\cdot83} \text{ millivolts} = 106 \text{ mV}$$

## (ii) Current measurement

The method here is to measure the PD across the resistor carrying the current, and hence to calculate the current using Ohm's law. The value of the resistor must be known accurately.

## Example 11.11

The current in a circuit is to be measured using a CRO, which is connected as shown in Figure 11.49. The value of the resistor $R$ is $2\cdot5$ $\Omega$, and the vertical trace on the screen of the CRO is 8 cm high with a Y amplifier setting of 5 V/cm. Assuming that the current is sinusoidal, calculate its RMS value.

Peak-to-peak PD = 8 cm × 5 V/cm = 40 V

$$\text{RMS PD} = \frac{40}{2\sqrt{2}} \text{ volts} = \frac{40}{2\cdot83} \text{ volts} = 14\cdot1 \text{ V}$$

$$\text{RMS current } I = \frac{V}{R} = \frac{14\cdot1}{2\cdot5} \text{ amperes} = 5\cdot65 \text{ A}$$

## (iii) Frequency measurement

The timebase control of a CRO consists usually of two knobs, one for coarse and the other for fine variation of timebase frequency. The coarse control, adjustable in steps, is always calibrated in the time for one sweep in fractions of a second; expensive oscilloscopes have the fine control similarly calibrated.

To measure the periodic time, and hence the frequency (see Section 10.1 of 'Electrical Craft Principles' Volume 1) of a voltage applied to the Y plates, the synchronising facility of the timebase is switched off, and the timebase controls adjusted until a display of one cycle of the voltage is stationary on the screen. The periodic time of the voltage can then be read off from the timebase controls.

## Example 11.12

When one cycle of a voltage applied to the Y plates of a CRO is shown as a stationary display on the screen, the timebase controls are set at 2·5 ms/sweep. Calculate the frequency of the applied voltage.

The periodic time $T$ is 2·5 ms.

$$f = \frac{1}{T} = \frac{1}{2\cdot5 \times 10^{-3}} \text{ hertz} = \frac{10^3}{2\cdot5} \text{ herz} = 400 \text{ Hz}$$

## Double-beam CRO

A cathode-ray tube with two electron guns or with one electron gun and a beam-splitting plate, producing two separate electron beams, is the basis of the double-beam CRO. Each beam passes through its own set of Y plates for separate vertical deflection before passing through the single set of X plates which gives common horizontal deflection. There are therefore two separate inputs, $Y_1$ and $Y_2$, each channel being complete with its own calibrated amplifiers.

A double-beam CRO is useful for simultaneous display, and hence comparison, of two waves, either of which may represent a current or a voltage. Relative magnitude and phase can easily be seen in this case. Figure 11.51 shows the connections of a double-beam CRO to display supply voltage on channel $Y_2$ and current on channel $Y_1$. Figure 11.52 is a photograph of a typical double-beam CRO.

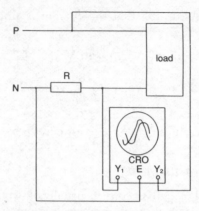

**Figure 11.51    A double-beam CRO connected to display waveforms of both current and voltage for a circuit**

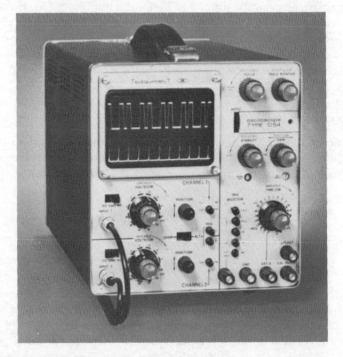

**Figure 11.52    A double-beam CRO**

## 11.22    Summary of formulas for Chapter 11

Wheatstone bridge:

$$R_X = \frac{R_A}{R_B} \times R_C$$

The formula must be read in conjunction with Figure 11.25.

Slidewire bridge:

$$R_X = R \times \frac{l_1}{l_2}$$

The formula must be read in conjunction with Figure 11.26.

Murray-loop test:

$$l = 2l_1 \times \frac{Q}{P + Q}$$

The formula must be read in conjunction with Figures 11.27 and 11.28.

## 11.23　Exercises

1　Draw a clearly labelled sketch to show the construction of a permanent-magnet moving-coil instrument. List the advantages and disadvantages of the device.

2　Using sketches to illustrate your answer, explain the principles of operation of a permanent-magnet moving-coil instrument. Why is the scale of such an instrument linear?

3　Name the important disadvantage of a permanent-magnet moving-coil instrument. Show how this disadvantage is overcome in
(*a*) the rectifier instrument, and
(*b*) the thermal (or thermocouple) instrument.

4　(*a*) Make a sketch or diagram showing the construction of a moving-iron ammeter and describe its principle of operation.
(*b*) Why is the instrument scale uneven?

5　Draw a labelled sketch to show the construction of a dynamometer wattmeter. Show the methods of control and damping. Clearly indicate the current and voltage coils. Draw a circuit diagram showing how the instrument coils are connected to the circuit under test.

6　A permanent-magnet moving-coil instrument has a full-scale deflection at 1 mA and has a resistance of 20 Ω. What is the resistance of a shunt to convert it for use up to 1 A?

7　The voltage drop across an instrument of resistance 4 Ω is 40 mV at full-scale deflection. What is the resistance of the shunt required to convert the instrument for use as a 0–40 A ammeter?

8　The instrument of Exercise 6 is to be converted for use as a 0–250 V voltmeter. What is the resistance of the multiplier required?

9　The instrument of Exercise 7 is to have a multiplier connected in series with it to convert it for use as a voltmeter to read up to 50 V. What will be the multiplier resistance?

10　A permanent-magnet moving-coil instrument has a resistance of 3 Ω, but has a 27 Ω swamping resistor connected in series with it. The movement requires a current of 15 mA for full-scale deflection. Calculate the resistance of a shunt required to convert it for use as a 0–10 A ammeter. What will be the resistance of a multiplier to convert the unit for use as a 0–300 V voltmeter if the swamping resistor remains in circuit?

11  Make a clearly labelled diagram or sketch showing the construction of a moving-coil meter. Describe its action, and explain why it has a uniform scale.

A moving-coil meter of resistance 5 $\Omega$ gives full-scale deflection at a potential difference of 0·075 V. Show how you could adapt the instrument to read up to 240 V. Calculate the value of the extra resistance needed.

12  The resistance material used for meter shunts is chosen for its very low value of temperature coefficient of resistance. Why is this so?

13  State briefly with the aid of circuit diagrams:
   (a) how to extend the range of a 0–110 V AC voltmeter to read (i) 220 V and (ii) 22 kV,
   (b) how to extend the range of a 0–5 A AC ammeter to read 500 V, and
   (c) how to extend the range of a 0·5 A DC ammeter to read 500 A.

14  Draw a circuit diagram showing a single-phase wattmeter measuring the power in a single-phase 6·6 kV 500 A circuit by means of current and voltage transformers.

15  Calculate the sensitivity in ohms/volt of voltmeters giving full-scale deflection with currents of (a) 10 mA (b) 0·5 mA, and (c) 50 μA.

16  A 0–250 V voltmeter with a sensitivity of 2000 $\Omega$/V is used to measure the terminal voltage of an equipment with an internal voltage of 220 V and an internal resistance of 30 k$\Omega$. Calculate the voltmeter reading.

17  When the value of a resistor is measured using the ammeter and voltmeter method, the instrument readings are: (a) 4 A, 20 V (b) 10 mA, 60 V, and (c) 5 A, 3 V.

Neglecting the effects of instrument resistances, calculate the value of the resistor in each case.

18  The resistance of an earth-continuity conductor is measured using the ammeter and voltmeter method with a wander lead as shown in Figure 11.23. When the wander lead is connected to point Y the readings are 5 A and 6 V, while connection to point X gives readings of 6 A, 6 V. Calculate the earth-continuity-conductor resistance.

19  To find the resistance of a resistor, various currents were passed through the resistor and the corresponding voltage drops across it were measured. The results obtained were:

| Current (A) | 1 | 2 | 3 | 4 | 5 | 6 |
|---|---|---|---|---|---|---|
| Voltage drop (V) | 2·4 | 5·1 | 7·4 | 9·0 | 12·3 | 15·0 |

   (a) Draw to a suitable scale the graph connecting these quantities, with current values horizontal, and from the graph estimate the resistance value.
   (b) One serious reading error has been made; estimate what the voltage reading should be.

20  A resistor R is measured by the ammeter-and-voltmeter method, as in Figure 11.53. The ammeter and voltmeter readings are shown on the figure and the resistance of the voltmeter is 400 $\Omega$. Find the apparent and the true values of R.

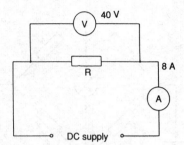

**Figure 11.53   Diagram for Exercise 20**

21 The resistance $R$ of a circuit is measured by the ammeter-and-voltmeter method with the instruments connected as shown in Figure 11.53. The voltmeter, which has a range of 0–100 V and a sensitivity of 5 $\Omega$/V, reads 65 V, whilst the ammeter reads 7 A. Calculate the apparent and true values of the circuit resistance $R$.

22 Calculate the value of resistor $R_X$ in the Wheatstone-bridge circuit of Figure 11.25 if the values of the other resistors at balance are:
(a) $R_A = 50\ \Omega$    $R_B = 50\ \Omega$    $R_C = 36 \cdot 63\ \Omega$
(b) $R_A = 100\ \Omega$    $R_B = 10\ \Omega$    $R_C = 19 \cdot 76\ \Omega$
(c) $R_A = 10\ \Omega$    $R_B = 1000\ \Omega$    $R_C = 75 \cdot 43\ \Omega$
(d) $R_A = 1000\ \Omega$    $R_B = 50\ \Omega$    $R_C = 46 \cdot 19\ \Omega$

23 (a) Sketch a simple diagram showing a 'Wheatstone-bridge' arrangement for measuring the resistance of a length of wire.
(b) In a balanced Wheatstone-bridge circuit ABCD, AB = 1000 $\Omega$, BC = 100 $\Omega$, CD = 80 $\Omega$. Find DA.

24 Explain the principle on which the Wheatstone bridge works. Show how it could be used to measure the resistance of a busbar joint by comparison with a standard 0·1 $\Omega$ resistor.

25 Draw a schematic circuit diagram of a Wheatstone bridge arranged for resistance measurement. Prove the relationship between the resistances at 'balance'. Set out resistances, labelling each with its value, representing a bridge at balance when the measured resistance is 1·752 $\Omega$. The variable-resistor dials give units, tens, hundreds and thousands, 0–10 of each; the ratio arms can each be set at 10, 100, 1000 or 10 000 $\Omega$.

26 Figure 11.54 shows a Wheatstone-bridge circuit. The ratio arms P and Q can each be set at 10, 100, 1000 or 10 000 $\Omega$ and the variable resistor $R$ can be varied from 1 $\Omega$ to 10 000 $\Omega$ in steps of 1 $\Omega$.
(a) Describe briefly how the bridge would be used to determine the value of the resistor $X$.
(b) Calculate the values of $P$, $Q$ and $R$ if the bridge is at balance and the resistor $X$ is 2·864 $\Omega$.

27 A slidewire bridge has a wire 1 m long and is arranged as shown in Figure 11.26. What is the value of the unknown resistor when the value of the known resistor and the distance of the slider from the left-hand end of the wire for balance are, respectively,
      (a) 10 $\Omega$, 27 cm          (b) 2 $\Omega$, 68 cm          (c) 60 $\Omega$, 42 cm?

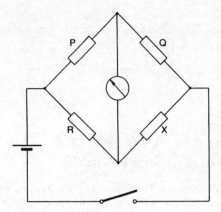

**Figure 11.54   Diagram for Exercise 26**

28 Describe how you would find the resistance of a coil of insulated wire using:

either (i) a slide-wire bridge
or     (ii) an ammeter and voltmeter.

29 A twin-core PVC-insulated and armoured cable is known to have a low-resistance fault to earth on one core only. The second core is in sound condition.

Describe, with diagrams, the Murray-loop method of finding the position of the fault, and explain how the test results would be used for this purpose.

30 Figure 11.55 shows the various pieces of apparatus to be used in determining the position of an earth fault on a metal-sheathed two-core cable AB, by means of a Murray-loop test.

Complete the diagram to show the apparatus connected to the cable ready for the test.

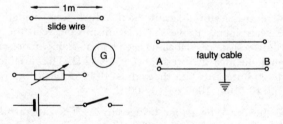

**Figure 11.55   Diagram for Exercise 30**

31 A three-core cable 1·3 km long is found to have a fault between two cores. A Murray-loop test is carried out using the circuit shown in Figure 11.28, and at balance the resistors P and Q gave values of 17·4 Ω and 15 Ω, respectively. Calculate the distance of the fault from the test end of the cable.

32 Describe, with the aid of a diagram, an instrument for the measurement of insulation resistance. The insulation resistances to earth of three separate lighting subcircuits are 75 MΩ, 40 MΩ and 4 MΩ, respectively. Calculate the combined insulation resistance of the three circuits.

33 (*a*) Illustrate by sketches connections of the following instruments when used to measure the input to a single-phase motor:
(i) voltmeter    (ii) ammeter    (iii) wattmeter
(*b*) Discuss the use of 'moving-coil' and 'moving-iron' instruments on AC and DC circuits.

34 Draw a labelled diagram to show the construction of an electrostatically deflected cathode-ray tube. Describe how the device operates to produce a spot on the screen.

35 Explain briefly the principles of both electrostatic deflection and of electromagnetic deflection in a cathode-ray tube. Indicate the use of a tube for which each type of deflection is most suitable.

36 (*a*) Describe briefly the principle of operation of a cathode-ray tube.
(*b*) Explain how the tube can be used to show the trace of an alternating-current waveform.

37 Explain with the aid of sketches the voltage waveform to be applied between the X plates of a CRO so that the waveform of a voltage applied to the Y plates may be displayed. What must be the frequency of the X-plate voltage to give a steady trace?

38 Calculate the RMS values of a sinusoidal alternating voltage applied to the Y plates of a CRO if it gives the following vertical lengths of trace, the Y amplifier control setting for each display being shown after the trace length:

(*a*) 5 cm, 10 V/cm,
(*b*) 9 cm, 50 V/cm,
(*c*) 4 cm, 100 mV/cm.

39 The current in a load is displayed on a CRO by passing it through a $1 \cdot 5 \, \Omega$ resistor and applying the PD across the resistor to the Y amplifier of the CRO. The peak-to-peak amplitude of the display is $7 \cdot 5$ cm when the amplifier setting is 200 mV/cm. Assuming the current waveshape to be sinusoidal, calculate the RMS value of the current.

40 A double-beam cathode-ray oscilloscope is used to measure the voltage and the current in a circuit. The supply voltage is applied to the $Y_1$ amplifier, which is set at 50 V/cm, while the current passes through a $0 \cdot 1 \, \Omega$ resistor, the PD across which is fed to the $Y_2$ amplifier with its controls set at $0 \cdot 2$ V/cm. Calculate the peak and RMS values of voltage and current if the peak-to-peak trace measurements are 6 cm and 3 cm, respectively.

# 11.24   Multiple-choice exercises

11M1 The biggest disadvantage of the permanent-magnet moving-coil instrument is that it:
(*a*) has very low power consumption
(*b*) will not read alternating values
(*c*) has a linear scale
(*d*) has built-in damping.

11M2 The biggest disadvantage of the rectifier permanent-magnet moving-coil instrument is that it:
(a) cannot read direct-current values
(b) can be adapted as a multirange instrument
(c) is extremely large and heavy
(d) is in error if the applied waveshape is not sinusoidal.

11M3 The biggest advantage of the moving-iron instrument is that it:
(a) reads true RMS values    (b) is inaccurate
(c) needs special damping    (d) has a nonlinear scale.

11M4 The most usual application of the dynamometer instrument is as:
(a) an ammeter    (b) a voltmeter
(c) a wattmeter    (d) a frequency meter.

11M5 An additional resistor is often connected in series with an instrument across a shunt to compensate for changes in the resistance of the instrument as temperature changes. This resistor is known as a:
(a) PTC resistor    (b) series resistor
(c) swamping resistor    (d) NTC resistor.

11M6 A voltmeter movement gives full-scale deflection when carrying a current of 25 µA. Its sensitivity will be:
(a) 40 kΩ/V    (b) 25 kΩ/V    (c) 40 Ω/V    (d) 40 kV/Ω.

11M7 A clamp-type ammeter has the advantage that:
(a) it is less expensive than other types
(b) it has a clearer scale than other types
(c) it is more accurate than other types
(d) no circuit disconnection is required.

11M8 A three-and-a-half digit display will give a maximum reading of:
(a) 2999    (b) 1999    (c) 9999    (d) 3599.

11M9 The resistance of an earthing conductor is measured using a long wander lead, a battery, an ammeter and a voltmeter. When the wander lead is connected across the battery the measured current is 400 mA at $1 \cdot 4$ V. The readings for the wandering lead and the earthing conductor in series are 220 mA and $1 \cdot 44$ V. The resistance of the earthing conductor is thus:
(a) $3 \cdot 5$ Ω    (b) $15 \cdot 9$ Ω    (c) $6 \cdot 55$ Ω    (d) $3 \cdot 05$ Ω.

11M10 The Wheatstone bridge is a particularly accurate method for measuring resistance because:

(a) it is a 'null-deflection' method, so the indicating instrument accuracy does not affect the result
(b) three resistors are required as well as that to be measured
(c) a sensitive centre-zero galvonometer is used
(d) the slide-wire method can be used.

11M11 The Murray-loop test can be used to:
(a) measure the sweetness of a sugar solution
(b) find the position of a cable fault
(c) measure the combined resistance of both cores of a two-core cable
(d) take measurements from which cable length can be calculated.

11M12 A megohm-meter used to test insulation resistance must apply a direct voltage so that:
(a) it is safer to use
(b) the reading is in megohms rather than in kilohms
(c) the scale is linear
(d) it measures insulation resistance rather than insulation impedance.

11M13 Because of the heavy current it carries, the earth-fault-loop tester must not be connected for more than:
(a) 1 s      (b) 50 ms      (c) 30 ms      (d) 40 ms.

11M14 A tachometer is an instrument used to measure:
(a) the prospective short-circuit current
(b) the linear speed of a vehicle
(c) rotational speed
(d) the force needed to drive home an earth spike.

11M15 Most modern temperature-measuring systems rely on:
(a) a low ambient temperature
(b) the thermocouple principle
(c) radiant heat from the subject
(d) a passive infrared detector.

11M16 A Schmitt-trigger circuit is used in an electronic frequency-measuring device to:
(a) bring the measured frequency and the crystal output into phase
(b) convert the measured waveform to a square wave
(c) start the two signals at exactly the same instant
(d) act as an AND logic gate.

11M17 The assembly in a cathode-ray tube which fires a stream of electrons at the screen is called the:
(a) X plates      (b) deflection assembly
(c) electron gun      (d) cathode.

11M18 A sinusoidal voltage applied to the input of a CRO with the timebase turned off will result in:
(a) a vertical line on the screen
(b) no screen display at all
(c) a sinusoidal display
(d) the display of the positive halfcycle of the input waveform.

11M19 If two complete cycles of the alternating signal applied to the X input of a CRO are to be displayed, the frequency of the timebase signal applied to the Y input must be:
(a) the same as the X input frequency
(b) twice the X input frequency
(c) any value at all
(d) half the X input frequency.

11M20 A sinusoidal alternating-voltage wave is applied to a CRO and the peak-to-peak measurement of the trace is $4 \cdot 2$ cm. If the Y amplifier setting is 100 V/cm, the RMS value of the displayed voltage is:
(a) 420 V      (b) 296 V      (c) 210 V      (d) 148 V.

11M21 When one complete cycle of an alternating voltage is displayed on the screen of a CRO, the timebase control is set at 50 μs/sweep. The frequency of the displayed value is:

(a) 50 MHz      (b) 50 Hz          (c) 20 kHz          (d) 200 Hz.

*Chapter 12*
# Lighting

## 12.1   Introduction

Lighting is an essential feature of our civilisation.

Natural light, in the form of electromagnetic radiation from the sun, is variable and is only with us for approximately half of the time. We augment and replace this with artificial light from lamps which almost invariably are electrically powered.

The object of this chapter is to indicate the nature of light, and to show how its correct use can be calculated. The different types of electrical lamp will then be considered, with operating circuits and data to show relative outputs and efficacies. The chapter will conclude with a short introduction to the special requirements of circuits feeding electric lamps.

## 12.2   The nature of light

Light is a very small portion of the complete *electromagnetic spectrum*. This family of waves enables energy to be transmitted through space at the constant velocity of $3 \times 10^8$ metres each second. Included in the spectrum are radio, television, radar, radiant heat, ultra-violet, X-rays etc. These phenomena differ from each other in terms of the wavelength or frequency at which they occur. As for alternating current, the wavelengths (or periodic time) and frequency are inversely proportional (see Section 10.1 of 'Electrical Craft Principles' Volume 1) and

$$f = \frac{v}{\lambda}$$

where   $f$ = frequency (Hz)
$v$ = velocity of waves ($3 \times 10^8$ m/s)
$\lambda$ = wavelength (m)

## Example 12.1

If Radio 1 broadcasts on a wavelength of 275 m, calculate its frequency

$$f = \frac{v}{\lambda} = \frac{3 \times 10^8}{275} \text{ hertz} = 1 \cdot 09 \text{ MHz}$$

Light is a very small part of the total spectrum, sandwiched between infra-red (heat) and ultraviolet radiations. It is usual to express the visible range in terms of wavelength rather than frequencies. The wavelengths in this region of the spectrum are very small, and are usually measured in thousandths of millionths of a metre, called nanometres (nm). The visible part of the spectrum is from about 380 nm (violet) to about 720 nm (red) (see Figure 12.1).

Between these two extremes the eye sees the various wavelengths in the different colours of the visible spectrum, commonly called 'the colours of the rainbow'. When these colours are mixed together in the correct proportions the eye sees them as white light.

The eyes are the 'windows' through which we see, but little is known about the way that they transmit their 'picture' to the brain. With the aid of eyelids and an adjustable aperture (the pupil), bounded by the iris, the amount of light entering the eye is adjusted so that we normally make the best use of the available light. In adapting itself for various levels of light, the sensitivity of the eye changes. The curve shown in Figure 12.1 is an average one; it will have a maximum sensitivity at a slightly shorter wavelength when dark-adapted, and at a longer wavelength when light-adapted.

## 12.3   Light units

**Luminous intensity** (symbol $I$) of a light source is a measure of the light given out by a particular source (lamp). It is measured in **candelas** (symbol cd).

**Luminous flux** (symbol $F$) is the rate of flow of light energy from the source, and is measured in **lumens** (symbol lm). A light source of one candela emitting equally in all directions radiates $4\pi$ lumens.

**Efficacy**: the input to a lamp is measured in its electrical power unit, the watt. It is very difficult to measure the light output in watts, but relatively simple in lumens. Because of this, the term 'efficiency' is seldom used for lamps, and is replaced by 'efficacy', which gives the number of lumens output for each watt of input.

$$\text{efficacy (lm/W)} = \frac{\text{light output (lm)}}{\text{electrical input (W)}}$$

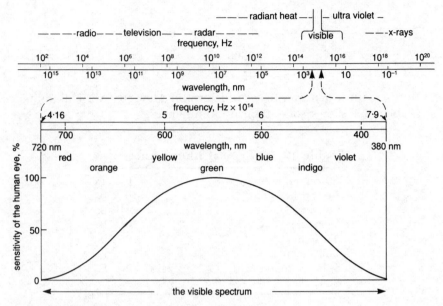

**Figure 12.1  The electromagnetic spectrum**

## Example 12.2

The efficacy of a 2400 mm, 125 W white tubular fluorescent lamp averaged over its normal life is 61·6 lm/W. Calculate the average output of luminous flux

$$\text{efficacy} = \frac{\text{light output}}{\text{electrical input}}$$

$$\text{light output} = \text{efficacy} \times \text{electrical input}$$

$$= 61 \cdot 6 \times 125 \text{ lumens} = 7700 \text{ lm}$$

**Illuminance** (symbol $E$) is a measure of the amount of light received by each part of a surface. Its unit is the lumen per square metre, which is called the **lux** (symbol lx).

$$\text{average illuminance (lx)} = \frac{\text{total luminous flux (lm)}}{\text{surface area illuminated (m}^2)}$$

or

$$E = \frac{F}{A}$$

## Example 12.3

A room is $2 \cdot 5$ m $\times$ 4 m, and the total luminous flux reaching the working plane is 2300 lm. Calculate the average illuminance.

$$E = \frac{F}{A}$$

$$= \frac{2300}{2 \cdot 5 \times 4} \text{ lux} = 230 \text{ lx}$$

The ease with which we can see is determined largely by the illuminance. Table 12.1 shows some typical illuminances.

**Table 12.1: Typical illuminances**

| | |
|---|---|
| Direct sunlight | $10^5$ lx |
| Bright daylight | $10^4$ lx |
| Overcast day | $10^3$ lx |
| Very dark day | $10^2$ lx |
| Twilight | 10 lx |
| Full moon | $10^{-1}$ lx |
| Quarter moon | $10^{-2}$ lx |
| Starlight | $10^{-3}$ lx |

## 12.4   'Point-source' calculations

This section will use the basic physical laws of light to calculate the illuminance of surfaces due to lamps in particular situations. These laws assume that lamps provide 'point sources' of light, i.e. that the light source has no volume. Clearly this is not true.

It is very difficult to take account of reflections from surfaces such as ceilings and walls using this method. For this reason, the 'point-source' method is limited to outside lighting and very large interiors where reflections are unimportant. The distance between the lamps and the illuminated surfaces in such situations is usually large, so the size of the light source can be neglected.

Calculations of this type are based on two laws.

### (i)  The inverse-square law

As light spreads out from a point source, it covers an area which increases as the square of the distance from the source (see Figure 12.2).

Thus, provided the light falls normally (at right angles) to the surface

$$E = \frac{I}{d^2}$$

where   $E$ = surface illuminance (lx)
   $I$ = source intensity (cd)
   $d$ = distance, source to surface (m)

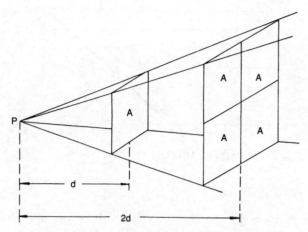

**Figure 12.2  To illustrate the inverse-square law**

## Example 12.4

A 1200 cd lamp is hung above a level surface. Calculate the illuminance of the surface directly below the lamp if the height of the lamp above the surface is (*a*) 3 m, and (*b*) 5 m.

(*a*) $E = \dfrac{I}{d^2} = \dfrac{1200}{3^2}$ lux = 133 lx

(*b*) $E = \dfrac{I}{d^2} = \dfrac{1200}{5^2}$ lux = 48 lx

## (ii) The cosine law

If the light reaching a surface is not normal (at right angles) to it, a given amount of light spreads over an increasing area as the angle between the surface and the incident light increases. Figure 12.3 shows an angled screen AB intercepting a beam of light. The angle between the normal to the surface and the incident light is $\theta$ (Greek 'theta'). If the screen were normal to the beam, a width of only BC would be needed to intercept it. Thus, the light in the beam is spread over a larger area in the ratio AB:BC, so the illuminance must have reduced in the ratio BC:AB. However,

$$\frac{BC}{AB} = \cos \theta$$

so light making an angle of $\theta$ to the normal gives an illuminance $\cos \theta$ times that which fell normally onto the surface.

## Example 12.5

An 800 cd lamp is 2 m from a surface. Calculate the illuminance of the

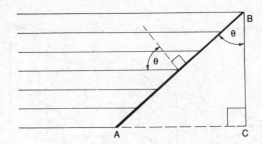

**Figure 12.3   To illustrate the cosine law**

surface if it is normal to the lamp. What will be the new illuminance if the screen is tilted through (*a*) 20°, and (*b*) 45°?

$$E = \frac{I}{d^2} = \frac{800}{2^2} \text{ lux} = 200 \text{ lx}$$

(*a*) $\theta = 20°$, so $\cos \theta = 0 \cdot 9397$

$$E_a = 200 \times 0 \cdot 9397 \text{ lux} = 188 \text{ lx}$$

(*b*) $\theta = 45°$, so $\cos \theta = 0 \cdot 7071$

$$E_b = 200 \times 0 \cdot 7071 \text{ lux} = 141 \text{ lx}$$

Combining the two laws we obtain the expression

$$E = \frac{I \cos \theta}{d^2}$$

where   $E$ = surface illuminance (lx)
   $I$ = source intensity (cd)
   $d$ = distance, source to surface (m)
   $\theta$ = angle between incident light and normal to surface

## Example 12.6

A 1200 cd lamp is suspended 3 m above a level roadway. Calculate the illuminance on the road surface

   (*a*) directly beneath the lamp, and
   (*b*) 4 m from the point beneath the lamp.

Figure 12.4 shows the arrangement.

(*a*) $E = \dfrac{I}{d^2} = \dfrac{1200}{3^2} \text{ lux} = 133 \text{ lx}$

(*b*) $\cos \theta = \dfrac{\text{adjacent}}{\text{hypotenuse}} = \dfrac{3}{\sqrt{(3^2 \times 4^2)}}$

$$E = \frac{I \cos \theta}{d^2}$$

$$= \frac{1200 \times 3}{(3^2 + 4^2) \times \sqrt{(3^2 + 4^2)}}$$

$$= \frac{1200}{25} \times \frac{3}{5} \text{ lux} = 28 \cdot 8 \text{ lx}$$

Note that the illuminance is the number of lumens per unit area. If the light comes from more than one direction, the total flux is the sum of the individual fluxes. It follows that the total illuminance of a surface is the sum of the illuminances if more than one source is involved.

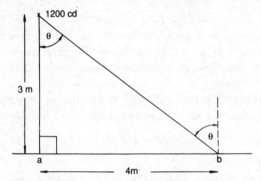

**Figure 12.4   Diagram for Example 12.6**

## Example 12.7

A short, level driveway is illuminated by three 800 cd lamps mounted in a straight line 6 m apart and $2 \cdot 5$ m above the road surface. Calculate the illuminance of the driveway halfway between two lamps.
The arrangement is shown in Figure 12.5.

$$\text{total illuminance} = E_A + E_B + E_C$$

Lamps B and C will give the same illuminance at the point concerned because it is halfway between them, so

total illuminance $E = E_A + 2E_C$

$$E = \frac{I_A \cos \theta_A}{d_A^2} + \frac{2I_C \cos \theta_C}{d_C^2}$$

$$= \left( \frac{800}{9^2 + 2 \cdot 5^2} \right) \times \frac{2 \cdot 5}{\sqrt{(9^2 + 2 \cdot 5^2)}} + 2 \left( \frac{800}{3^2 + 2 \cdot 5^2} \right) \times \frac{2 \cdot 5}{\sqrt{(3^2 + 2 \cdot 5^2)}} \text{ lx}$$

$$= \left( \frac{800 \times 2 \cdot 5}{87 \cdot 25 \times 9 \cdot 34} \right) + 2 \left( \frac{800 \times 2 \cdot 5}{15 \cdot 25 \times 3 \cdot 91} \right)$$

$$= 2 \cdot 45 + (2 \times 33 \cdot 54) \text{ lux} = 69 \cdot 6 \text{ lx}$$

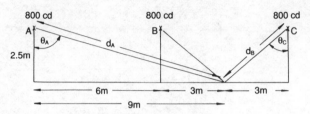

**Figure 12.5   Diagram for Example 12.7**

## 12.5   'Luminous-flux' calculations

Point-source calculations of the type considered in the previous section are perfectly satisfactory where light only reaches surfaces directly from lamps. When light 'bounces' from reflecting surfaces, such as ceilings and walls before reaching the area to be illuminated, such calculations become impossibly complicated.

The luminous-flux method starts by calculating the number of lumens required to illuminate a given area to a given illuminance.

$$F = E \times A$$

where   $F$ = luminous flux (lm)
$E$ = required illuminance (lx)
$A$ = area to be illuminated (m$^2$)

If every lumen emitted by every lamp arrived at the surface concerned, this calculation would give the lumen output required from the lamps. In practice, only a proportion of the emitted lumens reach the surface, and this proportion is called the **coefficient of utilisation** (symbol $U$). Practical values for this coefficient vary widely between about $0 \cdot 1$ (10% of emitted light reaches working surface) and about $0 \cdot 95$ (95%).

Its value depends on a number of factors. These are

(i)   Type of luminaire (lighting fitting). Enclosed luminaires will allow less light to escape than open types.
(ii)   Colour and texture of walls and ceiling. Dark, matt surfaces reflect less light than light-coloured, smooth surfaces.
(iii)   Room size. A large room allows more light to be received directly from luminaires without reflection.
(iv)   Number and size of windows. An uncurtained window reflects virtually no light.
(v)   Mounting height of luminaires. The inverse-square law still applies, although not used in this method.

It will be clear that reflection from luminaire surfaces, walls, ceilings, floor etc. play an important part in the efficiency of an interior-lighting system. If such surfaces become dirty, or discolour with age, less light will be reflected. This is taken into account by applying a **maintenance factor** (symbol $M$), which is the illuminance for an average dirty condition compared with that when the situation is completely clean. Values vary from about $0\cdot8$ for an office which is regularly cleaned to about $0\cdot2$ for a dirty industrial situation.

Extending the formula given near the beginning of this section,

$$F = \frac{E \times A}{U \times M}$$

where  $F$ = luminous-flux output from lamps (lm)
$E$ = required illuminance (1x)
$A$ = area to be illuminated ($m^2$)
$U$ = coefficient of utilisation
$M$ = maintenance factor

The assessment of values for the coefficient of utilisation and the maintenance factor are complex processes which are beyond the scope of this book. The maintenance factor takes into account the reduction in light output as lamps age, together with the loss due to the collection of dirt on lamps, reflectors, walls, ceiling etc. If required, reference should be made to the CIBSE *Code for Interior Lighting*, published by the Chartered Institution of Building Services Engineers, which also recommends illumination values for various interiors.

## Example 12.8

A workshop 12 m × 18 m is to be illuminated to a level of 300 lx by fluorescent luminaires each containing two 2400 mm, 125 W lamps which have an efficacy of 60 lm/W. Given a coefficient of utilisation of $0\cdot6$ and a maintenance factor of $0\cdot8$, calculate the number of luminaires required.

$$F = \frac{E \times A}{U \times M}$$

$$= \frac{300 \times 12 \times 18}{0\cdot6 \times 0\cdot8} \text{ lm} = 135\ 000 \text{ lm}$$

Each 125 W lamp gives 60 lumens for each watt, and each luminaire provides

$$2 \times 125 \times 60 \text{ lumens} = 15\ 000 \text{ lm}$$

Number of luminaires required

$$= \frac{\text{total lumens}}{\text{lumens per luminaire}}$$

$$= \frac{135\ 000}{15\ 000} = 9 \text{ luminaires}$$

In most cases, the number calculated will not work out to a whole number. The number chosen then should be more than the calculated figure, and should provide for a sensible layout.

## Example 12.9

It is estimated that the lamps in an office 6 m × 8 m emit 19 000 lm. Assuming a coefficient of utilisation of $0 \cdot 65$ and a maintenance factor of $0 \cdot 9$, estimate illuminance

$$E = \frac{F \times U \times M}{A} = \frac{19\ 000 \times 0 \cdot 65 \times 0 \cdot 9}{6 \times 8} \text{ lux} = 232 \text{ lx}$$

Recommended illuminances for situations of most types can be found in the CIBSE *Code for Interior Lighting*.

### Glare

Glare can be defined as discomfort or difficulty in seeing the task in hand when parts of the field of view are excessively bright compared with their surroundings. Advanced lighting calculations use the glare-index system, which provides a numerical value for a particular system and enables discomfort glare to be quantified.

In practical terms, it is important to ensure that a lighting installation is designed so that bright light sources are not directly visible by the user. This can usually be achieved by careful positioning and the use of diffusers or baffles.

### Computer calculations

The calculations made earlier in this section are not complicated, but are extremely tedious and time consuming. Very many lighting manufacturers and software specialists produce computer programs which will give the results of a particular lighting arrangement very quickly after entry of data such as luminaire type and output, position, spacing and so on. Such programs have the advantage of showing very quickly the results of changes in the chosen light source and have revolutionised lighting design calculations.

However, it is of the greatest importance that the lighting designer understands the basis of the calculations being made by the computer and its program and any approximations or simplifications, so a thorough understanding of the calculating methods is essential. There are too many programs for their uses to be considered here in detail.

## 12.6   Filament lamps

If a material is made white hot it emits light. This is the principle of operation of the lamp which uses a tungsten filament. Tungsten has both a high melting temperature (3380°C) and the ability to be drawn out in fine wire. To prevent the filament oxidising and failing prematurely, all oxygen must be removed from the enclosing glass bulb. Small lamps are evacuated, but larger lamps are filled with argon, which reduces filament evaporation at high temperatures. The efficacy of filament lamps is relatively low but increases with the larger sizes ($8 \cdot 0$ lm/W for a 25 W up to $11 \cdot 6$ lm/W for a 100 W lamp). Coiling the already coiled filament (coiled-coil lamp) retains heat and improves efficacy to $12 \cdot 6$ lm/W for the 100 W size. The construction of a general-lighting-service (GLS) lamp is shown in Figure 12.6*a*. The nominal rated life of a tungsten-filament lamp is 1000 hours.

Filament lamps are made in a bewildering range of sizes, colours and shapes. Reference should be made to the catalogue of a good manufacturer if details are required. Lamps are often used in fittings called **luminaires** which are designed to provide the required light control. If the luminaire is enclosed, it will be marked to indicate the rating of the lamp for which it was designed; using a higher-power lamp will cause overheating and fire may result. Incandescent lamps with integral reflectors are often used for display work, and result in a great deal of heat, as well as light, reaching the subject. To reduce this effect, some lamps have dichroic reflectors, which reflect the visible light but transmit a high proportion of the heat. Luminaires for use with such lamps must be specially designed because of the additional energy retained.

## Tungsten–halogen lamp

If a few drops of halogen such as iodine are included with the gas filling, a curious effect occurs. A clear gaseous sheath forms round the filament, almost entirely preventing evaporation and allowing the lamp to run much hotter, resulting in a higher efficacy (up to about 23 lm/W). The halogen condenses at about 300°C, so the lamp bulb must be kept above this temperature. It is made of quartz glass; such bulbs must not be handled or the sodium from the skin will form white opaque patches on the glass when it becomes hot. Tungsten–halogen lamps are widely used for floodlighting and vehicle-lighting applications. Typical lamps of this type are shown in Figure 12.6*b*. A small, low-voltage tungsten–halogen lamp has been produced for display lighting. Usually operating at 12 V, the filaments of these lamps are small, and thus extremely precise light output can be achieved with the integral reflector (see Figure 12.6*c*). With ratings of 15 W, 20 W, 35 W and 50 W, these lamps are very widely used for display lighting in shops, museums and similar situations. Care must be taken in selecting cable sizes or lighting track ratings on the low-voltage side of a supply transformer. With mains-voltage low-power lamps, we have become used to thinking of lamp current as very small. Remember that at low voltages the current is correspondingly higher; for example, a 50 W 12 V amp takes a current of more than 4 A. These lamps are provided with either bipin or small-bayonet-cap (SBC) connections, and usually have a rated life of 2000 hours, twice that of the standard filament

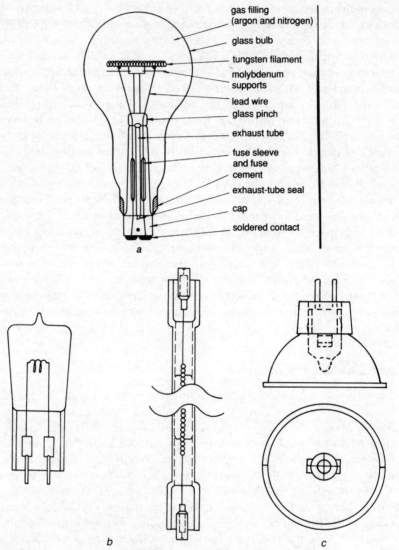

gas filling
(argon and nitrogen)

glass bulb

tungsten filament

molybdenum
supports

lead wire

glass pinch

exhaust tube

fuse sleeve
and fuse

cement

exhaust-tube seal

cap

soldered contact

*a*

*b*

*c*

**Figure 12.6   Filament lamps**

    *a* General-lighting service (GLS) lamp
    *b* Mains-voltage tungsten–halogen lamps
    *c* A low-voltage tungsten–halogen display lamp

lamp. There may be problems when dimming tungsten–halogen lamps. If the lamp temperature is reduced by lowering the voltage, it is possible that the bulb temperature will fall low enough to cause halogen condensation and hence reduced life.

    Filament lamps are inefficient, and the increased awareness of the need to conserve energy has resulted in the much wider use of discharge lamps.

The low first cost and the ease with which they can be handled by unskilled personnel suggests, however, that filament lamps will be used for very many years.

## 12.7   Discharge-lamp principles

It is easy to understand how a heated filament will produce light; understanding how radiation can be emitted from an electric current flowing in a gas or vapour is more difficult. Electrons driven through the gas or vapour by the tube voltage collide with atoms as they go. Some collisions are severe enough to break a loosely held electron from an atom, leaving behind a positively charged **ion**. This type of collision needs a fairly high tube voltage, and results in an **electron avalanche**. Each detached electron has further collisions in its path down the tube, so that very large numbers arrive at the positive electrode (the anode). The positive ions move more slowly towards the negative electrode (the cathode) collecting electrons and becoming complete atoms as they go. The mixture of electrons and ions is called a **plasma**, and has very low resistance so that the tube current must be limited by an external **ballast** in the form of a resistor (DC supplies) or a choke (AC supplies).

The heavy current in the tube when it has **struck** (the electron avalanche begins) results in a very large number of less severe collisions, which **excite** the atom. An electron is lifted to a higher orbit. In this condition the atom is unstable, and the electron quickly drops back to its correct position, giving off the energy received in the collision as a flash of electromagnetic radiation. There are tremendous numbers of such flashes, so the tube emits radiation continuously at a wavelength (and hence at a colour) depending on the type of gas filling and on its pressure. For lamps used on AC supplies an iron-cored inductor (choke) is preferred to a resistor because the necessary voltage drop is achieved without such a large power loss, so the circuit is more efficient. The inductor causes current to lag voltage, resulting in low power factor, so the circuit is provided with a capacitor for power-factor-improvement purposes. The inductor usually used as a ballast is large and heavy, particularly when used in conjunction with high-power lamps, and, although the power loss is a great deal lower than for a resistive ballast, it is usually significant. Electronic ballasts use thyristors and triacs to provide phase control of the lamp voltage, and thus to replace the ballast. Size and losses are usually much reduced and the need for power-factor correction is removed.

## 12.8   Low-pressure mercury-vapour (fluorescent) lamps

These are the most common of the discharge lamps because they are very efficient and are suitable for internal applications. The lamps are made in a variety of lengths from 150 mm up to 2400 mm (nominal lengths) and are filled with mercury vapour at low pressure. The radiation is almost all in the invisible ultraviolet part of the spectrum, but is converted to visible light by a coating of fluorescent powder on the inner tube surface.

By using differing combinations of fluorescent powders, it is possible to control the colour of the light. White lamps are most efficient, but the appearance of coloured surfaces illuminated by them may be altered (poor colour rendering). Lamps emitting light of more pleasing appearance (warm white) or giving good colour matching (northlight) have traditionally used halophosphate phosphors and had lower efficiency (see Table 12.2).

However, this is not the case with modern triphosphor lamps. A high voltage is needed to cause the tube to strike, and there are several methods for providing it.

## Switch-start circuit

The circuit is shown in Figure 12.7, together with the construction of the glow-type switch. When the supply is switched on, the switch is open so mains voltage appears across its contacts. The switch is enclosed in a small glass tube containing neon gas, which glows and produces heat. The switch contacts are bimetals, and the heat causes them to bend so that they touch.

A circuit is thus made through the choke and through both lamp filaments, which are warmed and emit electrons. Meanwhile, the switch contacts cool and open, breaking the circuit. The EMF induced in the choke by the current change is added to the supply voltage and applies to the tube to cause it to strike. The current in the tube causes a voltage drop in the choke, so that the lamp PD applied to the switch contacts is too low to cause a glow in the neon, and the starter switch takes no further part in the operation of the circuit.

If the tube fails to strike first time, the process repeats until striking is achieved. This type of circuit is inexpensive, but the delay in starting and the flashing of the tube during starting may cause annoyance.

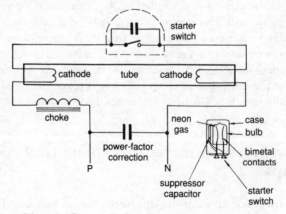

**Figure 12.7   Circuit for switch-start fluorescent lamp**

## Quick-start circuit

Figure 12.8 shows that an autotransformer is used to heat the lamp electrodes and sometimes to apply an increased voltage across the tube. The presence of earthed metal near the tube helps ionisation and the tube strikes. In very dirty situations, a metal earthing strip along the outside of the tube may be necessary to ensure starting.

## Semi-resonant-start circuit

The place of the choke is taken by a special transformer (Figure 12.9). Primary and secondary are reverse connected so that the voltages in them are almost 180° out of phase, increasing the tube voltage so that it strikes as soon as the electrodes emit electrons. When the tube has struck, the primary winding acts as a choke and limits the current. The circuit is very successful with the longer tubes which are difficult to strike by other methods, and when temperatures are low (down to – 5°C).

The inductive choke or transformer in these circuits results in a low lagging power factor which is corrected by a parallel capacitor (see Section 5.12). In the semi-resonant-start circuit, the series capacitor assists starting and improves power factor.

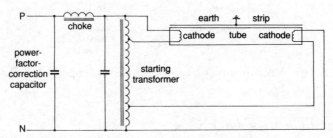

**Figure 12.8  Circuit for quick-start fluorescent lamp**

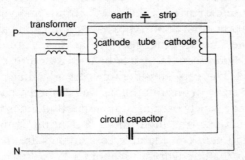

**Figure 12.9  Circuit for semi-resonant-start fluorescent lamp**

The nominal rated life of fluorescent lamps is 7000 hours. Data for the efficacy of fluorescent lamps are given in Table 12.2.

### Electronic control gear

Modern low-cost and compact electronics have made it possible to dispense with the inductive choke ballast as well as the starter switch and to limit running current and provide a high-voltage pulse for starting. Such systems are lighter than the traditional control gear, but are more expensive and have not so far been used very widely.

### High-frequency fluorescent lamps

Research has shown that a fluorescent lamp operated at a frequency much higher than that available from the public supply has many advantages. These include higher efficacy, the absence of flicker (which may cause headaches or, in a very few cases, the onset of epileptic attacks) and the absence of the stroboscopic effect (see Section 12.13).

A typical high-frequency fluorescent lamp will operate at 35 kHz. This frequency is cheaply and easily derived using electronic means, as shown in block-diagram form in Figure 12.10.

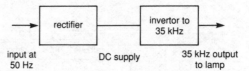

**Figure 12.10   Production of high-frequency supply for fluorescent lamp**

### Compact fluorescent lamps

The low efficacy of tungsten-filament lamps has led to intense development of low-pressure mercury-vapour (fluorescent) replacements for them. A number of versions has been available for some time (see Figure 12.11) which will replace filament lamps with counterparts giving the same light output for about one-quarter of the power input, and with rated lives of 5000 instead of 1000 hours. Some types have an integral ballast and starting switch, whilst others use external gear, the advantage of the former being that the compact fluorescent lamp can be used as a direct replacement for the filament type. The very considerable cost advantage of the new type of lamp (reduced energy consumption as well as increased life) means that the lamps have been widely used in commercial and industrial installations. For domestic situations, the high initial cost of the lamp has slowed its adoption, although very considerable cost savings are made over the life of the lamp. Lamp loadings vary from about 5 W to 36 W, replacing 25 W to 150 W filament types.

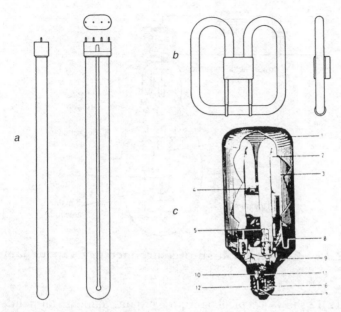

**Figure 12.11   Compact fluorescent lamps**
    *a* Philips PL lamp
    *b* GE 2D lamp
    *c* Philips SL lamp

| | | |
|---|---|---|
| 1 outer prismatic bulb | 5 tube cathode | 10 thermal cutout |
| 2 discharge tube | 6 & 7 starter switch | 11 starter-switch capacitor |
| 3 internal fluorescent coating | 8 mounting plate | 12 lamp cap (ES or BC) |
| 4 inductive choke | 9 outer housing | |

## 12.9   High-pressure mercury-vapour lamps

When the mercury vapour pressure is greater than that of the fluorescent lamp, the wavelength of the output radiation increases, so that the output is visible without the need for fluorescent powders. Many types of high-pressure lamp are available, but all have short arc tubes made of quartz or other materials which will withstand the high temperatures and pressures involved.

The mercury lamp gives most output in the violet, green and yellow parts of the spectrum, so the basic colour rendering is not good. The light output is often modified by adding metallic halides to the gas, or by using a fluorescent coating inside the outer bulb.

At normal temperatures mercury exists as a liquid. In consequence, the arc must be struck in the argon gas filling; the heat produced vaporises the mercury, increasing the pressure and the light output. The arc is struck by using a secondary electrode close to a main electrode so that mains voltage will cause a discharge which spreads down the tube. The secondary electrode takes little part after the initial discharge because of a high-value series-connected resistor. Operating current is limited by the choke.

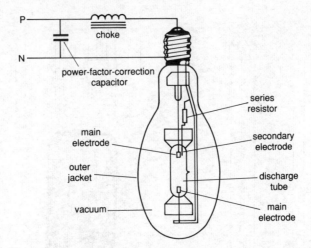

**Figure 12.12   Circuit for high-pressure mercury-vapour lamp**

Figure 12.12 shows a typical operating circuit. Some metal-halide lamps need an additional pulse starter rather like that used for the high-pressure sodium lamp (see Figure 12.16). Because the operating pressure of the lamp is often several times that of the atmosphere, it will not normally restrike after switching off until it has cooled and the pressure has fallen. Where it is important that lighting should be maintained for safety or security purposes, restriking may be forced using a high-voltage electronic-pulse restarting system.

High-pressure mercury-vapour lamps of the various types are made (see Figure 12.13) with electrical loadings from 50 W to 1 kW. Typical details are shown in Table 12.2. Rated nominal life is 7000 hours.

A variant on the mercury-vapour lamp is the mercury blended (MBT) lamp which combines a mercury-arc tube and a tungsten filament in the same envelope. The filament is connected in series with the arc tube and acts as a ballast. The predominantly red light output from the filament also improves the overall colour of the output of the lamp.

## 12.10   Low-pressure sodium lamps

These lamps have a much longer arc path than high-pressure lamps, and are started by applying a high voltage across the electrodes by means of a 'leaky' autotransformer. Sodium is solid at normal temperatures, so the arc is struck in neon gas, which gives a characteristic red glow until the sodium vaporises and provides its output. This light is almost all at one wavelength (monochromatic) in the orange part of the visible spectrum, so the colour-rendering properties of the lamp are poor. Its efficacy, however, is very high, so it is widely used for road lighting.

**Table 12.2: Lamp data**

| Type | Description | Power (W) | Nominal life (hours) | Output lm | Efficacy (lm/W) |
|---|---|---|---|---|---|
| E40 | tungsten filament | 40 | 1000 | 385 | 9·6 |
| E150 | tungsten filament | 150 | 1000 | 1950 | 13 |
| M50 | tungsten halogen miniature 12V | 50 | 3000 | 1030 | 20·6 |
| K4 | tungsten halogen linear 240 V | 1000 | 2000 | 21 000 | 21 |
| D7BC4 | single ended compact fluorescent | 7 | 8000 | 400 | 57 |
| D20BC4 | single ended compact fluorescent | 20 | 8000 | 1200 | 60 |
| L6520 | fluorescent 1500 mm cool white | 65 | 8000 | 4500 | 69 |
| L5830 | fluorescent 1500 mm white | 65 | 8000 | 4800 | 74 |
| L7023 | fluorescent 1800 mm white | 70 | 8000 | 5800 | 83 |
| MBFT | high pressure mercury blended (tungsten ballast) | 160 | 4000 | 2500 | 16 |
| H400SN | high pressure mercury metal halide | 400 | 8000 | 35 000 | 87 |
| S250P | high pressure sodium | 250 | 8000 | 31 250 | 125 |
| S1000 | high pressure sodium | 1000 | 8000 | 120 000 | 120 |
| SX135 | low pressure sodium (U tube) | 55 | 8000 | 7800 | 142 |

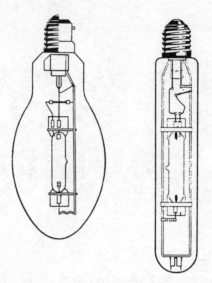

**Figure 12.13   High-pressure mercury-vapour lamps**

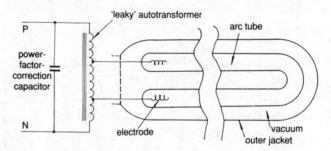

**Figure 12.14   Circuit for low-pressure sodium lamp**

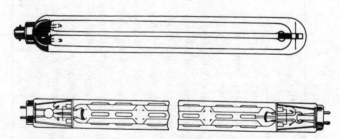

**Figure 12.15   Low-pressure sodium lamps**

Figure 12.14 shows a typical low-pressure-sodium-lamp circuit, and Figure 12.15 shows the construction of this type of lamp; the various types have electrical loadings in the range 18 W to 180 W. Typical details are shown in Table 12.2. Rated nominal life is 7000 hours.

## 12.11 High-pressure sodium lamps

A sodium lamp runs at a higher pressure, like the mercury lamp, and emits radiation over a wider range of wavelengths than its low-pressure counterpart. This gives the high-pressure sodium lamp much better colour rendering properties, the output being a pleasant golden colour. The short arc tube is usually made of sintered aluminium oxide, which is one of the only materials known to stand up to the attack of hot ionised sodium vapour at a pressure approaching one-half of an atmosphere. A choke is again used for current control, but starting is almost always be means of an electronic high-voltage pulse starter. The high-pressure sodium lamp is available in a range of sizes from 50 W to 1 kW. They are used for large interiors such as gymnasia, factories, swimming pools etc., as well as for road lighting and floodlighting. Data for these lamps are given in Table 12.2. Figure 12.16 shows a circuit for a high-pressure sodium lamp, and Figure 12.17 shows its construction. Rated nominal life is 7000 hours.

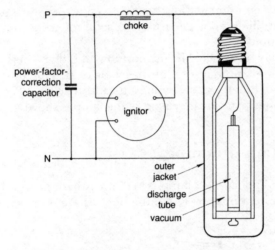

**Figure 12.16   Circuit for high-pressure sodium lamp**

## 12.12   Induction lamp

This is a recent introduction to the lamp market. One type consists of a glass bulb with a gas filling which is excited into an ultraviolet light-emitting state by the application of a very high-frequency magnetic field, typically of the order of 2 MHz, which is provided from a built-in electronic circuit. A fluorescent coating on the inside of the glass bulb converts this invisible radiation to visible light as in the low-pressure mercury-vapour (fluorescent) lamp. Efficacy is about four times that of the incandescent lamp of similar light output, so that, together with a claimed life of 10 000 hours an

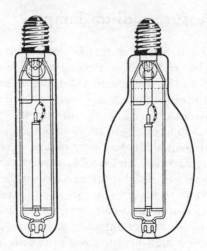

**Figure 12.17  High-pressure sodium lamps**

economic case can be made for this relatively expensive lamp. The long life means that the lamp is particularly useful for installation in difficult-to-reach situations.

The lamp may be installed in any position and, due to its high-frequency operation, there is no question of flicker or of stroboscopic effect. Light output starts within 1 s of switching on, and reaches 80% of maximum level in 8 minutes. The arrangement of the lamp is shown in Figure 12.18.

## 12.13   Lamp dimming

As the voltage applied to a lamp is reduced, there is also a reduction in current and in light output. Dimming devices are thus voltage-reduction

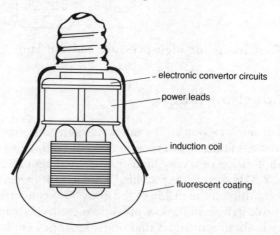

**Figure 12.18   Arrangement of one type of induction lamp**

systems, but it must be appreciated that light output is not proportional to the applied voltage. Figure 12.19 shows a graph which relates the light output of a filament lamp to the applied voltage. It shows that, at about 40% of normal voltage, there is no light output at all. While the curve shown is correct for the filament lamp, it does not follow that this relationship applies to all lamps, and it is important that manufacturers are consulted before systems are designed.

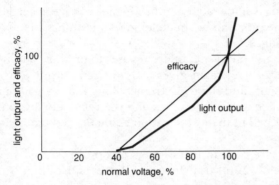

**Figure 12.19   Variation of light output and efficacy with voltage**

Lamp dimming is a useful design tool, because the same lighting scheme, if dimmed, can fulfil more than one purpose. For example, lighting suitable for reading is likely to be too bright for comfortable viewing of television, so dimming circuits are often used in the home and in educational situations. The change in psychological mood with a change in lighting intensity is also often of importance.

The lighting designer must be aware of the fact that lamp dimming is often accompanied by a change in the colour of the light output; a filament lamp, for example, will provide light which is redder in colour as it is dimmed. A reduction in efficiency and in efficacy usually accompanies the process of dimming.

Most filament lamps are suitable for dimming, although there is a question mark over the prolonged use of tungsten–halogen lamps at low voltages where the envelope temperature may fall to a level where correct operation is unlikely. Many fluorescent lamps may be dimmed, but there are exceptions and manufacturers should be consulted. When fluorescent lamps are dimmed, special circuitry is necessary to ensure that the filaments are fed at their full voltage throughout the operation to prevent sudden loss of light due to discharge failure during voltage reduction, or delayed striking on voltage increase.

Thyristor and triac circuits are usually used for voltage control in light-dimming circuits, and are considered in Chapter 6. Great care in design is necessary if excessive electromagnetic emissions are to be avoided which would break the EMC (Electro-Magnetic Compatibility) regulations.

## 12.14   Electric signs

All the discharge lamps so far considered are 'hot-cathode' lamps. The cathode is heated either by a current flowing through it for the purpose (fluorescent lamp) or due to bombardment by electrons and ions (other lamps) and it is given a coating of special emitting material such as barium or thorium. The life of the discharge lamp is limited by the wearing away of the coating.

If the cathode is not heated, a much higher voltage is needed to maintain the arc, and the tube can be much longer and thinner than is the case with the lower voltages.

Some lamps use fillings of different gases to produce outputs of differing colours. For example, neon gives a bright red output, carbon dioxide a white/buff output, and so on. Extremely long life is an advantage (20 000 hours), but the length of the tubes and the relatively low output per unit length has resulted in their application being limited mainly to advertising signs.

Glass tubes are bent to form letters or other required shapes, and are fed from double-wound transformers. Tubes must often be connected in series, when nickel–chromium single-strand wire, enclosed in a very thin capillary tube, is used. Since a high voltage is present, the tubes and their interconnections must always be arranged so that they cannot be touched, such as high up on buildings (see Figure 12.20). Signs often run unattended at night and, in the event of a fire, the fire-fighters could be in danger if the sign were still live. The IEE Wiring Regulations require that an external **fireman's switch** allows easy disconnection. Similarly, to prevent danger to maintenance personnel if the circuits are switched on while being maintained, a locked switch must also be provided. This enables the person working on the circuit to remove the operating handle of the switch so that it cannot inadvertently be switched on.

The IEE Wiring Regulations require that no lamp of this type may have a PD to earth exceeding 5 kV. However, the use of a transformer with its secondary centre-tapped to earth enables voltages up to 10 kV to be used.

## 12.15   Installation of discharge lamps

The foregoing sections have made it clear that all discharge lamps use transformers or inductors (chokes) to control or limit the current. Such inductive circuits will operate at a lagging power factor, so that the current needed for a given power output is greater than that calculated simply from dividing power by applied voltage. In addition, whenever an inductive circuit is broken, an EMF is induced which tries to keep the current flowing and results in severe arcing at the switch contacts.

For this reason, the IEE Wiring Regulations demand that switches controlling discharge lamps should, unless specially designed to withstand

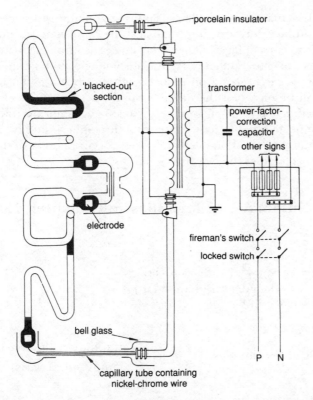

**Figure 12.20   Arrangement of a high-voltage cold-cathode electric sign**

arcing, have twice the current rating of the load that they control. Similarly, the circuit rating in voltamperes (used for calculating cable and fuse sizes) should not be less than 1·8 times the power rating in watts.

## Stroboscopic effect

The current in an AC circuit falls to zero twice in every cycle, so a discharge lamp fed from a 50 Hz supply actually goes out and comes on 100 times every second. This rapid flicker is normally invisible to the human eye, which cannot usually follow separate variations at a rate exceeding about 20 each second.

The rapid flicker is known as the stroboscopic effect, and it can present dangers under particular circumstances. For example, if a rotating machine has a speed which is a submultiple of 100 revolutions each second, its position will be the same each time it is lit by a flash from the lamp, and it may appear to be stationary. Many accidents have occurred because of this effect.

Modern discharge lamps are more efficient than their older counterparts, and stay on for longer in each halfcycle, reducing the stroboscopic effect. If, however, it becomes troublesome, there are two methods of overcoming it.

(i) Lead–lag circuit: Two discharge lamps are mounted close together (e.g. a twin-tube fluorescent luminaire). One lamp is fed from a normal circuit, so current lags voltage. The other is fed from a circuit including a series capacitor so that current leads voltage. Both lamps still go out 100 times each second on a 50 Hz supply, but are 'out of step'. So one will be on when the other is off.

(ii) Connection to a three-phase supply: Successive lamps, or rows of lamps, are connected to different phases of a three-phase supply, so they are 120° out of phase. Possible dangers will occur because of the presence of line voltage (415 V) between lamps.

## 12.16   Summary of formulas for Chapter 12

$$f = \frac{v}{\lambda} \qquad v = f\lambda \qquad \lambda = \frac{v}{f}$$

where   $f$ = frequency of radiation (Hz)
   $\lambda$ = wavelength of radiation (m)
   $v$ = velocity of radiation (m/s)

(the velocity of light is $3 \times 10^8$ m/s)

$$\text{Efficacy (lumens per watt)} = \frac{\text{lamp output (lumens)}}{\text{lamp input (watts)}}$$

$$E = \frac{F}{A} \qquad F = EA \qquad A = \frac{F}{E}$$

where   $E$ = illuminance (lx)
   $F$ = luminous flux (lm)
   $A$ = area (m$^2$)

$$E = \frac{I \cos \theta}{d^2} \qquad I = \frac{Ed^2}{\cos \theta} \qquad \cos \theta = \frac{Ed^2}{I} \qquad d = \sqrt{\frac{I \cos \theta}{E}}$$

where   $E$ = illuminance (lx)
   $I$ = luminous intensity (cd)
   $\theta$ = angle between incident light and normal (degrees or radians)
   $d$ = distance, source to surface (m)

$$F = \frac{EA}{UM} \qquad E = \frac{FUM}{A} \qquad A = \frac{FUM}{E} \qquad U = \frac{EA}{FM} \qquad M = \frac{EA}{FU}$$

where   $F$ = luminous flux (lm)
   $E$ = illuminance (lx)
   $A$ = area (m$^2$)
   $U$ = coefficient of utilisation
   $M$ = maintenance factor

## 12.17  Exercises

1 Maximum sensitivity of the human eye is in the green part of the visible spectrum where the frequency is about $5 \cdot 5 \times 10^{14}$ Hz. Calculate the wavelength of green light.

2 A red traffic sign provides light at a wavelength of 670 nm. What is the frequency of the red light?

3 A 135 W low-pressure sodium (SOX) lamp provides an output of 21 200 lm. Calculate its efficacy.

4 A mercury blended (MBT) lamp provides 11 500 lm at an efficacy of 23 lm/W. Calculate its power input.

5 A linear tungsten–halogen lamp is rated at 300 W and provides 5000 lm. Calculate its efficacy.

6 An area 10 m × 25 m is to be illuminated to a level of 150 lx. How many lumens must reach the area?

7 A lamp provides 6875 lm in the direction of a surface, resulting in an illuminance of 550 lx. Calculate the area of the surface.

8 A lighting system provides 15 000 lm to an area 8 m × 7·5 m. Calculate the average illuminance.

9 A light source mounted 8 m above a level surface is required to give an illuminance of 60 lx on the surface directly beneath the source. What must be the luminous intensity of the source?

10 A 2200 cd lamp mounted directly in front of an advertising hoarding gives a maximum illuminance on the hoarding of 180 lx. Calculate the distance between the hoarding and the lamp.

11 Lamps of 800 cd and 1200 cd are mounted 4 m apart on a level surface. A vertical screen is arranged between the lamps so that its surfaces are normal to the lamps. The screen is 2·5 m from the 800 cd lamp. Calculate the illuminance of both sides of the screen.

12 Two lamps, each of intensity 2400 cd, are mounted 6 m above a level roadway and 20 m apart. Calculate the illuminance of the road surface,
   (a) directly beneath one lamp,
   (b) halfway between the points beneath the lamps, and
   (c) 4 m from the point beneath one lamp, and between them.

13 Three 1500 cd lamps are mounted on 3·5 m-high columns 12 m apart in a straight line above a level road. Calculate the illuminance of the road surface,
   (a) beneath the centre lamp, and
   (b) beneath the end lamp.

14 Two lamps are each of intensity 750 cd and are mounted 6 m apart above a level path. One is 3 m above the path, and the other 4 m high. Calculate the illuminance on the path beneath the 3 m-high lamp. To what level will the illuminance fall if the 3 m-high lamp fails?

15 A level area 15 m square is illuminated by four 1400 cd lamps, one above each corner at a height of 4 m. Calculate the surface illuminance
   (*a*) at the centre of the area,
   (*b*) directly beneath one lamp, and
   (*c*) at the centre of one side of the area.

16 A swimming pool 120 m × 60 m is to be illuminated to a level of 400 lx using 400 W high-pressure sodium lamps, each providing 41 000 lm. Given a maintenance factor of 0·7 and a coefficient of utilisation of 0·75, how many lamps will be required?

17 A theatre foyer is 10 m square. As a feature it is to be illuminated to a level of 150 lx with 40 W GLS lamps which have an efficacy of 8 lm/W. The maintenance factor is 0·8 and the coefficient of utilisation 0·7. How many lamps are required?

18 The lighting system in a rolling mill has a coefficient of utilisation of 0·5 and a maintenance factor of 0·4. It is 20 m wide and 150 m long and is illuminated by 50 high-pressure mercury lamps, each of which provides 32 000 lm. Estimate the average illuminance.

19 A gymnasium 90 m × 45 m is lit with triple-tube 1800 mm 65 W fluorescent-lamp luminaires, each lamp emitting 4200 lm. Assuming a maintenance factor of 0·7 and a coefficient of utilisation of 0·5, estimate the power consumption of the lighting scheme which provides an average illuminance of 250 lx.

20 Draw a labelled sketch to show the construction of a general-lighting-service (GLS) filament lamp. Indicate a typical range of values for the efficacy of these lamps.

21 Why is the tungsten–halogen lamp more efficient than a normal tungsten-filament lamp?

22 Explain in simple terms how electromagnetic radiation is emitted from a low-pressure gas or vapour when an electric current is flowing in it. Indicate how current can flow in the gas or vapour.

23 Why is a ballast needed for the operation of a discharge lamp? Explain why the ballast used will be more efficient if it takes the form of a choke rather than a resistor.

24 Indicate why a majority of discharge-lamp circuits have a power-factor-correction capacitor included.

25 Draw circuit diagrams for
   (*a*) switch-start, and
   (*b*) quick-start fluorescent lamps,
   and describe their operation.

26 Write a short account of the effect of the colour of light output on the efficacy of fluorescent lamps. Indicate the circumstances in which various colours of the lamp light output may be used.

27 Draw a circuit diagram for a high-pressure mercury-vapour lamp and explain its operation. Suggest where this type of lamp may be used, and indicate how the manufacturers can improve its colour-rendering properties.

28 With the aid of a diagram, explain the construction and operation of the low-pressure sodium lamp.

29 The low-pressure sodium lamp is the most efficient of all the electric lamps. Explain why it is not widely used.

30 Use a diagram to assist an explanation of the operation of the high-pressure sodium lamp. Why is it sometimes preferred to its low-pressure counterpart, even though it is less efficient?

31 Explain why an electric sign may operate at 10 kV even though the IEE Wiring Regulations limit the maximum voltage to earth to 5 kV. Describe the arrangement of a typical sign with the aid of a diagram.

32 What is
(a) a locked switch, and
(b) a fireman's switch,
and why are they necessary in a high-voltage electric-sign circuit?

33 Explain why cables and switches used in conjunction with discharge lamps must often be overrated.

34 What is 'stroboscopic effect'? Describe why it may be dangerous, and explain how it may be overcome.

35 A hospital corridor is lit by twelve 100 W GLS lamps. As an economy measure, the GLS lamps are to be replaced with SL18 integral compact-fluorescent lamps, each of which consumes 18 W and provides a similar lumen output to the 100 W GLS types. If the GLS lamps have an expected life of 1000 hours at a cost of 34p each and SL18s are rated 5000 hours and cost £5·25 each, calculate the cost saving over a burning period of 10 000 hours. Ignore the savings in labour costs for lamp changing, and assume that the energy cost over the period will be 5·2p/kWh.

# 12.18   Multiple-choice exercises

12M1   Light is one form of electromagnetic radiation, which has a constant velocity in free space of:
(a) 186 000 m/s          (b) $3 \times 10^8$ m/s
(c) 186 000 mile/h       (d) infinity.

12M2   The frequency of electromagnetic radiation and its wavelength are:
(a) of no interest to the lighting designer
(b) proportional to one another
(c) the same thing
(d) inversely proportional.

12M3   The visible part of the electromagnetic spectrum:
(a) lies between the infrared and the ultraviolet sections
(b) is the green part of the spectrum
(c) cannot be seen by the human eye
(d) is at a wavelength of 700 nm.

12M4   The illuminance of a surface is measured in units which are called:
(*a*) lumens          (*b*) lux          (*c*) candle power  (*d*) candelas.

12M5   A 2000 cd lamp is hung above a level surface, which has an illuminance of 125 lx directly below the lamp. The distance from the lamp to the surface is:
(*a*) 16 m          (*b*) 6·25 m          (*c*) 4 m          (*d*) 4 cm.

12M6   If the surface of Example 12M5 were tilted through 25° the illumination of the surface directly below the lamp would become:
(*a*) 113 lux          (*b*) 22 lux          (*c*) 53 lux          (*d*) 138 lux.

12M7   The formula relating surface illuminance $E$, luminous intensity $I$, source to surface distance $d$ and the angle between the normal to the surface and the incident light $\phi$ is:

$$(a)\ E = \frac{Id^2}{\cos \phi} \qquad\qquad (b)\ I = \frac{E \cos \phi}{d^2}$$

$$(c)\ E = Id^2 \cos \phi \qquad\quad (d)\ E = \frac{I \cos \phi}{d^2}$$

12M8   Four 1000 cd lamps are mounted 4 m apart on 3 m-high poles in a straight line above a level surface. The lighting level on the surface directly beneath the second lamp in the line will be:
(*a*) 111 lux          (*b*) 161 lux          (*c*) 24 lux          (*d*) 1·6 lux.

12M9   The single factor in the following list which does **not** affect the coefficient of utilisation is:
(*a*) the mounting height of the luminaires
(*b*) the room size
(*c*) the colours and textures of the wall and ceiling finishes
(*d*) the required illuminance.

12M10  The number of 65 W fluorescent luminaires which will be required to provide an illuminance of 500 lx for an office 6 m by 15 m if the coefficient of utilisation is calculated at 0·55, the maintenance factor can be ignored and the luminaire efficacy is 60 lm/W is:
(*a*) 90          (*b*) 21          (*c*) 10          (*d*) 15.

12M11  The usual gas filling for a GLS filament lamp is:
(*a*) argon and nitrogen     (*b*) oxygen
(*c*) helium                 (*d*) air.

12M12  A lamp reflector with the ability to reflect visible light but to transmit heat is referred to as:
(*a*) overheated             (*b*) silvered
(*c*) dichroic               (*d*) tungsten–halogen.

12M13  A lighting track feeds seven 20 W 12 V tungsten–halogen lamps. The current rating of a suitable transformer would be:
(*a*) 1·7 A          (*b*) 42 A          (*c*) 11·7 A          (*d*) 7 A.

12M14  A manufacturer controls the colour of a tubular fluorescent lamp by:
(*a*) varying the lamp voltage
(*b*) choice of the fluorescent powder coating the inside of the tube

(c) choice of choke rating

(d) filament current and temperature.

12M15 The fluorescent-lamp circuit shown in Figure 12M1 is for a:

(a) semi-resonant system   (b) quick-start system

(c) compact lamp      (d) switch-start system.

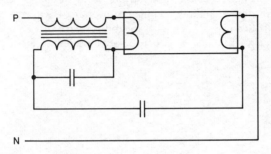

**Figure 12.21   Circuit for Exercise 12M15**

12M16 The compact fluorescent lamp is much less expensive to operate than the filament lamp because:

(a) its purchase cost is cheaper

(b) the light output is whiter

(c) it is heavier

(d) efficacy is greater and life is longer.

12M17 The electromagnetic output of the low-pressure mercury discharge in a fluorescent lamp is:

(a) pure white light

(b) a mixture of the colours of the visible spectrum

(c) likely to give a loud mains hum on any nearby radio receiver

(d) almost entirely invisible but is converted to visible light by the fluorescent powder.

12M18 The light output of a high-pressure mercury-vapour lamp is:

(a) very low considering the high power input usually required

(b) largely in the green and yellow parts of the visible spectrum

(c) like evening sunlight

(d) pure red, so that the lamp is used mainly for signs.

12M19 The discharge lamp with the highest efficacy but the poorest colour rendering is the:

(a) high-pressure sodium lamp

(b) fluorescent lamp

(c) low-pressure sodium lamp

(d) high-pressure mercury lamp.

12M20 The high-pressure sodium lamp

(a) provides a pure white light output

(b) usually requires a high-voltage pulse to start it

(c) is never used for high-bay industrial lighting

(d) is the least efficient of all the discharge lamps.

12M21 The discharge lamps most used in outdoor advertising signs are:
    (*a*) high-voltage cold-cathode types
    (*b*) high-pressure mercury types
    (*c*) low-voltage hot-cathode types
    (*d*) low-pressure mercury-vapour types.

12M22 The rapid flicker in the light output from discharge lamps which may make rotating machinery appear stationary is called:
    (*a*) rapid flicker        (*b*) disability glare
    (*c*) epilepsy           (*d*) the stroboscopic effect.

*Chapter 13*

# Supply Economics

## 13.1 Introduction

We must all pay for the electrical energy we use, and this chapter will be concerned with why we pay at the rate we do, which is called the **tariff**. Even the small domestic consumer now has a choice of tariffs, since (s)he can opt for the normal rate based on consumption, or on a two-part tariff which depends on when the energy is used as well as on how much is used. As well as an energy charge, (s)he will also have to pay a standing charge. For the larger commercial or industrial consumer things are more complicated, since maximum demand and power factor have to be considered.

This chapter will endeavour to explain the principles behind the energy charges made to consumers by the electricity companies.

## 13.2 Load curves

Perhaps the biggest single problem faced by the power-supply companies is that the demand for electrical energy varies so much, from minute to minute and from day to day. All electricity consumers expect that they can take electrical energy from the supply system at any time; it seldom occurs to them that equipment costing several thousands of pounds will be in use for every kilowatt of power they demand, and that this equipment has to be available 24 hours of every day of the year whether they use it or not.

Figure 13.1 shows typical national load curves. The first thing to notice is that the demand in the early hours of the morning is typically less than one-third of that in the middle of the day. This is because most people are asleep at this time, the load being accounted for by space and water heating, appliances such as freezers and refrigerators which are connected all the time, the emergency services and so on. It is usual for the curve

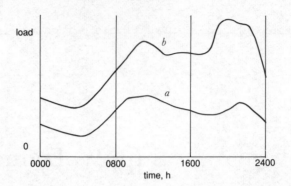

**Figure 13.1   Typical daily load curves for Great Britain**
  *a* A warm summer day
  *b* A cold winter day

to change from day to day. For example, if a cold winter day is followed by a warmer one, the demand is likely to be less.

We should next notice the very large difference between the summer and winter curves. Demand is about twice as high in the winter owing to the extra heating and lighting loads which are not required when the weather is warmer and brighter. The total cost of the generating system (power stations), the transmission system (grid lines) and the distribution system (local company's overhead and underground supply lines, substations etc.) is very high indeed. An analysis of the curves suggests that, on average, more than half of the capacity is unused for more than half of the time, clearly a very inefficient use of expensive resources.

It is in the interests of all concerned with the public electricity-supply system to reduce the 'peaks' and to fill in the 'valleys' of the demand curve. The ideal curve would be a straight horizontal line, but this is an impossibility because it would mean that on average all consumers take a constant load all the time. The tariff for the electrical energy provided is arranged to attempt to make the consumers change their power-consumption habits.

## Maximum demand (MD)

In an attempt to make consumers reduce their maximum demand, or to make them contribute to the costs if they are unable to do so, most supply companies apply a maximum-demand tariff to larger consumers. The cost of the special metering involved makes MD charging uneconomic for the supply company except for large industrial or commercial users. In these cases, a maximum-demand meter is used, which indicates the highest demand during any 30 minute period of the charge interval, which is typically one month. The maximum demand is expressed in kilovolt-amperes (kVA), and thus takes power factor into account. The tariff is likely to include an amount for each kVA of MD during the monthly accounting period. Often this amount will vary with the time of year. For example, the MD charge may be 50p per kVA per month in the summer months, rising to £3·50 in January and February.

There are two ways in which the user can seek to reduce maximum-demand payments.

The first is to curb MD. This may be done by making arrangements to stagger heavy loads of short duration so that they do not coincide. In most cases it is unlikely that all connected loads will be in use simultaneously; this fact is usually taken into account in the installation design, and is known as **diversity**. In the extreme, an MD alarm may be arranged to sound after, say, 20 minutes of the 30 minute period of operation of the MD meter to indicate that, if demand continues at its present rate, the predetermined target-maximum demand will be exceeded. Heavy loads could then be switched off to ensure that there is no increase in the reading. It must be stressed that the disruption to production caused by this method may result in greater financial losses than the extra cost of the MD tariff, especially if an artificially low figure for maximum demand has been assumed.

The second is to ensure that power factor is corrected to a reasonably high level. Power-factor correction will be considered in Section 13.4. If power factor is corrected, the kVA will reduce towards the true power (kW).

The maximum-demand part of the tariff will usually apply only to the larger consumer who has entered into a contract with the supplier, who need not be the local electricity supply company. There are so many types of contract available that there is no typical type which can be described here. All will, however, tend to follow the basic principles set out above.

## *Fixed or standing charges*

This is the name (sometimes called the standard rate) given to the charge made regardless of the amount of energy used. With the high energy user, the cost is usually very small when compared with the energy cost, but it can be a significant proportion of the total when energy use is low. The charge is often missing from the tariffs applied to industrial and commercial customers because their accounts are high.

It is commonly assumed by consumers, especially of the small energy-using domestic kind, that this charge is some kind of iniquitous tax. In fact, it is a sensible method of asking the consumer to make a contribution to the very high cost of providing a supply. Other methods of making the charge, such as increasing the unit cost, have been proposed, but have been rejected because they would be less fair generally than the present system.

## 13.3 Tariffs

The tariff is the system of charging the consumer for the electrical energy used. Tariffs will follow the general principles outlined above i.e. the consumer will pay a rate which takes account of the demand on the supply network and whether that demand takes place during a period when many other consumers are also requiring energy. There are many tariffs which vary depending on the likely usage of the consumer and on the geographical area in which the demand is situated. The examples given below are those usually found, but are not claimed to be exhaustive. All tariffs use the kilwatt-hour (kWh) as the basic unit of energy, that is, a power load of one kilowatt used for one hour. The kilowatt-hour is always referred to in this context as a 'unit'. The SI unit of energy is the joule (J) which is the same as the watt second. $1 \text{ kWh} = 3 \cdot 6 \times 10^6 \text{J}$ or $3 \cdot 6$ MJ.

For all tariffs, charges are slightly lower where an agreement exists for payment by direct debit — in these cases, the account is paid automatically by the customer's bank and the supply company will never be kept waiting for payment.

## Domestic tariffs

These tariffs apply to energy used in the home. Since the typical consumer, is unlikely to use large amounts of energy, it is not economic for the supply companies to install kVA maximum-demand metering, and simple kWh meters are used.

The general domestic tariff consists of a standing charge plus a unit charge to account for the energy used. In some cases a coin or token meter is installed, and this attracts an additional charge. Value Added Tax (VAT) is payable on the whole of the domestic electrical energy account, currently at a rate of 8%.

Off-peak tariffs are also available for energy used at night as described below.

## Off-peak tariff

To encourage the use of electrical energy at night, when the demand on the power stations is at its lowest, most supply companies offer an off-peak tariff. Two meters are used with a changeover switch operated by a timing device so that the energy used during the off-peak period, which is often seven hours during the night (known as 'Economy 7'), can be charged at a lower rate than the balance of the energy used during the day. In some cases, a ten-hour reduced rate applies (known as 'Economy 10'), five hours at night and two two-and-a-half hour periods during the day. The extra cost of metering is recouped by a higher standing charge than for the single-tariff consumer. This tariff can be beneficial to those who use a significant amount of energy during the off-peak period, such as those who use storage heating.

The standing charge for the off-peak tariffs is higher than that for the normal tariff, but the daytime rate is usually the same for both, with cheaper rates for units used during the economy periods. These rates are typically 37% of the day rate for Economy 7 and 45% for Economy 10.

The off-peak tariffs may also show significant savings for the consumer who heats water overnight for daytime use, where refrigerators or freezers are used, when a washing machine or dishwasher with a timer is arranged to operate at night, and so on. Most consumers can save by sensible energy use during off-peak periods. The savings will occur when the reduction in energy cost on the off-peak tariff exceeds the extra cost of the standing charge.

## Business and industrial tariffs

The tariff usually applicable to the small business has a standing charge with a variable unit charge, typically the first thousand units used each month being charged at a slightly higher rate than the balance. Where part of the premises is used as living accommodation, the first block of units (typically about 500 per month) is charged at a slightly lower rate than the balance.

The Economy 7 tariff (but not usually Economy 10) is also available to the business user on payment of a higher standing charge, although daytime units used in the business (but not in the living accommodation) are at a slightly higher rate.

Tariffs will differ depending on the size and kind of business involved as well as on which supply company provides the energy. For example, special tariffs are available for businesses such as clubs and public houses which mainly use energy during the hours when there is little demand from retail businesses or from industry. In some cases, an 'availability charge' is made for each kVA of supply capacity or of substation capacity, and special higher rates may apply for energy used during the early evening when general demand is heavy.

Larger consumers are free to 'shop around' for the best tariffs from all the supply companies as well as from the generating companies. The fact that their energy is supplied through the distribution system of their local supply company does not prevent their technically being supplied by, and charged by, another. At the time of writing this applies only to larger consumers, but it is intended that the limit will fall, so that free choice eventually applies to all, including domestic consumers. As mentioned above, special tariffs are negotiated with the larger consumers which are likely to vary very widely. The user with a low power factor is likely to be penalised by making a monthly charge for each kilovolt-ampere-hour (kVAh) measured by the maximum-demand meter.

## 13.4   The effect of power factor on the tariff

In the ideal world, where all power factors are unity, there would be no need for power-factor correction, and the current supplied to the consumer would generate true power measured in kilowatts (kW). Unfortunately, the widespread use of reactive loads, such as motors and chokes for discharge lighting, means that current is often out of phase with voltage. Thus, more current must be provided by the supply company than would be the case at unity power factor, leading to greater losses in its supply system. A study of Section 5.8 of this book will provide the necessary understanding of power factor and its effects.

Figure 13.2 shows phasor and power diagrams for a single-phase system operating at a lagging power factor. The power factor of the load shown is power divided by voltamperes.

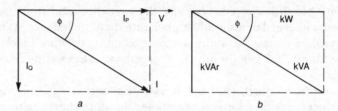

**Figure 13.2   Power factor for a single-phase system**

    *a* Phasor diagram for single-phase load with a lagging power factor

    *b* Power diagram for the load at Figure 13.2*a*

$$\text{power factor} = \frac{P}{VA} = \frac{I_P}{I} = \frac{\text{kW}}{\text{kVA}} = \cos \phi$$

The total current taken at this power factor is $I$, but the only part providing useful power is $I_P$, the component $I_Q$ being reactive current which contributes no power. Unless the power factor is unity (1) $I$ will always be larger than $I_P$. Similarly, the circuit voltamperes (kVA) will always exceed the power (kW). The reactive component of current $I_Q$ provides the reactive voltamperes (kVAr) and provides no useful power. The diagrams will make it clear why the supply company will penalise the consumer whose power factor is lower than unity, and why it may pay the consumer to improve the power factor. If this is done, the cost of improvement may well be repaid by the lower cost of electrical energy within a year or two, after which savings will continue to be made.

## Power-factor improvement

A study of Section 5.12 of this book will give the necessary details of the principles of power-factor improvement (often called power-factor correction). As indicated, the power factor encountered in many installations is lagging, since inductive loads such as motors and chokes are common. This means that the standard method of correction is the connection of capacitors.

## Bulk or group correction

If a capacitor bank is connected at the supply-intake position to correct the power factor of the complete installation, this is known as bulk correction or group correction. There is no doubt that this is the cheapest method of improving power factor for the purpose of reducing the cost of electrical energy, but unfortunately it has drawbacks.

The first of these is that while the total supply current to the complete installation is reduced to a minimum, there is no current reduction in individual circuits within the installation. Thus it is possible that the cables installed are larger than they would need to be if power-factor-correction capacitors were installed closer to the loads concerned; alternatively, cables might be overloaded if they were sized on the assumption that power-factor correction at the loads was to be used. There will also be greater power losses within the installation where currents are higher.

The second difficulty is the possibility of the installation operating at a leading power factor if loads are switched off while the correcting capacitors are still connected. The power diagrams of Figure 13.3 show how this could happen. It could be argued that a consumer with a leading power factor is helping the supply company because the installation is offsetting the lagging power factor of other consumers. However, the meters used do not differentiate between leading and lagging power factors, so the consumer with a low leading power factor will pay the same financial penalty as if the power factor were lagging.

A method of preventing this overcorrection is to have individual capacitor banks which can be switched in and out as load rises and falls. This could be done manually, but is more likely to be carried out using automatic sensing and switching systems. These arrangements are expensive, increasing the overall cost of correction and making the payback time longer.

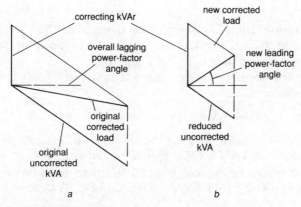

**Figure 13.3   Lagging and leading power factor**
  *a* Full load corrected to a high lagging power factor
  *b* Reduced load overcorrected to a leading power factor

## *Individual or load correction*

The problems associated with a single central capacitor bank will be overcome if each individual load which operates at a lagging power factor is provided with its own correction capacitor. This is known as individual or load correction. The cost of buying a large number of individual correction units will be much higher than for a single bulk unit, and when the labour costs for fitting numerous units is added there is usually a much higher first cost for load correction than for bulk correction. Against this, however, must be set the advantages of continuous optimum power-factor correction and reduced current in supply cables.

The most usual approach is to strike a happy medium between the two methods, correcting all large loads with individual local capacitors and providing a central correcting bank at the incoming mains position to correct the individual small loads (see Figure 5.18).

The calculations for power-factor correction of individual loads are covered in Section 5.12. For tariff purposes, it is uncommon for capacitance to be expressed in microfarads ($\mu F$), the kilovoltampere reactive (kVAr) being universally used. A formula applied to calculate the rating of a capacitor bank is:

$$kVAr = kW(\tan \phi_1 - \tan \phi_2)$$

where   kVAr is the kilovoltampere reactive rating of the required capacitor bank

kW is the load power

$\tan \phi_1$ is the tangent of the uncorrected power-factor angle $\phi_1$

$\tan \phi_2$ is the tangent of the corrected power-factor angle $\phi_2$

The formula assumes that both the initial (uncorrected) and the final (corrected) power factors are lagging.

## Example 13.1

An electrical installation has a maximum demand of 450 kVA at a lagging power factor of $0 \cdot 78$. Calculate the rating of the capacitor bank needed to improve the power factor to $0 \cdot 97$ lagging.

We first need to calculate the maximum demand in terms of power rather than voltamperes.

$$power = kVA \times PF = 450 \times 0 \cdot 78 = 351 \text{ kW}$$

Since initial power factor ($\cos \phi_1$) is $0 \cdot 78$, then from our calculator $\phi_1$ is $38 \cdot 7°$ (key in 0·78 INV COS). Hence $\tan \phi_1$ is $0 \cdot 802$.

Since final power factor ($\cos \phi_2$) is $0 \cdot 97$, then from our calculator $\phi_2$ is $14 \cdot 1°$ (key in 0·97 INV COS). Hence $\tan \phi_2$ is $0 \cdot 251$.

$$capacitor \ rating = kW(\tan \phi_1 - \tan \phi_2) = 351(0 \cdot 802 - 0 \cdot 251)$$

$$= 351 \times 0 \cdot 551 = 193 \cdot 4 \text{ kVAr}$$

## Example 13.2

A consumer has a maximum demand of 240 kVA at a power factor of $0 \cdot 86$ lagging. His tariff includes a figure of £16 per annum for each kVA of maximum demand. If he wishes to improve his power factor to $0 \cdot 98$ lagging, what maximum price should he pay for each kVAr of correcting capacitors if the correction cost is to be repaid by tariff savings within two years?

First we must calculate the rating of the necessary correcting capacitors.

$$\text{power rating} = kVA_1 \times PF = 240 \times 0 \cdot 86 = 206 \cdot 4 \text{ kW}$$

Since $\cos \phi_1$ is $0 \cdot 86$, then $\phi_1$ is $30 \cdot 7°$ and $\tan \phi_1$ is $0 \cdot 593$.

Since $\cos \phi_2$ is $0 \cdot 98$, then $\phi_2$ is $11 \cdot 5°$ and $\tan \phi_2$ is $0 \cdot 203$.

$$\text{capacitor rating} = kW(\tan \phi_1 - \tan \phi_2) = 206 \cdot 4 \, (0 \cdot 593 - 0 \cdot 203)$$

$$= 206 \cdot 4 \times 0 \cdot 390 = 80 \cdot 5 \text{ kVAr}$$

$$\text{new kVA rating} \quad kVA_2 = \frac{kW}{PF_2} = \frac{206 \cdot 4}{0 \cdot 98} = 211 \text{ kVA}$$

$$\text{reduction in kVA} = kVA_1 - kVA_2 = 240 - 211 = 29 \text{ kVA}$$

Tariff saving per annum $= 29 \times 16 = £464$

Cost saving over two-year period $= 464 \times 2 = £928$

Maximum cost of each kVAr of correction system to give two-year payback

$$= \frac{928}{80 \cdot 5} = £11.53 \text{ per kVAr}$$

## 13.5 Exercises

1 Use sketches to explain the meaning of the term 'load curves'. Draw a typical load curve for a large industrial area for a cold winter day and account for the shape of the curve. Why may the curve show significant changes from one day to the next?

2 Explain the meaning of the term 'maximum demand' as applied to the consumption of electrical energy. Why should a consumer with a large maximum demand expect to pay more for his electricity than another who uses a similar amount of energy but has a lower maximum demand? Describe the device which is used to measure maximum demand, and explain how its reading may have a considerable effect on the charge for electrical energy.

3 Why do electricity supply companies include fixed charges in their accounts for electricity?

4 What is an 'off-peak tariff'? Explain the structure of the Economy 7 and Economy 10 tariffs, and describe the pattern of demand from a domestic consumer who could be expected to opt for such tariffs.

5 Why may an electricity supply company or a consumer decide that a coin or token prepayment meter should be used in a particular installation?

6 An industrial consumer has a charge for the availability of supply capacity included in his account for electricity. Explain why this charge is made.

7 Explain why it is usually cost effective for an industrial consumer to spend money on correcting the power factor of his load.

8 Why is the maximum demand of an industrial consumer measured in kilovoltamperes (kVA) rather than in kilowatts (kW)?

9 The maximum-demand part of the tariff for an industrial consumer costs £18·00 per annum per kVA of MD. If the consumer assesses his maximum demand at 120 kW at a lagging power factor of 0·75, and the cost of correction capacitors fitted is £22·00 per kVAr, what will be the payback time for correcting his power factor to 0·95 lagging?

10 Explain the relative advantages of bulk and load correction of power factor. Why may a consumer find himself operating at a leading power factor if he has installed bulk power-factor-correction equipment?

11 An electrical installation has a maximum demand of 945 kVA at 0·88 power factor lagging. Calculate the kVAr rating of a capacitor bank to correct the power factor to 0·97 lagging.

12 An electrical installation is to be power-factor corrected. One large motor has a maximum rating of 125 kW at 0·78 power factor lagging. Calculate the kVAr rating of a capacitor to correct the motor power factor to 0·99 lagging.

## 13.6   Multiple-choice exercises

13M1  The electricity supply company rules which decide how much we pay for our electrical energy is called the:
(*a*) account    (*b*) tariff    (*c*) charge    (*d*) bill.

13M2  The demand for electrical energy is almost always much lower during the early hours of the morning than during the day. This is because:
(*a*) of the way that the off-peak tariffs are arranged
(*b*) power stations are shut down at night
(*c*) electricity costs are very much higher in the early hours of the morning
(*d*) most consumers are asleep and not using electricity.

13M3  The national demand on the electrical system on a warm and bright day is likely to be:
(*a*) low    (*b*) unchanged    (*c*) high    (*d*) zero.

13M4  An industrial maximum-demand meter indicates the demand during periods of:
(*a*) one month    (*b*) 30 minutes    (*c*) one hour    (*d*) a quarter.

13M5  The maximum-demand meter gives its reading in terms of:
(*a*) kW    (*b*) A    (*c*) kVA    (*d*) kVAr.

13M6  A formula from which the kVA rating of a capacitor bank needed to correct power factor can be calculated is:

(*a*) $\cos \phi = \dfrac{\text{kVA}}{\text{kW}}$          (*b*) $\text{kVAr} = \text{kW}(\tan \phi_1 - \tan \phi_2)$

(c) $kVAr = kVA - kW$     (d) $\cos \phi = \dfrac{kW}{kVA}$

13M7   The fixed or standing charge part of a domestic tariff is:
(a) intended to pay for the total energy used
(b) a contribution towards the cost of providing the supply
(c) a government tax on the consumer
(d) the value-added tax.

13M8   An off-peak tariff is:
(a) a low energy rate for electricity used at off-peak times
(b) a special tariff for large consumers
(c) never applied to domestic consumers of electricity
(d) a high energy rate for electricity used at off-peak times.

13M9   An electricity consumer with a demand of 500 kW:
(a) can arrange to take his supply from any of the electricity supply or generating companies
(b) will have difficulty in obtaining an electricity supplier for such a large load
(c) must take his supply from the local electricity supply company
(d) must take his supply from one of the generating companies.

13M10   The industrial electricity consumer whose load is at a low power factor will:
(a) have difficulty in obtaining a supply
(b) take a higher current than is necessary to provide his loads
(c) pay a higher rate for his electrical energy than he would if his power factor were improved
(d) pay a low rate for the electrical energy he consumes.

13M11   A single central power-factor correction capacitor bank at the incoming supply position provides:
(a) no power-factor correction
(b) load power-factor correction
(c) poor power-factor correction
(d) bulk power-factor correction.

13M12   The advantages of load power-factor correction include:
(a) overall reduced cost of correction
(b) reduced power losses in installation supply mains
(c) reduced tariff costs when compared with group correction
(d) lower currents within individual loads.

13M13   A bulk power-factor-correction system which is left connected to an installation where the load has reduced considerably may result in:
(a) damage to the installation cables
(b) a reduced supply tariff
(c) operation at a leading power factor
(d) considerably reduced maximum demand.

# Numerical answers to exercises

## Chapter 1
### Exercises (Section 1.14)

| | | | |
|---|---|---|---|
| 1  5 $\mu$F | 8  2·5 mm | 14 (a) 0·12 $\mu$F | 17  6 pF |
| 2  250 V | 9 (b) 2·4 mC | (b) 0·5 $\mu$F | 18  101 |
| 3  2 ms | 11 (c) (i) 3·75 $\mu$F | 15 (a) 1·98 $\mu$F | 19  1·66 mm |
| 5  8 A | (ii) 16 $\mu$F | (b) 38 $\mu$F | 20  0·346 J |
| 6  400 V/mm | 13 (a) 0·067 $\mu$F | 16 (i) 2·55 $\mu$F | 21  25 $\mu$F |
| 7  2·8 kV | (b) 0·3 $\mu$F | (ii) 24 $\mu$F | 22  200 V |

### Multiple-choice answers (Section 1.15)

| | | | |
|---|---|---|---|
| 1M1 (c) | 1M2 (d) | 1M3 (a) | 1M4 (b) |
| 1M5 (c) | 1M6 (d) | 1M7 (a) | 1M8 (c) |
| 1M9 (d) | 1M10 (b) | 1M11 (b) | 1M12 (d) |
| 1M13 (c) | 1M14 (a) | 1M15 (c) | 1M16 (d) |
| 1M17 (a) | 1M18 (c) | | |

## Chapter 2
### Exercises (Section 2.10)

| | | | |
|---|---|---|---|
| 1  500 V | 8  1 H | 15  0·12 ms | 24  0·1 H |
| 2  400 V | 9  50 $\mu$Wb | 17  0·5 V | 25  40 turns |
| 3  160 V | 10  6 turns | 19  1·2 H | 26  0·75 A |
| 4  80 mWb | 11  21 A | 20  2·4 V | 29 (b) 0·5 H |
| 5  0·3 Wb | 12  2·2 V | 21  150 A | 30  3·5 J |
| 6  2 s | 13  125 mH | 22  0·5 ms | 31  2 H |
| 7  500 turns | 14  0·75 Wb | 23  0·24 mWb | 32  10 A |

### Multiple-choice answers (Section 2.11)

| | | | |
|---|---|---|---|
| 2M1 (c) | 2M2 (d) | 2M3 (b) | 2M4 (b) |
| 2M5 (d) | 2M6 (c) | 2M7 (a) | 2M8 (d) |
| 2M9 (c) | 2M10 (b) | 2M11 (a) | 2M12 (c) |
| 2M13 (b) | 2M14 (a) | 2M15 (c) | 2M16 (b) |
| 2M17 (a) | 2M18 (c) | 2M19 (b) | 2M20 (d) |

## Chapter 3

### Multiple-choice answers (Section 3.10)

| | | | |
|---|---|---|---|
| 3M1 (*b*) | 3M2 (*a*) | 3M3 (*b*) | 3M4 (*c*) |
| 3M5 (*d*) | 3M6 (*a*) | 3M7 (*c*) | 3M8 (*d*) |
| 3M9 (*b*) | 3M10 (*d*) | 3M11 (*b*) | 3M12 (*c*) |
| 3M13 (*a*) | | | |

## Chapter 4

### Exercises (Section 4.10)

1  240 V
2  86·6 V
3  15 V
4  20 Ω, 43·6 Ω, 48 Ω, 0·139 H, 65·4°
5  92 V, 69 V, 79·6 mH, 36·9°
6  50 Ω
7  10 Ω, 8·66 Ω, 3·45 mH, 60°
8  6 Ω, 60·7 mH
9  8·24 mH
10  (*a*) (i) 15 Ω
    (ii) 19·6 Ω
    (iii) 0·04 H
11  19·75 Ω, 15·7 Ω, 0·05 H, 12 Ω
12  (*a*) 6·44 A
    (*b*) 129 V, 202 V

13  25 V
14  17·3 Ω
15  240 V
16  66·1 V
17  133 V
18  750 Ω, 661 Ω, 1000 Ω, 0·241 μF, 41·4°
19  (i) 1594 Ω
    (ii) 0·260 A
    (iii) 26 V
    (iv) 415 V
    (v) 86·4° lead
20  15 Ω, 177 μF, 36·9° lead
21  (*a*) 15·5 A
    (*b*) 49·7° lead
    (*c*) $V_R$ = 155 V, $V_L$ = 146 V,

$V_C$ = 329 V
22  (i) 390 Ω
    (ii) 1·06 A
    (iii) 106 V
    (iv) 1693 V
    (v) 59·2° lead
23  (*a*) 2·24 A, 134 V, 63·4° lead
    (*b*) 4·47 A, 134 V, 26·6° lag
27  2 A, $V_R$ = 24 V, $V_L$ = $V_C$ = 224 V
28  24Ω, 1570 V
29  (i) 178 Ω,
    (ii) 67·4 mA
    (iii) 63·3° lagging
    (iv) 21·4 V
30  15·2 A
31  28·9 A

32  1·5 A, 0°, $V_R$ = 30 V, $V_L$ = $V_C$ = 597 V
33  5·4 A, 27° leading
34  10 A, 20 A, 5 A
35  45·6 A
36  (*a*) 50 V
    (*b*) 8·94 A
37  3·92 A, 18·9° leading
41  10 A
42  11·5 A
43  13·9 A
44  21·8 A
45  10·4 Ω, 19·1 mH
46  34·6 A, 60 A
47  26·1 A, 45·2 A
48  7·27 A

### Multiple-choice answers (Section 4.11)

| | | | |
|---|---|---|---|
| 4M1 (*b*) | 4M2 (*d*) | 4M3 (*c*) | 4M4 (*a*) |
| 4M5 (*c*) | 4M6 (*a*) | 4M7 (*d*) | 4M8 (*b*) |
| 4M9 (*b*) | 4M10 (*d*) | 4M11 (*a*) | 4M12 (*c*) |
| 4M13 (*b*) | 4M14 (*d*) | 4M15 (*a*) | 4M16 (*c*) |
| 4M17 (*c*) | 4M18 (*a*) | 4M19 (*b*) | 4M20 (*d*) |
| 4M21 (*a*) | 4M22 (*c*) | 4M23 (*b*) | 4M24 (*b*) |
| 4M25 (*d*) | 4M26 (*a*) | 4M27 (*c*) | 4M28 (*d*) |
| 4M29 (*b*) | 4M30 (*c*) | | |

## Chapter 5

### Exercises (Section 5.15)

1  0·417 A
2  4·17 A, 57·6 Ω
3  833 W
4  1·3 A, 150 VAr
5  50 μF

16  69·5 Ω, 71·9 Ω
17  3·33 kW, 2·66 kW
18  6·35 A
19  2·40 kW
20  (*a*) 100 W

32  (*a*) 100 W, 100 VA
    (*b*) zero, 100 VA
    (*c*) 60 W, 100 VA
33  85% 0·729 lagging
34  84%

42  (*a*) 140 A
    (*b*) 107 A
    (*c*) 140 A
43  (i) 21·2 A
    (ii) 17·2 A

6  14·1 Ω, 14·1 A,
   45° leading,
   2·00 kW
7  88·5 W
8  (a) 250 V
   (b) 283 V
   (c) 450 W
9  77·6 μF
10 6 Ω, 331 μF
11 0·197 A, 7·73 W
12 11·2 A, 63·4°
   lagging, 500 W
13 2·88 kW,
   5·76 kWh
14 22·4 W
15 69·4 mH

(b) 400 W
21 48 W
22 0·707 leading
23 0·514 leading
24 0·600 leading
25 0·785 lagging
26 0·446 lagging
27 0·866 lagging
28 (a) 0·448 leading
   (b) 0·894 lagging
29 (b) 3·84 kW
30 9·6 A, 0·600
   lagging
31 (a) 21·9 A
   (b) 0·820 lagging
   (c) 4·31 kW

35 (a) 18 kW
   (b) 6 kVAr
   (c) 19 kVA
   (d) 0·948 lagging
   (e) 79·3 A
36 20 kW, 6 kVAr,
   20·9 kVA, 0·957
   lagging
37 279 kW, 149 kVAr,
   316 kVA, 0·882
   lagging
39 (a) 131 kW
   (b) 142 kVA
40  (i) 106 A
    (ii) 19·1 kW
    (iii) 25·4 kVA

(iii) 14·9 A
44 3 A
45 (a) 47·6 A
   (b) (i) 31·5 A
       (ii) 418 μF
46 1·68 A, 23·3 μF
47 22·5 A, 299 μF
48 (a) 225 kVAr
   (b) 150 kVAr
49 (a) 440 kVAr
   (b) 280 kVAr
50 (a) 44·0 kVAr
   (b) 27·7 kVAr
51 7·28 A, 1·59 kW
52 21·8 A, 4·77 kW

## Multiple-choice answers (Section 5.16)

| | | | |
|---|---|---|---|
| 5M1 (b) | 5M2 (d) | 5M3 (c) | 5M4 (b) |
| 5M5 (a) | 5M6 (d) | 5M7 (a) | 5M8 (c) |
| 5M9 (b) | 5M10 (a) | 5M11 (c) | 5M12 (d) |
| 5M13 (b) | 5M14 (a) | 5M15 (c) | 5M16 (d) |
| 5M17 (d) | 5M18 (b) | 5M19 (b) | 5M20 (b) |
| 5M21 (c) | 5M22 (a) | 5M23 (d) | |

## *Chapter 6*
## Exercises (Section 6.10)

14 108 V, 3·61 A
15 108 V, 3·61 A
16 216 V, 7·22 A

17 (a) 222 V
   (b) 111 V
18 8·1 A

19 13·4 V
29 2 × 10$^5$
30 33·3 μs

36 1
37 AND

## Multiple-choice answers (Section 6.11)

| | | | |
|---|---|---|---|
| 6M1 (b) | 6M2 (d) | 6M3 (a) | 6M4 (c) |
| 6M5 (a) | 6M6 (b) | 6M7 (d) | 6M8 (b) |
| 6M9 (a) | 6M10 (c) | 6M11 (a) | 6M12 (d) |
| 6M13 (b) | 6M14 (b) | 6M15 (c) | 6M16 (a) |
| 6M17 (a) | 6M18 (d) | 6M19 (b) | 6M20 (d) |
| 6M21 (b) | 6M22 (d) | 6M23 (c) | 6M24 (a) |

## *Chapter 7*
## Exercises (Section 7.13)

1  48 V
2  6:25
3  2400 V
4  9·1 A, 250 A
5  10 A

11 125 A (primary
   line), 72·3 A
   (primary winding)
   2283 V, 22·8 A
   (secondary line),

(b) 2·02 A
(c) 3·5 A
(d) 28·6:1
14 (a) 415 V
   (b) 7·87 A (line),

(b) 180 A
22 (a) 1200 V
   (b) 16·7 A
   (c) 83·3 A
   (d) 66·6 A

6  0·1 A, 240 V
7  990 V, 1042 V,
   1100 V, 1165 V
8  387 turns
9  80·0 V, 74·7 V,
   69·3 V, 64·0 V
10 92 turns, 48·1 A,
   105 A

13·2 A (secondary
   winding)
12 (a) 38·1 kV
   (b) 11 kV
   (c) 17·5 A
   (d) 105 A
   (e) 60·6 A
13 (a) 57·7 A

4·55 A
   (winding),
   (c) 209 A (line and
   winding)
16 1900:150
18 220 V, 9·1 A,
   136 A
19 (a) 2250:150

25 (a) 18 kW, 97·1%
   (b) 13·5 kW, 96·7%
   (c) 10·9 kW, 94.8%
26 4·5 kW
27 (b) 461 V
28 6·56%
29 22·1 V
30 3300 V

## Multiple-choice answers (Section 7.14)

| | | | |
|---|---|---|---|
| 7M1 (c) | 7M2 (a) | 7M3 (b) | 7M4 (a) |
| 7M5 (d) | 7M6 (b) | 7M7 (a) | 7M8 (d) |
| 7M9 (c) | 7M10 (a) | 7M11 (b) | 7M12 (a) |
| 7M13 (c) | 7M14 (b) | 7M15 (d) | 7M16 (c) |
| 7M17 (b) | 7M18 (a) | 7M19 (b) | 7M20 (c) |
| 7M21 (d) | 7M22 (c) | 7M23 (a) | 7M24 (b) |
| 7M25 (d) | 7M26 (c) | 7M27 (b) | 7M28 (a) |
| 7M29 (c) | 7M30 (b) | | |

## *Chapter 8*
## Exercises (Section 8.7)

1  150 V
2  66·7 mWb
3  50 r/s, 100 Hz
4  300 V
5  3840 r/min
6  200 V
7  (a) 30 Hz
   (b) 48 Hz
   (c) 75 Hz

8  (b) 192 V
   (c) 300 V
9  (a) 4 poles
   (b) 8 poles
   (c) 2 poles
10 16·7 r/s, 264 V,
   60 Hz
11 400 Hz
12 20 poles

13 900 r/min
14 (a) 16 kV, 18 500 A
   (b) 9·24 kV,
   32 000 A
15 (a) 720 V, 300 A
   (b) 415 V, 519 A
16 50 Hz, 18 r/s,
   167 A
17 1000 r/min, 167 V

18 360 V, 30 Hz
19 267 V
20 (a) 750 r/min
   (b) 254 V
   (c) 500 r/min
   (d) 169 V
21 (a) 415 V
   (b) 50 A
   (c) 8·3 Ω

## Multiple choice answers (Section 8.8)

| | | | |
|---|---|---|---|
| 8M1 (b) | 8M2 (d) | 8M3 (c) | 8M4 (b) |
| 8M5 (a) | 8M6 (d) | 8M7 (b) | 8M8 (a) |

## *Chapter 9*
## Exercises (Section 9.15)

1  672 V
2  25 r/s
3  36 mWb
4  800
5  250 V
6  242 V
7  2·94 A
8  27 r/s

14 225 A
15 150 V
16 0·25 Ω
17 332 V
18 300 A
19 415 V
20 (a) 245 V
   (b) 260 V

26 (i) 442 V
   (ii) 66·7 A
27 (a) 58·5 A
   (b) 41·5 A
28 250 V
29 232·5 V
30 200 V
31 0·15 Ω

34 (b) (i) 154 A, 4 A
       (ii) 146 A, 4 A
35 (b) 7·5 A
36 (b) (i) 137·7 A,
          2·3 A
       (ii) 142·3 A,
          2·3 A
38 7·25 Ω

9 53·3 Nm

10 24 A

11 121 Nm

12 28·8 Nm

13 17·1 mWb

21 280 V

22 192 V

23 298 V

24 1000 A

25 0·4 Ω

32 (i) 88 A

   (ii) 253 V

33 (a) 100 A

   (b) 257 V

43 (b) (i) 240 V

    (ii) 97·9% of

      first speed

46 600 r/min (10 r/s)

## Multiple-choice answers (Section 9.16)

| 9M1 | (c) | 9M2 | (a) | 9M3 | (d) | 9M4 | (a) |
|---|---|---|---|---|---|---|---|
| 9M5 | (b) | 9M6 | (b) | 9M7 | (d) | 9M8 | (a) |
| 9M9 | (c) | 9M10 | (a) | 9M11 | (d) | 9M12 | (b) |
| 9M13 | (d) | 9M14 | (b) | 9M15 | (c) | 9M16 | (a) |
| 9M17 | (d) | 9M18 | (a) | 9M19 | (b) | 9M20 | (d) |
| 9M21 | (c) | 9M22 | (a) | 9M23 | (c) | 9M24 | (b) |
| 9M25 | (d) | | | | | | |

## *Chapter 10*

## Exercises (Section 10.10)

2 10 r/s

3 8 poles

4 60 Hz

7 0·06, 6%

8 1176 r/min

9 50 Hz

11 (b) 50 Hz

12 3 Hz, 1·2 Hz, 2 Hz

13 (b) 950 r/min,

   2·5 Hz

14 (a) 50 Hz

   (b) 2 Hz

15 940 r/min,

   1500 r/min

## Multiple-choice answers (Section 10.1)

| 10M1 | (b) | 10M2 | (d) | 10M3 | (a) | 10M4 | (c) |
|---|---|---|---|---|---|---|---|
| 10M5 | (b) | 10M6 | (c) | 10M7 | (a) | 10M8 | (b) |
| 10M9 | (d) | 10M10 | (c) | 10M11 | (b) | 10M12 | (d) |
| 10M13 | (c) | 10M14 | (d) | 10M15 | (a) | 10M16 | (a) |
| 10M17 | (c) | 10M18 | (b) | 10M19 | (c) | 10M20 | (d) |

## *Chapter 11*

## Exercises (Section 11.23)

6 0·02002 Ω

7 0·00100025 Ω

8 249 980 Ω

9 4996 Ω

10 0·04507 Ω

  19 970 Ω

11 15 995 Ω

15 (a) 100 Ω/V

  (b) 2 kΩ/V

  (c) 20 kΩ/V

16 207 V

17 (a) 5 Ω

  (b) 6 kΩ

  (c) 0·6 Ω

18 0·2 Ω

19 (a) 2·46 Ω

  (b) 9·84 V

20 5 Ω, 5·06 Ω

21 9·29 Ω, 9·46 Ω

22 (a) 36·63 Ω

(b) 197·6 Ω

(c) 0·7543 Ω

(d) 923.8 Ω

23 (b) 800 Ω

25 1752 Ω, 10 000 Ω,

  10 Ω

26 (b) $P$ = 10 000 Ω,

  $Q$ = 10 Ω,

  $R$ = 2864 Ω

27 (a) 27·0 Ω

(b) 0·941 Ω

(c) 82·9 Ω

31 1204 m

32 3·47 MΩ

38 (a) 17·7 V

  (b) 159 V

  (c) 0·141 V

39 0·354 A

40 150 V, 106 V, 3 A,

  2·12 A

## Multiple-choice answers (Section 11.24)

| | | | |
|---|---|---|---|
| 11M1 *(b)* | 11M2 *(d)* | 11M3 *(a)* | 11M4 *(c)* |
| 11M5 *(c)* | 11M6 *(a)* | 11M7 *(d)* | 11M8 *(b)* |
| 11M9 *(d)* | 11M10 *(a)* | 11M11 *(b)* | 11M12 *(d)* |
| 11M13 *(d)* | 11M14 *(c)* | 11M15 *(b)* | 11M16 *(b)* |
| 11M17 *(c)* | 11M18 *(a)* | 11M19 *(d)* | 11M20 *(d)* |
| 11M21 *(c)* | | | |

## *Chapter 12*

## Exercises (Section 12.17)

| | | | |
|---|---|---|---|
| 1  545 nm | 8  250 lx | *(c)* 41·3 lx | *(c)* 20·4 lx |
| 2  $4·48 \times 10^{14}$ Hz | 9  3840 cd | 13 *(a)* 128 lx | 16  134 lamps |
| 3  157 lm/W | 10  3·5 m | *(b)* 125·5 lx | 17  84 lamps |
| 4  500 W | 11  128 lx, 533 lx | 14  91·3 lx, 8·0 lx | 18  107 lx |
| 5  16·7 lm/W | 12 *(a)* 68·3 lx | 15 *(a)* 15·4 lx | 19  44·8 kw |
| 6  37 500 lm | *(b)* 18·2 lx | *(b)* 91·1 lx | 35  £426·48 |
| 7  12·5 m$^2$ | | | |

## Multiple-choice answers (Section 12.18)

| | | | |
|---|---|---|---|
| 12M1 *(b)* | 12M2 *(d)* | 12M3 *(a)* | 12M4 *(b)* |
| 12M5 *(c)* | 12M6 *(a)* | 12M7 *(d)* | 12M8 *(b)* |
| 12M9 *(d)* | 12M10 *(b)* | 12M11 *(a)* | 12M12 *(c)* |
| 12M13 *(c)* | 12M14 *(b)* | 12M15 *(a)* | 12M16 *(d)* |
| 12M17 *(d)* | 12M18 *(b)* | 12M19 *(c)* | 12M20 *(b)* |
| 12M21 *(a)* | 12M22 *(d)* | | |

## *Chapter 13*

## Exercises (Section 13.5)

| | | |
|---|---|---|
| 9  2·37 years | 11  240 kVAr | 12  82·5 kVAr |

## Multiple-choice answers (Section 13.6)

| | | | |
|---|---|---|---|
| 13M1 *(b)* | 13M2 *(d)* | 13M3 *(a)* | 13M4 *(b)* |
| 13M5 *(c)* | 13M6 *(b)* | 13M7 *(b)* | 13M8 *(a)* |
| 13M9 *(a)* | 13M10 *(c)* | 13M11 *(d)* | 13M12 *(b)* |
| 13M13 *(c)* | | | |

# Index